Lecture Notes in Mathematics 2071

Editors:
J.-M. Morel, Cachan
B. Teissier, Paris

Editors *Mathematical Biosciences Subseries:*
P.K. Maini, Oxford

W0192919

For further volumes:
http://www.springer.com/series/304

Mark A. Lewis
Philip K. Maini
Sergei V. Petrovskii
Editors

Dispersal, Individual Movement and Spatial Ecology

A Mathematical Perspective

 Springer

Editors
Mark A. Lewis
Mathematical and Statistical Sciences
University of Alberta
Edmonton
Alberta, Canada

Philip K. Maini
Mathematical Institute
Centre for Mathematical Biology
University of Oxford
Oxford, United Kingdom

Sergei V. Petrovskii
Department of Mathematics
University of Leicester
Leicester, United Kingdom

ISBN 978-3-642-35496-0   ISBN 978-3-642-35497-7 (eBook)
DOI 10.1007/978-3-642-35497-7
Springer Heidelberg New York Dordrecht London

Lecture Notes in Mathematics ISSN print edition: 0075-8434
                              ISSN electronic edition: 1617-9692

Library of Congress Control Number: 2013932344

Mathematics Subject Classification (2010): M31000, M13003, L19147, L19007, M13090, M14068

Printed on acid-free paper

Springer is part of Springer Science+Business Media (www.springer.com)

Foreword

The study of biological populations is one of the oldest and most successful areas in mathematical biology, dating back at least a century to the work of Vito Volterra, the Italian mathematician equally famous for his contributions to the theory of integral equations. Indeed, there are examples even earlier of the use of mathematics in population biology, especially in demography and population growth; even Fibonacci dabbled in this subject largely to illustrate how the sequence of numbers that bears his name could arise easily in a population model. But Volterra's foray into mathematical ecology was an event of significance, because it demonstrated not only how sophisticated mathematics could contribute to biology but also that serious attention to biology could stimulate advances in mathematics. Both aspects are illustrated in this volume, which provides further evidence of the irresistible appeal of population problems for mathematicians.

Volterra's investigations focused on the dynamics of well-mixed populations and did not consider the spatial dimension, though his contributions to integral equations would certainly have put him in a position to advance the subject of spatial population biology. The first major efforts in that direction actually came from population genetics, where Fisher, Haldane and Wright all made major contributions in the 1930s and later. Fisher, in particular, was the first to note that the asymptotic speed of propagation of an advantageous allele would be twice the square root of the product of the intrinsic rate of increase and the diffusion coefficient, a result profound enough once again to attract leading mathematicians to provide formal analysis [10]. Indeed, attention to that rich problem has continued to be of interest to mathematicians [2, 4, 7], including those in this volume.

In ecology, the landmark paper was undoubtedly Skellam's 1951 treatise [18], which developed a broader framework for the consideration of spread, including those in response to climate change, and furthermore addressed the problem of critical patch size for persistence. These topics have remained of continuing interest for more than half a century, both for practical reasons [1] and because of their inherent mathematical richness. Skellam's framework allowed easily for the consideration of long-distance transport and was followed by papers such as Mollison's [12] and later work [15, 20] that explicitly dealt further with long-distance movements.

The consideration of spatial clines [9] and more general patterns [11], in addition to the problems mentioned earlier, has spawned a rich mathematical literature and one that has close contact with the biology [5, 13, 16, 17].

As this volume provides evidence, problems in dispersal, movement and spatial ecology continue to attract the attention of serious mathematicians and continue to grow in ecological importance [19]. On the biological side, we have seen the birth of a new sub-discipline called movement ecology [14]; and from many directions, interest in anomalous diffusion has grown [3, 21]. Conservation biology has raised many new problems, including those associated with the design of nature reserves, and the fascinating subject of collective motion has attracted the attention of biologists, mathematicians and physicists alike [6, 21]. Substantive mathematical problems remain, like the problem of scaling from the microscopic to the macroscopic, marrying the Lagrangian and Eulerian perspectives [8]. All of these issues are evident in the broad scope of the papers in this volume.

This collection is a welcome addition to the literature, illustrating once again the mathematical richness that underlies the movement problems as well as the ecological importance.

Princeton, New Jersey   Simon Levin
May 26, 2012

References

1. D.A. Andow, P.M. Kareiva, S.A. Levin, A. Okubo, Spread of invading organisms. Landscape Ecol. **4**(2/3), 177–188 (1990)
2. D.G. Aronson, H.F. Weinberger, Nonlinear diffusion in population genetics, combustion, and nerve pulse propagation. In *Partial Differential Equations and Related Topics*, vol 446, ed. by J.A. Goldstein (Springer, Berlin, 1975), pp. 5–49
3. F. Bartumeus, P. Fernandez, M.G.E. da Luz, J. Catalan, R.V. Sole, S.A. Levin, Superdiffusion and encounter rates in diluted, low dimensional worlds. Eur. Phys. J. Spec. Top. **157**, 157–166 (2008). doi: Doi 10.1140/Epjst/E2008-00638-6
4. M. Bramson, *Convergence of Solutions of the Kolmogorov Equation to Travelling Waves*. Paper presented at the (Mem. Am. Math. Soc., Providence, Rhode Island, 1983)
5. R.S. Cantrell, C. Cosner, *Spatial Ecology via Reaction-Diffusion Equations* (Wiley, Chichester, West Sussex, England; Hoboken, NJ, 2003)
6. I.D. Couzin, J. Krause, N.R. Franks, S.A. Levin, Effective leadership and decision making in animal groups on the move. Nature **433**, 513–516 (2005)
7. P.C. Fife, J.B. McLeod, The approach of solutions of nonlinear diffusion equations to travelling wave solutions. Arch. For Rat. Mech. Anal. **65**, 335–361 (1977)
8. G. Flierl, D. Grünbaum, S. Levin, D. Olson, From individuals to aggregations: the interplay between behavior and physics. J. Theor. Biol. **196**, 397–454 (1999)
9. J.B.S. Haldane, The theory of cline. J. Genet. **48**, 227–284 (1948)
10. A.N. Kolmogorov, I. Petrovsky, N. Piscounov, Etude de l'equation de la diffusion avec croissance de la quantite de matiere et son application a un problema biologique. Moscow University Bull. Ser. Internat. Sect. A **1**, 1–25 (1937)
11. S.A. Levin, Dispersion and population interactions. Am. Nat. **108**(960), 207–228 (1974)

12. D. Mollison, Spatial contact models for ecological and epidemic spread (with discussion). J. R. Stat. Soc. B **39**, 283–326 (1977)
13. J.D. Murray, *Mathematical Biology*, 2nd ed., vol. 19 (Springer, Heidelberg, 1990)
14. R. Nathan, Movement ecology. Special Feature Proc. Natl. Acad. Sci. **105**, 19050–19125 (2008)
15. M.G. Neubert, M. Kot, M.A. Lewis, Invasion speeds in fluctuating environments. Proc. R. Soc. Lond. B **267**, 1603–1610 (2000)
16. A. Okubo, S.A. Levin, *Diffusion and Ecological Problems: Modern Perspectives,* 2nd ed., vol. 14 (Springer, New York, 2001)
17. N. Shigesada, K. Kawasaki, *Biological Invasions: Theory and Practice* (Oxford University Press, Oxford, 1997)
18. J.G. Skellam, Random dispersal in theoretical populations. Biometrika **38**, 196–218 (1951)
19. D. Tilman, P. Kareiva, *Spatial Ecology* (Princeton University Press, Princeton, 1997)
20. F. van den Bosch, J. Metz, O. Diekmann, The velocity of spatial population expansion. J. Math. Biol. **28**(5), 529–565 (1990)
21. G.M. Viswanathan, M.G.E. da Luz, E.P. Raposo, H.E. Stanley, *The Physics of Foraging: An Introduction to Random Searches and Biological Encounters* (*Cambridge University Press, Cambridge,* 2011)

Preface

It has long been recognized that ecological dynamics is essentially spatial. Population aggregation that can be either self-organized or induced by heterogeneity in the environment is a commonly observed phenomenon. Spatial patterning has a variety of implications for biodiversity, harvesting, pest control, species extinction, and nature conservation. Dispersal is the process that results in a coupling between local populations and thus integrates them at a global level into an ecological entity. The properties of the entity can be very different from the properties of its parts. Thus it is important for us to know how to correctly interpret at the macroscopic level behavior at the local level if we are to determine how the entity behaves.

The approaches to study dispersal can differ greatly in terms of their focus and the level of detail involved. According to a commonly accepted definition, dispersal is the movement of organisms away from their parent source. The primary focus of dispersal is therefore on individual animal movement. Correspondingly, the focus of research is on individual movement paths and the most detailed description of dispersal should include all necessary information about the individual movement pattern.

However, this comprehensive description of dispersal is neither always possible nor always necessary. Once the state of the system is described by mean-field variables, e.g., by the population densities, information about individuals is lost. In fact, it is not required: Once the dispersal kernel is known, mathematical models are capable of grasping essential features of the population dynamics. Biological invasion is one example where application of population-level models has been particularly successful. One of the advantages of the population models is that they appear to be analytically more tractable than individual-based models allowing a fuller classification of different types of behavior in parameter space.

The most interesting part of the story is probably the bridge between the two "extremes." How can we derive the equations of the spatiotemporal population dynamics from the properties of the individual animal movement? Can we combine the benefits of the two approaches? What pattern of individual movement is behind a particular population dynamics model? One should recall here that population models are usually obtained from empirical or heuristic arguments rather than derived from first principles. Mathematical rigor is often lacking in this approach and, as a result, the empirical models may have hidden pitfalls and caveats that are difficult to identify. For example, implicitly assumptions may have been made that are erroneous or inconsistent with each other.

The structure of this book follows the general logic of dispersal studies outlined above. Part I (Chaps. 1–3) is concerned with individual animal movement. This subject has been increasingly controversial, sometimes even resulting in rather heated debates. Classical studies assumed that the individuals move around in a diffusive manner, i.e., a random walk process known as Brownian motion where the step length/size is described by a normal or exponential distribution effectively suppressing long steps. However, over the last two decades there has been increasing evidence that this might not always be the case. Indeed, field and laboratory data often show a rate of decay in the step size distribution which is much slower than exponential, e.g., as a power law. Correspondingly, stochastic processes such as Levy flights and/or Levy walks were introduced to take into account the long jumps in order to describe and analyze data on animal movement. However, the biological relevance of the Levy statistics still remains a controversial issue as it is not always clear whether it is a genuine pattern of the individual movement or an artifact of data collection and processing. The chapters in Part I contribute to this discussion and partially reflect this controversy by providing different points of view of the subject.

Part II (Chaps. 4–8) considers how the properties of individual movement can be scaled up to the population level. It starts with a review of mathematical models of self-organized population patterning with an emphasis on interaction and communication between the individuals (Chap. 4). Chapter 5 gives an overview of hybrid approaches that attempt to incorporate individual-based description to population-level models by considering movement of discrete objects (e.g., animals) in a continuous environment, chemotaxis being used as a paradigm. A different type of hybrid model is studied in Chap. 6 where foraging behavior is described as a space- and time-continuous process but transition between consequent generations (multiplication) is described as a time-discrete map.

The analysis of Chaps. 4–6 is mostly focused on self-organized behavior in a homogeneous and isotropic environment. This assumption is relaxed in Chaps. 7 and 8. In particular, Chap. 7 considers population models when individual movement is anisotropic, e.g., occurring in an environment with a directional bias. The population dynamics of wolves in a forest with seismic lines is used as an instructive example. Chapter 8 considers complex foraging behavior of zooplankton in a prey–predator (e.g., phyto-zooplankton) system in a vertically stratified water column. Interestingly, the behavioral response to stratification can result in a change of the predator function response, so that the Holling type II response assumed in local grazing gives way to type III after averaging over water column height.

Part III (Chaps. 9–13) considers dispersal and its implications on the level of populations and communities. One of the main objectives here is to understand how the population abundance, e.g., as quantified by the population density, changes in space and time because of the interplay between dispersal and the local population dynamics. The two phenomena that are essentially attributed to this interplay are biological invasion and population range shift (Chaps. 9 and 10). The properties of dispersal may affect the rate of species spread significantly. For instance, it is well known that fat-tailed dispersal can increase the invasion rate considerably. It therefore becomes important to develop analytical approaches which allow us to reveal the properties of the dispersal kernel (Chap. 9) and to better understand how the population behavior depends on the kernel used.

Another major issue is population dynamics on a fragmented habitat. Dispersal coupling results in the possibility of re-colonization of empty patches. Chapter 11 shows that the effect of re-colonization can be subtle and counterintuitive depending on how much detail of the food web is taken into account.

With the spatiotemporal complexity of dispersal in mind, perhaps it is not surprising that dispersal has not only ecological but also evolutionary implications. Chapter 12 considers interaction between the processes going on different temporal scales and concludes that dispersal coupled with non-local resource consumption can be a crucial factor resulting in speciation.

Finally, Chap. 13 considers the implication of dispersal—regarded here as diffusion—for the pest population size estimation commonly required in pest control programs. Somewhat counterintuitively, it shows that a pest with a lower diffusivity may be more difficult to monitor than a highly mobile one.

The idea of this book emerged and was eventually shaped into its final form during a series of meetings, in particular at the conference *Models in Population Dynamics and Ecology 2010* (Leicester, September 1–3, 2010) and the MBI Workshop: *Ecology and Control of Invasive Species* (Columbus, February 21–25, 2011). Obviously, considerable progress has been made over the last two decades in understanding all aspects of dispersal, as can be traced from the references provided with the chapters. Appreciation of the diversity of studies focused on or closely related to dispersal led to the feeling that an account of the state of the art in this field may be timely and useful. It is for the reader to decide whether this goal has been achieved and how comprehensive is the account. Whichever is the case, we hope that this book is going to be stimulating for future research.

Mark A. Lewis
Philip K. Maini
Sergei V. Petrovskii

Contents

Part I
Individual Animal Movement

Stochastic Optimal Foraging Theory

Frederic Bartumeus, Ernesto P. Raposo, Gandhi M. Viswanathan,
and Marcos G.E. da Luz

Abstract We present here the core elements of a stochastic optimal foraging theory
(SOFT), essentially, a random search theory for ecologists. SOFT complements
classic optimal foraging theory (OFT) in that it assumes fully uninformed searchers
in an explicit space. Mathematically, the theory quantifies the time spent by a
random walker (the forager) on a spatial region delimited by absorbing boundaries
(the targets). The walker starts from a given initial position and has no previous
knowledge (nor the possibility to gain knowledge) on target/patch locations.
Averages on such process can describe the dynamics of an uninformed forager
looking for successive targets in a diverse and dynamical spatial environment. The
framework provides a means to advance in the study of search uncertainty and
animal information use in natural foraging systems.

F. Bartumeus (✉)
Center for Advanced Studies of Blanes, Accés Cala Sant Francesc 14, 17300, Blanes,
Girona, Spain
e-mail: fbartu@ceab.csic.es

E.P. Raposo
Laboratório de Física Teórica e Computacional, Departamento de Física, Universidade Federal de
Pernambuco, Recife, Brazil
e-mail: ernesto@df.ufpe.br

G.M. Viswanathan
Departamento de Física Teórica e Experimental, Universidade Federal do Rio Grande do Norte,
Natal, Brazil
e-mail: gandhi.viswanathan@pq.cnpq.br

M.G.E. da Luz
Departamento de Física, Universidade Federal do Paraná, Curitiba, Brazil
e-mail: luz@fisica.ufpr.br

M.A. Lewis et al. (eds.), *Dispersal, Individual Movement and Spatial Ecology*,
Lecture Notes in Mathematics 2071, DOI 10.1007/978-3-642-35497-7_1,
© Springer-Verlag Berlin Heidelberg 2013

1 Introduction

Classic optimal foraging theory (OFT) assumes fully informed foragers. Hence, animals can recognize a patch instantaneously, knowing in advance the expected patch quality as well as the average travel time between patches [19]. Stephens and Krebs (1986) called such conceptual framework the *complete information* assumption [47, 48].

Based on simple cases, theoreticians have addressed the problem of *incomplete information* [47, 48], acknowledging the presence of environmental uncertainty in foraging processes. The key questions are related to how animals obtain information about the environment while foraging [1, 20, 21, 31, 34]. The use of information to both discriminate the properties of a given patch and to figure out large-scale environmental properties have been shown to modify patch-exploitation and patch-leaving strategies [48]. Simple memory rules based on previous environment exploration experiences [32] and potential acquaintance with the travel times between patches [13, 14, 17, 24] also impact on the foraging strategy.

Here we introduce a theoretical framework to study aspects of foraging processes rooted on the assumption of *complete lack of knowledge* and with the virtue of being spatially explicit (here we address the one-dimensional case). In its core formulation, SOFT quantifies the distance traveled (or equivalently time spent) by a random walker that starts moving from a given initial position within a spatial region delimited by absorbing boundaries. Each time the walker reaches the boundaries, the process starts all over again. Averages on the properties of many walks realizations are aimed to reproduce the dynamics of a forager looking for successive targets in a diverse and dynamical environment. This modeling approach differs from classic theory in a very important point: it switches the patch-encounter problem of foraging theory from the *traveling salesman* [1] to the *random search* optimization problem [4, 16, 49, 51].

While useful as analytic simplifications, classic theoretical studies on foraging usually lack the explicit inclusion of space and are not focused on the search optimization problem, in which a forager with limited information explores a landscape to find scarce cues [4, 16, 51]. In OFT patch locations are known in advance and the goal is to find the shortest path connecting them. In SOFT, the locations and travel distances between patches are unknown, and thus the task is to determine an uninformed exploration strategy (which necessarily use some element of randomness), maximizing the number of patch encounters [4, 51]. Out of doubt, the theory described here is at the far end of the spectrum that begins with the mean-field and full-knowledge assumptions of classic OFT [19, 47, 48].

It does not escape to us that the assumption of a foraging animal as a "brainless" random walker (i.e., with no internal states nor sensory or memory capabilities) should be viewed as a first-order approximation to the actual dynamics. Hence it does not represent the ultimate description of animal information use and movement complexity. Nevertheless, memory-less models can be realistic when the searcher looks for dynamic targets that move away from their original location on time scales shorter than typical revisiting times by the searcher. In any case, limiting models are good starting points to think on complex problems and have an extraordinary success in making general scientific predictions. Importantly, they play a complementary role to biologically more specific models and shed light on different aspects of movement phenomena [51]. In this chapter, we hope to demonstrate that a spatially explicit random search theory can serve as the seed for more realistic (yet still simple) models [15] to advance in the study of information use in natural foraging systems. New ideas and results on random searching [2,9,29,41,49] clearly show that random walk and diffusion theory [35,43,44,51] can better fit the concepts of search and uncertainty in behavioral ecology. Routes to integrate both theories, the classical OFT and the recent SOFT, will be needed in order to properly answer questions about efficiency and uncertainty of animal foraging strategies [3,4,51].

2 Some Preliminary Assumptions of the Model

We begin by considering a random searcher looking for point-like target sites in a one-dimensional (1D) search space. We consider a lattice of targets separated by the distance λ, i.e. the targets positions are $x = j\lambda$, with j integer. Suppose, initially, that the walker starts from a distance x_0 to the closest target. The walker thus searches for the two nearest (boundary) targets by taking steps of length ℓ from a probability density function (pdf) $p(\ell)$, which is kept the same for all steps. In Sects. 3–6, every time an encounter occurs the search resets and restarts over again from the same distance x_0 to the last target found. For example, if the position of the n-th target found is, say, $x = 10\lambda$, then the next starting point will be $10\lambda + x_0$ or $10\lambda - x_0$. In this sense, the search for any target is statistically indistinguishable from the search for the very first target: in both cases, the closest and farthest targets are, respectively, at initial distances x_0 and $\lambda - x_0$ from the searcher, and the pdf $p(\ell)$ of step lengths is the same. Therefore, without loss of generality we can restrict our analysis to the region $0 \le x \le \lambda$, with the targets at $x = 0$ and $x = \lambda$ being the system absorbing boundaries. This is actually possible since leaps over targets without detection are not allowed in this study. For an interesting account of leapover statistics in the context of Lévy flights, see [27]. As a consequence, in the present framework the overall search trajectory can be viewed as the concatenated sum of partial paths between consecutive encounters. In Sect. 7, the constraint of always starting from the same distance x_0 to the last target found is relaxed, and searches in landscapes with targets heterogeneously distributed are considered (see below). In every case, averages over these partial paths will describe a random search process in an environment whose global density of targets is $\rho \sim 1/(\text{mean distance between targets}) = 1/\lambda$.

As commented above, at each starting process to find a new target we may or may not assume distinct initial positions of the searcher, x_0. The analysis presented in Sects. 3–6 assumes that the forager always restarts at a fixed $x_0 = a$. However, in the most general case x_0 can be drawn from a pdf $\pi(x_0)$. By considering a distribution

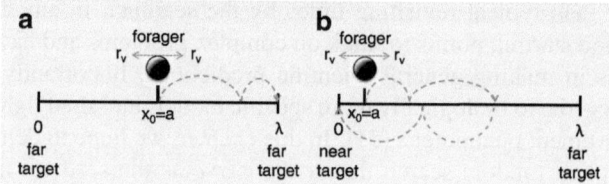

Fig. 1 Diagrams showing the two key initial conditions for the one-dimensional stochastic search model: (**a**) symmetric (destructive), (**b**) asymmetric (non-destructive). We denote by $x_0 = a$ the forager starting position at each search ($a = \lambda/2$ in the symmetric case and a $r_v + \epsilon$ in the asymmetric case). r_v denotes the forager's perceptive range or radius of vision

of x_0 values, the relative distances from the initial position of the searcher to the targets change at each search, thus describing an heterogeneous environment (but of global density $1/\lambda$). In Sect. 7 we consider various pdfs $\pi(x_0)$, so to address more realistic foraging situations in which the search landscape presents several degrees of heterogeneity.

In particular, for the case of fixed $x_0 = a$ two limiting situations are considered (see Fig. 1 and [23, 49]). The symmetric (or destructive) condition (i.e. $a = \lambda/2$) represents the situation in which, having located and consumed a food item, there are no other nearby food items available and the forager begins the next search positioned far away and relatively equidistant, on average, from the two closest food items (Fig. 1). The asymmetric (or non-destructive) condition represents the situation where, having located a food item, other nearby items exist, hence the forager begins the next search with a close and a faraway target (see Fig. 1). Non-destructive foraging, with a once-visited item always available for future searches, should be considered as the paradigmatic asymmetric condition. If the foraging dynamics is non-destructive but environmental uncertainty exists (such that the forager may repeatedly loose track of the items outside its perceptual range), it will systematically reproduce the asymmetric condition at each restarted search. Even though the idea of *non-destructive* stochastic search perfectly maps with the asymmetric condition, caution must be taken with the *destructive* searches, which can indeed accommodate both symmetric and asymmetric conditions, depending on the landscape structure (see Sect. 7). Importantly, in the context of foraging, the previous definitions of destructive/non-destructive search [49] have led to some misleading criticism [22, 36].

In our model the pdf $p(\ell)$ of step lengths ℓ is the same for each statistically independent step of the walk. The normalization condition imposes

$$\int_{-\infty}^{+\infty} p(\ell)d\ell = 1. \tag{1}$$

Notice that a "negative step length" just means that a step is taken to the left (negative) direction. We study the case in which it is equiprobable for the walker to go either to the left or to the right, so that $p(\ell) = p(-\ell)$. In addition, we consider the minimum step length as ℓ_0, resulting in $p(\ell) = 0$ for $|\ell| < \ell_0$. An important quantity is the radius of vision r_v, i.e. the walker's perceptive range. Whenever its distance to the nearest site is $\leq r_v$, it goes straight to the target. Events of finding a target actually lead to truncation of steps, as discussed below. In principle, ℓ_0 and r_v are independent parameters. However, in some of our calculations we set $r_v = \ell_0$. Here we are interested in the scarcity regime of low-food density, $\lambda \gg r_v$ and $\lambda \gg \ell_0$, with the forager's perception about the search landscape being limited. Hence, searches with stochastic character arise naturally.

We define the efficiency η of the search walk as the ratio between the total number of target sites found, N_{found}, and the total distance traveled by the walker, L_{tot}:

$$\eta = \frac{N_{\text{found}}}{L_{\text{tot}}}. \tag{2}$$

By writing $L_{\text{tot}} = N_{\text{found}}\langle L \rangle$, where $\langle L \rangle$ denotes the average distance traveled between two successive target sites found, we obtain

$$\eta = \frac{1}{\langle L \rangle}. \tag{3}$$

In the following, we work out a closed analytical expression for $\langle L \rangle$ for any probability density $p(\ell)$. Nevertheless, the focus of this contribution is on asymptotically power-law Lévy distributions [33] of step lengths. In particular, we focus on Lévy walk and not Lévy flight models. In the former models, jumps are not instantaneous but a time interval related to a finite velocity to complete the jump is involved (see Sect. 5).

3 Calculation of $\langle L \rangle$ and $\langle |\ell| \rangle$

We start by calculating the average distance $\langle L \rangle$ traversed by a walker starting at a fixed position $x_0 = a$ until reaching one of the borders located at $x = 0$ and $x = \lambda$. In the foraging process this quantity represents the distance traveled between two successively found target sites. Due to the perceptive range of the forager, we demand that $r_v < a < \lambda - r_v$. Here we follow the general method developed by Buldyrev et al. in [10, 11].

Let us consider a walker that finds either the boundary target at $x = 0$ or $x = \lambda$ after n steps. The distance traveled in this walk is

$$L_n = \sum_{i=1}^{n} |\ell_i|, \tag{4}$$

where $|\ell_i|$ denotes the length of the i-th step. Since the walker is not in free space, the possibility of truncation of steps makes $|\ell_i|$ dependent on the position x_{i-1} from which the step i starts. As a consequence, the last (n-th) step depends upon $x_0 = a$, since x_{n-1} depends on x_{n-2}, which, in turn, depends on x_{n-3}, and so on, all the way down to x_0. Therefore, we must have $L_n = L_n(a)$ as well.

By averaging over all possible walks that finds a target after n steps, we find

$$\langle L_n \rangle(a) = \sum_{i=1}^{n} \langle |\ell_i| \rangle. \tag{5}$$

Observe now that n can take any integer value, from 1 to ∞, meaning that the targets at $x = 0$ or $x = \lambda$ can be found just at the first step or after an infinitely long number of steps. We should also remark that the probability P_n of finding a target after n steps is not uniform, being, instead, dependent on n. Thus, when we average over all possible walks with the same starting point $x_0 = a$ in the interval of length λ, we must take into consideration the different weights of walks with distinct n's, so that

$$\langle L \rangle = \sum_{n=1}^{\infty} P_n \langle L_n \rangle. \tag{6}$$

The above equation implicity assumes the normalization condition $\sum_{n=1}^{\infty} P_n = 1$, so to assure that a target site, either at $x = 0$ or $x = \lambda$, is always found at the end. In this sense, we emphasize that $\langle L \rangle$ can be also interpreted as the average distance traversed by the searcher in the *first-passage-time* problem to find a boundary target at either $x = 0$ or $x = \lambda$. We return to this point in Sect. 6.

In order to calculate P_n we define $\rho_n(x_n)$ as the pdf to find the walker between x_n and $x_n + dx_n$ after n steps. Therefore, the probability that the walker has *not* yet encountered any of the targets after n steps is given by

$$P_n^{\text{not}} = \int_{r_v}^{\lambda - r_v} \rho_n(x_n) dx_n. \tag{7}$$

Conversely, the complementary probability of finding any of the targets in some step $n' \geq n + 1$ is thus

$$P_{n' \geq n+1} = 1 - P_n^{\text{not}}. \tag{8}$$

As a consequence, the probability of finding a target precisely after n steps reads

$$P_n = |P_{n' \geq n+1} - P_{n' \geq n}| = |P_n^{\text{not}} - P_{n-1}^{\text{not}}|, \tag{9}$$

which, by using (7) and dropping the subindexes in the dummy x_n and x_{n-1} variables of integration, leads to

$$P_n = \int_{r_v}^{\lambda - r_v} [\rho_{n-1}(x) - \rho_n(x)] dx. \tag{10}$$

Note that $\rho_{n-1}(x) > \rho_n(x)$, since the probability that the walker finds one of the targets grows with increasing n. From (10), we thus interpret $\rho_{n-1}(x) - \rho_n(x)$ as a pdf to encounter a target precisely after n steps.

By combining this fact with (6), we find

$$\langle L \rangle = \sum_{n=1}^{\infty} \int_{r_v}^{\lambda - r_v} dx [\rho_{n-1}(x) - \rho_n(x)] \langle L_n \rangle(x), \tag{11}$$

which can be conveniently broken into two sums:

$$\langle L \rangle = \sum_{n=1}^{\infty} \int_{r_v}^{\lambda - r_v} dx \rho_{n-1}(x) \langle L_n \rangle(x) - \sum_{n=1}^{\infty} \int_{r_v}^{\lambda - r_v} dx \rho_n(x) \langle L_n \rangle(x). \tag{12}$$

The integration from r_v to $\lambda - r_v$ takes into account all possible starting points x for the last n-th step. By changing the variable in the first sum, $m = n - 1$, and adding the $n = 0$ null term to the second sum (note that, by definition, $\langle L_{n=0} \rangle = 0$), we obtain

$$\langle L \rangle = \sum_{m=0}^{\infty} \int_{r_v}^{\lambda - r_v} dx \rho_m(x) \langle L_{m+1} \rangle - \sum_{n=0}^{\infty} \int_{r_v}^{\lambda - r_v} dx \rho_n(x) \langle L_n \rangle. \tag{13}$$

By using (5) above, we find

$$\langle L \rangle = \sum_{n=0}^{\infty} \int_{r_v}^{\lambda - r_v} dx \rho_n(x) \langle |\ell| \rangle(x). \tag{14}$$

To perform the integral (14), we need to work on $\rho_n(x)$ first. We note that, in general,

$$\rho_i(x_i) = \int_{r_v}^{\lambda - r_v} \rho_{i-1}(x_{i-1}) p(x_i - x_{i-1}) dx_{i-1}, \tag{15}$$

where we have recovered the subindexes to make explicit the positions of the walker after i and $i - 1$ steps, respectively x_i and x_{i-1}. The above expression sums over all the possibilities of reaching the site x_i from the site x_{i-1}, by performing a step of length $|x_i - x_{i-1}|$ with probability $p(x_i - x_{i-1}) dx_{i-1}$. By recursively applying (15) down to the very first step, we find n integrals, associated to $n - 1$ steps, from x_0 up to x_{n-1}, which denotes the starting point of the last n-th step:

$$\rho_n(x_n) = \int_{r_v}^{\lambda - r_v} \cdots \int_{r_v}^{\lambda - r_v} \left[\prod_{i=0}^{n-1} p(x_{i+1} - x_i) dx_i \right] \rho_0(x_0). \tag{16}$$

Since the initial position $x_0 = a$ is fixed, with $r_v < a < \lambda - r_v$, then from (16) for $n = 1$,

$$\rho_1(x_1) = \int_{r_v}^{\lambda - r_v} \rho_0(x_0) p(x_1 - x_0) dx_0 = p(x_1 - a). \tag{17}$$

Above, $\rho_0(x_0)$ is the pdf to find the walker at zero time steps. Since its initial position is a, then we have

$$\rho_0(x_0) = \delta(x_0 - a), \tag{18}$$

where δ denotes Dirac delta function.

Now, by substituting (16) into (14) we obtain

$$\langle L \rangle = \sum_{n=0}^{\infty} \int_{r_v}^{\lambda - r_v} \left\{ \int_{r_v}^{\lambda - r_v} \cdots \int_{r_v}^{\lambda - r_v} \left[\prod_{i=0}^{n-1} p(x_{i+1} - x_i) dx_i \right] \rho_0(x_0) \right\} \langle |\ell| \rangle (x_n) dx_n,$$

$$\tag{19}$$

where, once again, we have recovered the notation $x \rightarrow x_n$ from (14). This expression can be put in a much shorter form if one defines the following integral operator [10, 11]:

$$[\mathscr{L}\rho_n](x) = \int_{r_v}^{\lambda - r_v} p(x - x')\rho_n(x') dx', \tag{20}$$

so that, by comparing with (15), $\rho_1(x_1) = [\mathscr{L}\rho_0](x_1)$, $\rho_2(x_2) = [\mathscr{L}\rho_1](x_2) = [\mathscr{L}[\mathscr{L}\rho_0]](x_2) \equiv [\mathscr{L}^2\rho_0](x_2)$, and so on. Using this definition, we rewrite (19) as

$$\langle L \rangle = \sum_{n=0}^{\infty} \int_{r_v}^{\lambda - r_v} [\mathscr{L}^n \rho_0](x_n) \langle |\ell| \rangle (x_n) dx_n. \tag{21}$$

In formal analogy to Taylor's series expansion, we write

$$[(\mathscr{I} - \mathscr{L})^{-1}\rho_0](x) = \sum_{n=0}^{\infty} [\mathscr{L}^n \rho_0](x), \tag{22}$$

where $\mathscr{I}$ denotes the unitary operator: $[\mathscr{I}\rho](x) = \rho(x)$. Equation (21) thus becomes

$$\langle L \rangle = \int_{r_v}^{\lambda - r_v} [(\mathscr{I} - \mathscr{L})^{-1}\rho_0](x_n) \langle |\ell| \rangle (x_n) dx_n, \tag{23}$$

which, with the use of (18), leads to [10, 11]

$$\langle L \rangle (a) = [(\mathscr{I} - \mathscr{L})^{-1} \langle |\ell| \rangle](a). \tag{24}$$

This closed analytical expression is actually essential to determine the efficiency of the search, according to (3).

Now, in order to deal with (24), we need to calculate the average (modulus) length of a single step starting at $x_0 = a$ in the interval of length λ, $\langle|\ell|\rangle(a)$. As discussed, in the presence of target sites at $x = 0$ and $x = \lambda$ there is the possibility of truncation of steps. Thus, the usual average in free space, $\langle|\ell|\rangle = \int_{-\infty}^{\infty}|\ell|p(\ell)d\ell$, which does not depend on the starting position, must be replaced by

$$\langle|\ell|\rangle(a) = \int_{r_v}^{a-\ell_0}(a-x)p(x-a)dx + \int_{a+\ell_0}^{\lambda-r_v}(x-a)p(x-a)dx$$

$$+ (a-r_v)\int_{-\infty}^{r_v}p(x-a)dx + (\lambda-r_v-a)\int_{\lambda-r_v}^{\infty}p(x-a)dx, \quad (25)$$

valid for $r_v + \ell_0 \leq a \leq \lambda - r_v - \ell_0$. The meaning of this expression becomes clearer if we make the change of variable $\ell = x - a$ is all above integrals, to obtain

$$\langle|\ell|\rangle(a) = \int_{-(a-r_v)}^{-\ell_0}|\ell|p(\ell)d\ell + \int_{\ell_0}^{\lambda-r_v-a}|\ell|p(\ell)d\ell$$

$$+ (a-r_v)\int_{-\infty}^{-(a-r_v)}p(\ell)d\ell + (\lambda-r_v-a)\int_{\lambda-r_v-a}^{\infty}p(\ell)d\ell. \quad (26)$$

The first two integrals represent flights to the left and to the right which are not truncated by the encounter of a target. The third and fourth represent flights truncated by the encounter of the targets, respectively, at $x = 0$ and $x = \lambda$. In fact, due to the perceptive range or radius of vision, these sites are detected as soon as the walker reaches the respective positions $x = r_v$ and $x = \lambda - r_v$. In addition, since $p(\ell) = 0$ if $|\ell| < \ell_0$, then $\langle|\ell|\rangle(a)$ is given only by the second, third and fourth (first, third and fourth) integrals in the case $r_v < a \leq r_v + \ell_0$ ($\lambda - r_v - \ell_0 \leq a < \lambda - r_v$).

4 Discrete Space Calculation

The exact formal expression (24) can be numerically solved through a spatial discretization of the continuous range $0 \leq x \leq \lambda$. In order to accomplish it, we consider positions x which are multiple of some discretization length Δx, i.e. $x = j\Delta x$, with $j = 0, 1, \ldots, M$ and Δx much smaller than any relevant scale of the problem (ℓ_0, r_v, λ). In this case, the targets at $x = 0$ and $x = \lambda$ are respectively associated with the indexes $j = 0$ and $j = M = \lambda/\Delta x$ (M is the integer number of intervals of length Δx in which the range $0 \leq x \leq \lambda$ is subdivided). Similarly, we define $\ell_0 = m_0\Delta x$ and $r_v = m_r\Delta x$, with m_0 and m_r integers. The continuous limit is recovered by taking $\Delta x \to 0$ and $M \to \infty$, with $\lambda = M\Delta x$ fixed.

Our first aim is to write (16) in the discrete space. First, the set of continuous variables, $\{x_0, x_1, \ldots, x_{n-1}, x_n\}$, denoting, respectively, the position of the searcher after $\{0, 1, \ldots, n-1, n\}$ steps, must be replaced by the following set of discrete indices: $\{i_0, i_1, \ldots, i_{n-1}, i_n\}$, where $x_m = i_m \Delta x$. It thus follows that each integral over a continuous space variable must be changed to a sum over the respective discrete index. The probability $p(x_{m+1}-x_m)dx_m$ of reaching the site x_{m+1} from the site x_m by performing the $(i_m + 1)$-th step of length $|x_{m+1} - x_m| = |i_{m+1} - i_m|\Delta x$ should be replaced by the quantity a_{i_{m+1}, i_m}, to be determined below. With these considerations in mind, (16) can be discretized to

$$[\rho_n]_{i_n} = \sum_{i_0=m_r+1}^{M-m_r-1} \cdots \sum_{i_{n-1}=m_r+1}^{M-m_r-1} a_{i_n,i_{n-1}} a_{i_{n-1},i_{n-2}} \cdots a_{i_2,i_1} a_{i_1,i_0} [\rho_0]_{i_0}. \tag{27}$$

We observe above that $a_{i_m,i_m} = 0$ and $a_{i_{m+1},i_m} = a_{i_m,i_{m+1}}$, since the probabilities of step lengths $x_{m+1} - x_m$ and $x_m - x_{m+1}$ are the same. In addition, we have also taken into account that the lower and upper limits of each integral, respectively $x = r_v$ and $x = \lambda - r_v$, represent extreme positions which must not be considered in the above discrete summation, since at either of these sites the walker already detects a target and gets absorbed.

Notice that (27) has the structure of a sequence of matrix products. Indeed, we can regard the quantities $a_{k,j}$ as the matrix elements $[A]_{k,j}$ of a symmetric matrix A, with null diagonal elements and dimension $(M - 2m_r - 1) \times (M - 2m_r - 1)$ [note that $M - 2m_r - 1 = (M - m_r - 1) - (m_r + 1) + 1$]. Accordingly, $[\rho_m]_{i_m}$ denotes the i_m-th element of the column vector ρ_m of dimension $M - 2m_r - 1$. Equation (27) can thus be written in the form

$$[\rho_n]_{i_n} = \sum_{i_0=m_r+1}^{M-m_r-1} [A^n]_{i_n,i_0} [\rho_0]_{i_0}. \tag{28}$$

We further observe that, since the property $\int_{r_v}^{\lambda-r_v} \delta(x - a)dx = 1$ becomes $\sum_{j=m_r+1}^{M-m_r-1} \delta_{j,i_a} = 1$ in the discrete limit, with the initial position index defined as $i_a = a/\Delta x$, then the Dirac delta relates to the Kronecker delta via

$$\delta(x - a) \rightarrow \frac{\delta_{j,i_a}}{\Delta x}, \tag{29}$$

as

$$dx \rightarrow \Delta j \Delta x = \Delta x. \tag{30}$$

Observe now that by the same procedure (5) becomes, in the discrete limit,

$$[\langle L \rangle]_{i_a} = \sum_{n=0}^{\infty} \sum_{i_n=m_r+1}^{M-m_r-1} [\rho_n]_{i_n} [\langle|\ell|\rangle]_{i_n} \Delta x. \tag{31}$$

In this sense, each element $[\langle L \rangle]_{i_a}$ of the column vector $\langle L \rangle$ of dimension $M - 2m_r - 1$ represents the average distance traversed by the walker starting at a discrete position i_a until reaching one of the borders. By substituting (28) above, we obtain

$$[\langle L \rangle]_{i_a} = \sum_{n=0}^{\infty} \sum_{i_n=m_r+1}^{M-m_r-1} \sum_{i_0=m_r+1}^{M-m_r-1} [A^n]_{i_n,i_0} [\rho_0]_{i_0} [\langle |\ell| \rangle]_{i_n} \Delta x. \tag{32}$$

The assignment of the index i_a appears explicitly in (32) by using (18) and (29). Summing over i_0 and applying the symmetry property of matrix A we obtain

$$[\langle L \rangle]_{i_a} = \sum_{n=0}^{\infty} \sum_{i_n=m_r+1}^{M-m_r-1} [A^n]_{i_a,i_n} [\langle |\ell| \rangle]_{i_n}. \tag{33}$$

Finally, by summing over n we get the discrete equivalent of (24):

$$[\langle L \rangle]_{i_a} = \sum_{i=m_r+1}^{M-m_r-1} [(I - A)^{-1}]_{i_a,i} [\langle |\ell| \rangle]_{i}, \tag{34}$$

where we have renamed the dummy index i_n simply by i. I is the $(M - 2m_r - 1) \times (M - 2m_r - 1)$ unity matrix and $(I - A)^{-1}$ is the inverse of the matrix $(I - A)$.

In (34) we observe that $[\langle |\ell| \rangle]_i$ is determined by first calculating $\langle |\ell| \rangle(x)$ in continuous space from (25) or (26), and next applying the discretization of the parameters x, λ, ℓ_0 and r_v, according to the previous prescription.

At last, we also need to determine the matrix elements $[A]_{k,j}$. We observe that $[A]_{k,j}$ is the discrete representation of the probability $p(x - x')dx'$ of performing a step of length between $|x - x'| = |k - j|\Delta x$ and $|x - x'| + dx' = (|k - j| + 1)\Delta x$. Therefore, by considering

$$P(|x - x'| < |\ell| < |x - x'| + \Delta x) = \int_{|x-x'|}^{|x-x'|+\Delta x} p(\ell)d\ell, \tag{35}$$

its discrete limit implies

$$[A]_{k,j} = [A]_{j,k} = \int_{|k-j|\Delta x}^{(|k-j|+1)\Delta x} p(\ell)d\ell, \qquad k \neq j, \tag{36}$$

and where $[A]_{j,j} = 0$ and $[A]_{k,j} = 0$ if $|k - j| < m_0$ due to the minimum step length ℓ_0. After the matrix elements $[A]_{k,j}$ are calculated for a given pdf $p(\ell)$ of step lengths, one must invert the matrix $(I - A)$ so to determine the average distance $\langle L \rangle$ and the search efficiency η, (34) and (2), respectively, for a searcher starting from $x_0 = a = i_a \Delta x$. In the following we provide explicit calculations for Lévy random searchers.

5 Lévy Random Searchers

In this section we explicit the calculations for a (power-law) Lévy pdf of step lengths.

Our emphasis is on the the mentioned destructive and non-destructive cases, respectively corresponding to set symmetric and asymmetric initial conditions and identified with the starting positions $x_0 = a = \lambda/2$ and $x_0 = a = r_v + \Delta x$, as discussed.

For Lévy random walks in *free* space (i.e., with no a priori spatial truncations), the pdf of step lengths reads

$$p(\ell) = \mathscr{A} \frac{\Theta(|\ell| - \ell_0)}{|\ell|^\mu}, \tag{37}$$

where the theta function $\Theta(|\ell| - \ell_0) = 0$ if $|\ell| < \ell_0$ and $\Theta(|\ell| - \ell_0) = 1$ otherwise, assures the minimum step length ℓ_0. From (1) the normalization constant is given by:

$$\mathscr{A} = \frac{(\mu - 1)}{2} \ell_0^{\mu-1}, \qquad \mu > 1. \tag{38}$$

Actually, the power-law dependence of (37) represents the long-range asymptotical limit of the complete family of Lévy stable distributions of index $\alpha = \mu - 1$ [43,44]. Moreover, as the second moment of pdf (37) diverges for $1 < \mu \leq 3$, the central limit theorem does not hold, and anomalous superduffisive dynamics takes place, governed by the generalized central limit theorem. Indeed, Lévy walks and flights in *free* space are related to a Hurst exponent [43, 44] $H > 1/2$, whereas Brownian behavior (diffusive walks with $H = 1/2$) emerges for $\mu > 3$. In particular, for Lévy walks one finds $H = 1$ for $1 < \mu \leq 2$, with ballistic dynamics emerging as $\mu \to 1$ (the case $\mu = 2$ corresponds to the Cauchy distribution). For $\mu \leq 1$ the function (37) is not normalizable. Therefore, by varying the single parameter μ in (37) the whole range of search dynamics can be accessed (from Brownian to superdiffusive and ballistic).

We emphasize that these results for faster-than-diffusive dynamics hold in free space or, as in the present context, in the free part of the search path between consecutive target encounters. As one considers the total path as a whole, the truncation of steps by the finding of a statistically large number of target sites generates an effective *truncated* Lévy distribution [30], with finite moments and emergence of a crossover towards normal dynamics, as justified by the central limit theorem. This issue is discussed in more detail in Sect. 6.

By substituting (37) and (38) into (26), we find, for $r_v + \ell_0 \leq a \leq \lambda - r_v - \ell_0$,

$$\langle|\ell|\rangle(a) = \frac{(\lambda - a - r_v)}{2} + \frac{\ell_0(1-\mu)}{2(2-\mu)} \left[1 + \frac{((a - r_v)/\ell_0)^{2-\mu}}{1-\mu} \right], \quad 1 < \mu \leq 3, \mu \neq 2, \tag{39}$$

and

$$\langle |\ell| \rangle (a) = \frac{(\lambda - a - r_v)}{2} + \frac{\ell_0}{2} \left[1 + \ln((a - r_v)/\ell_0))\right], \quad \mu = 2. \quad (40)$$

Discrete space expressions associated with (39) and (40) are readily found by following the prescription of Sect. 4:

$$\langle |\ell| \rangle_{\iota_0} = \frac{(M - \iota_0 - m_r)\Delta x}{2}$$
$$+ \frac{m_0 \Delta x (1 - \mu)}{2(2 - \mu)} \left[1 + \frac{((\iota_0 - m_r)/m_0)^{2-\mu}}{1 - \mu}\right], \quad 1 < \mu \le 3, \mu \ne 2, \quad (41)$$

and

$$\langle |\ell| \rangle_{\iota_0} = \frac{(M - \iota_0 - m_r)\Delta x}{2} + \frac{m_0 \Delta x}{2} \left[1 + \ln((\iota_0 - m_r)/m_0))\right], \quad \mu = 2. \quad (42)$$

Moreover, as we mentioned in Sect. 3 (see discussion right after (26)), the results for the remaining intervals ($r_v < a \le r_v + \ell_0$ and $\lambda - r_v - \ell_0 \le a < \lambda - r_v$) can be obtained straightforwardly. Indeed, we quote them below in the continuous and discrete limits. For $r_v < a \le r_v + \ell_0$:

$$\langle |\ell| \rangle (a) = \frac{(a - r_v)}{2} + \frac{\ell_0 (1 - \mu)}{2(2 - \mu)} \left[1 + \frac{((\lambda - a - r_v)/\ell_0)^{2-\mu}}{1 - \mu}\right], \quad 1 < \mu \le 3, \mu \ne 2,$$
$$(43)$$

and

$$\langle |\ell| \rangle (a) = \frac{(a - r_v)}{2} + \frac{\ell_0}{2} \left[1 + \ln((\lambda - a - r_v)/\ell_0))\right], \quad \mu = 2, \quad (44)$$

and their discrete limits:

$$\langle |\ell| \rangle_{\iota_0} = \frac{(\iota_0 - m_r)\Delta x}{2} + \frac{m_0 \Delta x (1 - \mu)}{2(2 - \mu)}$$
$$\times \left[1 + \frac{((M - \iota_0 - m_r)/m_0)^{2-\mu}}{1 - \mu}\right], \quad 1 < \mu \le 3, \mu \ne 2, \quad (45)$$

and

$$\langle |\ell| \rangle_{\iota_0} = \frac{(\iota_0 - m_r)\Delta x}{2} + \frac{m_0 \Delta x}{2} \left[1 + \ln((M - \iota_0 - m_r)/m_0))\right], \quad \mu = 2. \quad (46)$$

For $\lambda - r_v - \ell_0 \le a < \lambda - r_v$:

$$\langle |\ell| \rangle (a) = \frac{\ell_0 (1 - \mu)}{2(2 - \mu)} \left[2 + \frac{((a - r_v)/\ell_0)^{2-\mu}}{1 - \mu}\right.$$
$$\left. + \frac{((\lambda - a - r_v)/\ell_0)^{2-\mu}}{1 - \mu}\right], \quad 1 < \mu \le 3, \mu \ne 2, \quad (47)$$

and

$$\langle|\ell|\rangle(a) = \ell_0[1 + \ln([(\lambda - a - r_v)(a - r_v)]^{1/2}/\ell_0)], \quad \mu = 2, \qquad (48)$$

and their discrete limits:

$$\langle|\ell|\rangle_{\iota_0} = \frac{m_0 \Delta x (1 - \mu)}{2(2 - \mu)} \left[2 + \frac{((\iota_0 - m_r)/m_0)^{2-\mu}}{1 - \mu} \right.$$

$$\left. + \frac{((M - \iota_0 - m_r)/m_0)^{2-\mu}}{1 - \mu} \right], \quad 1 < \mu \leq 3, \mu \neq 2, \qquad (49)$$

and

$$\langle|\ell|\rangle_{\iota_0} = m_0 \Delta x [1 + \ln([(M - \iota_0 - m_r)(\iota_0 - m_r)]^{1/2}/m_0)], \quad \mu = 2. \qquad (50)$$

These small intervals at the extremes of the search space generally only contribute in an important way when small steps are very frequent, as it happens for $\mu \to 3$.

Finally, the matrix A is determined by substituting (37) and (38) into (36), so that

$$A_{ij} = A_{ji} = \frac{1}{2} \left[\frac{1}{|i - j|^{\mu-1}} - \frac{1}{(|i - j| + 1)^{\mu-1}} \right], \quad i \neq j, \quad 1 < \mu \leq 3, \quad (51)$$

with $A_{ii} = 0$ and $A_{ij} = 0$ if $|i - j| < m_0$.

Substitution of the expressions for $\langle|\ell|\rangle_{\iota_0}$ in the respective intervals into (34), along with (51), leads to $\langle L \rangle_{\iota_0}$ and, therefore, also to the efficiency η, (3), in the case of Lévy searches.

Figure 2a, b display the efficiency of the symmetric (destructive) ($a = \lambda/2$ or $\iota_0 = M/2$) and asymmetric (non-destructive) ($a = r_v + \Delta x$ or $\iota_0 = m_r + 1$) cases, respectively.

It is striking the agreement between the analytical Eqs. (3) and (24) or (34) and numerical results. Obtained from simulations which closely resemble the features of the above search model. The optimal search strategy corresponds to ballistic ($\mu \to 1$) and superdiffusive ($\mu \approx 2$) dynamics, for the symmetric and asymmetric conditions respectively, in agreement with previous mathematical approximations [49].

6 Search Diffusivity

One way to characterize the dynamics generated by the search is by determining how the searcher's root-mean-square (r.m.s.) distance R, defined as function of averages of the forager's position x,

$$R \equiv [\langle(\Delta x)^2\rangle]^{1/2} = [\langle x^2 \rangle - \langle x \rangle^2]^{1/2}, \qquad (52)$$

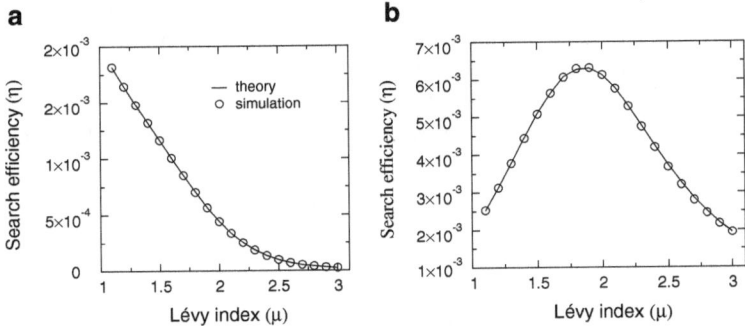

Fig. 2 Search efficiency, η, versus (power-law) Lévy exponent, μ, for both (**a**) symmetric (destructive) and (**b**) asymmetric (non-destructive) initial conditions. In each case, the optimal search strategy respectively corresponds to ballistic ($\mu \to 1$) and superdiffusive ($\mu \approx 2$) dynamics. Notice the striking agreement between the analytical Eqs. (3) and (24) or (34) and numerical results. Simulation parameters: $\Delta x = 0.2$, $r_v = \ell_0 = 1$, $\lambda = 10^3$, $a = \lambda/2$ (symmetric) and $a = 2r_v$ (asymmetric)

depends on time t, number of steps N and number of targets found N_{found}. The asymptotic scaling relation,

$$R \sim t^{\nu} \quad \text{or} \quad R \sim N^{\nu} \quad \text{or} \quad R \sim N_{\text{found}}^{\nu}, \tag{53}$$

implies normal (Brownian) diffusion for the diffusion exponent $\nu = 1/2$, superdiffusion with $\nu > 1/2$, and ballistic dynamics in the case of $\nu = 1$.

Due to the truncation of steps and the long-term prediction of the central limit theorem (see Sects. 3 and 5), we can anticipate that a crossover should occur between two dynamical regimes during a Lévy random search. There is an initial regime with superdiffusive character due to the Lévy pdf of single step lengths, occurring up to the encounter of the first targets [50]. Then, it follows a subsequent Brownian behavior for the overall search trajectory, which, as discussed, is viewed as the concatenated sum of partial paths between consecutive encounters. Indeed, the initial superdiffusive dynamics could not remain as such indefinitely, once the *truncated* Lévy pdf presents well-defined (finite) first and second moments.

At this point we should also observe that if the typical time scale of the search is smaller than the crossover time then the foraging process appears as effectively superdiffusive [7].

In the following we discuss the dynamics of a Lévy random searcher in both regimes, starting with the initial superdiffusive one.

6.1 Characterizing First-Passage-Time Diffusivity

As discussed in Sect. 3, since finding a target either at $x = 0$ or $x = \lambda$ is essentially a mean first-passage-time problem [38], we can initially ask about the r.m.s. distance associated with the encounter events. In other words, we start by considering the quantities $\langle x \rangle_{\text{fpt}}$ and $\langle x^2 \rangle_{\text{fpt}}$, which represent the average positions x and x^2 over all walks departing from $x_0 = a$ and ending either at $x = 0$ or $x = \lambda$ by an encounter event. In fact, by taking into account the radius of vision r_v, the detection of targets occurs at $x = r_v$ and $x = \lambda - r_v$, respectively, so that we can actually write

$$\langle x \rangle_{\text{fpt}} = r_v p_0 + (\lambda - r_v) p_\lambda \tag{54}$$

and

$$\langle x^2 \rangle_{\text{fpt}} = r_v^2 p_0 + (\lambda - r_v)^2 p_\lambda. \tag{55}$$

Above, $p_0(a)$ and $p_\lambda(a)$ denote, respectively, the probabilities for a walker starting at $x_0 = a$ to find the target site at $x = 0$ or $x = \lambda$. Notice that

$$p_0(a) + p_\lambda(a) = 1, \tag{56}$$

since an encounter always happens at the end of the process. By substituting (54)–(56) into the expression

$$R_{\text{fpt}} = [\langle x^2 \rangle_{\text{fpt}} - \langle x \rangle_{\text{fpt}}^2]^{1/2}, \tag{57}$$

we find the correspondent r.m.s. distance of the first time passage at positions $x = 0$ or $x = \lambda$:

$$R_{\text{fpt}} = (\lambda - 2r_v)(p_0 p_\lambda)^{1/2}. \tag{58}$$

It is clear now that the r.m.s. quantities R and R_{fpt} are not the same. In particular, there is no first-passage-time restriction in the calculation of R. Nevertheless, the dynamics of these two quantities are interrelated. As we show below for Lévy random searchers, the diffusion exponent ν is essentially the same for random search walkers restricted to the interval $r_v < x < \lambda - r_v$ and random walkers in free space, for which [18, 42]

$$\nu = \begin{cases} 1, & 1 < \mu < 2; \\ \dfrac{(4 - \mu)}{2}, & 2 < \mu < 3; \\ \dfrac{1}{2}, & \mu > 3. \end{cases} \tag{59}$$

We should stress, however, that no search activity takes place in free space, due to the absence of target sites.

The result of (58) still demands the knowledge of p_0 and p_λ. For such calculation, we consider initially a walker that starts at $x_0 = a$ and reaches the site $x = \lambda$ after n steps. Following the approach [10, 11] of the preceding sections, we write

$$p_{\lambda,n}(a) = \int_{r_v}^{\lambda - r_v} \rho_{n-1}(x_{n-1}) dx_{n-1} P(\ell \geq \lambda - r_v - x_{n-1}). \tag{60}$$

This expression can be understood as follows: first, $\rho_{n-1}(x_{n-1})dx_{n-1}$ represents the probability for the walker to be located in the interval $[x_{n-1}, x_{n-1}+dx_{n-1})$ after $n-1$ steps; since, up to this point, no target has been found yet, then $r_v < x_{n-1} < \lambda - r_v$. Second, we also have to multiply the probability that the next (n-th) step will reach the target at $x = \lambda$ and terminate the walk; so, $P(\ell \geq \lambda - r_v - x_{n-1})$ gives the probability that the n-th step has length $\ell \geq \lambda - r_v - x_{n-1}$, and thus certainly finds the target at $x = \lambda$ (recall that steps of length $\ell > \lambda - r_v - x_{n-1}$ end up truncated). Finally, the integral above sums over all possible values of x_{n-1}, consistently with this reasoning.

Since all walks are statistically independent, the total probability of walks with any number n of steps that start at $x_0 = a$ and terminate at $x = \lambda$ is simply a sum of $p_{\lambda,n}$ over all possibilities:

$$p_\lambda(a) = \sum_{n=1}^{\infty} p_{\lambda,n}(a), \tag{61}$$

that is,

$$p_\lambda(a) = \sum_{n=1}^{\infty} \int_{r_v}^{\lambda-r_v} \rho_{n-1}(x_{n-1})dx_{n-1} P(\ell \geq \lambda - r_v - x_{n-1}). \tag{62}$$

Now, by changing the variable $m = n - 1$, we obtain

$$p_\lambda(a) = \sum_{m=0}^{\infty} \int_{r_v}^{\lambda-r_v} \rho_m(x_m)dx_m P(\ell \geq \lambda - r_v - x_m). \tag{63}$$

Note that the above equation is similar to (14). Thus, from the same procedure detailed in Sect. 3, we find [10, 11]

$$p_\lambda(a) = [(\mathscr{I} - \mathscr{L})^{-1} P(\ell \geq \lambda - r_v - a)]. \tag{64}$$

We now take the discrete limit of (64) by following the general procedure described in Sect. 4. First of all, as before, we set $x = i\Delta x$, where $i = m_r + 1, \ldots, M - m_r - 1$. Equation (64) thus becomes

$$p_{\lambda,i_0} = [(I - A)^{-1} P_{i_0}], \tag{65}$$

where, now, p_{λ,i_0} and P_{i_0} are $(M - 2m_r - 1) \times 1$ column vectors.

To calculate P_{i_0} in (65), we write in the continuous limit

$$P(\ell \geq \lambda - r_v - a) = \int_{\lambda-r_v-a}^{\infty} p(\ell)d\ell, \tag{66}$$

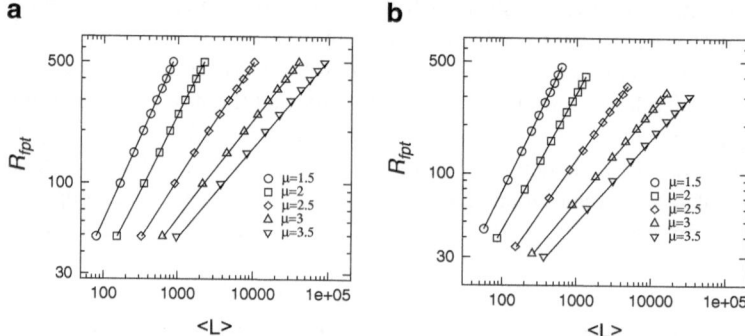

Fig. 3 R.m.s. distance related to first-passage-time diffusivity, R_{fpt}, defined in (57), versus average distance traveled by the searcher between consecutive encounters, $\langle L \rangle$, for both (**a**) symmetric (destructive) and (**b**) asymmetric (non-destructive) initial conditions. Notice the nice agreement between analytical (*solid lines*), Eq. (58), and numerical (*symbols*) results for all values of μ considered. Simulation parameters: $\Delta x = 0.2$, $r_v = \ell_0 = 1$, $\lambda = 10^3$, $a = \lambda/2$ (symmetric) and $a = 2r_v$ (asymmetric). The diffusion exponents $\nu(\mu)$, defined in (69), assume the values shown in (70), in close agreement with the theoretical prediction [18,42] for Lévy walks in free space, (59)

which, after integration, should go through the discretization process, leading to P_{ℓ_0} as function of the discrete settings for a, λ, r_v and ℓ_0. Analogously to the example of Lévy walks in Sect. 5, we obtain in the continuous and discrete limits, respectively,

$$P(\ell \geq \lambda - r_v - a) = \frac{1}{2}\left(\frac{\lambda - r_v - a}{\ell_0}\right)^{1-\mu} \tag{67}$$

and

$$P_{\ell_0} = \frac{1}{2}\left(\frac{M - m_r - \iota_0}{m_0}\right)^{1-\mu}, \tag{68}$$

if $r_v < a \leq \lambda - r_v - \ell_0$ (or $m_r < \iota_0 \leq M - m_r - m_0$), and $P(\ell \geq \lambda - r_v - a) = P_{\ell_0} = 1/2$ otherwise. The same protocol can also be used to calculate $p_0(a)$ (or p_{0,ι_0} in the discrete limit). However, we can always use (56), so that we actually only need to calculate either p_λ or p_0.

In the short-term regime (i.e. first-passage-time diffusivity), we must also comment on the possible validity of (58) to times (or number of steps) in which a boundary target has not been reached yet. In fact, although the calculation described in (54)–(58) explicitly refers to the encounter of extreme sites at $x = 0$ and $x = \lambda$, any two sites at positions $r_v < x_1 < a$ and $a < x_2 < \lambda - r_v$ can be assumed in the mean first-passage-time formulation. Thus, one can actually "follow" the dynamics of the searcher as it passes for the first time at positions x_1 or x_2, apart each other by a distance $(x_2 - x_1)$. In particular, if the ratio between the initial and total distances, $(a - x_1)/(x_2 - x_1) = (a - r_v)/(\lambda - 2r_v)$, is kept fixed, the evolution of R_{fpt} with the average distance traversed can be determined (see below).

In Fig. 3 we compare the prediction of (58) with results from numerical simulation. By considering unity velocity for the Lévy searcher, the average time to encounter a target is identical to the average distance traversed $\langle L \rangle$. Thus, we can actually probe the asymptotic relation,

$$R_{\text{fpt}} \sim \langle L \rangle^{\nu}, \tag{69}$$

for several distances $(x_2 - x_1)$, as discussed above. As Fig. 3 indicates, we have a nice agreement between the analytical and numerical results. Importantly, just as in the case of a Lévy walker in free space [18,42] (i.e. with no targets present, (59)), we identify ballistic, superdiffusive and Brownian short-term regimes in the respective ranges $1 < \mu < 2$, $2 < \mu < 3$ and $\mu > 3$, with (analytical and numerical) diffusion exponents:

$$\nu = \begin{cases} 0.99, & \mu = 1.5; \\ 0.85, & \mu = 2; \\ 0.67, & \mu = 2.5; \\ 0.51, & \mu = 3.5. \end{cases} \tag{70}$$

Observe that, in this case in which searches and encounters are actually being performed, the effect of hitting the boundaries are more pronounced for intermediate values of μ. Indeed, for $\mu \to 1$ and $\mu \to 3$ there is a fair agreement between the values of ν given by (59) and (70). On the other hand, for intermediate $\mu = 2.5$ the value of ν above should be compared with that of the free-space Lévy walker, $\nu = 0.75$.

6.2 Characterizing Search Dynamics Diffusivity

The dynamics of the long-term regime (i.e., after the encounter of a large number of targets, $N_{\text{found}} \gg 1$) can be worked through a suitable random-walk mapping. We describe below such approach for the asymmetric (non-destructive) case, in which the walker starts from a fixed distance $x_0 = a = r_v + \Delta x$ to the closest target, with $\Delta x \ll r_v \ll \lambda$. Generalization for any x_0 is possible.

We start by recalling that the set of targets are placed at positions $x = i\lambda$, where i is an integer (negative, null or positive) number. If the searcher begins the non-destructive walk at $x_0 = a = r_v + \Delta x$, then it can find either the target at $x = 0$ or $x = \lambda$. When the target at $x = \lambda$ is encountered, the forager can restart the search walk from $x = \lambda - r_v - \Delta x$ or $x = \lambda + r_v + \Delta x$ (in both cases, the distance to the closest site at $x = \lambda$ remains $a = r_v + \Delta x$; here we take any of these two possibilities with 50 % probability). After, say, a sequence of $N_{\text{found}} = 5$ encounters, one possible set of visited targets is $\{\lambda, \lambda, 0, -\lambda, -2\lambda\}$. Notice that after the first target (located at $x = \lambda$) is found, the searcher returns to it in the next (i.e. second) encounter, as allowed in the non-destructive case. By recalling that p_0 and p_λ denote, respectively, the probabilities to find the closest and farthest targets, and taking into account the radius of vision r_v, one generally has four possibilities after the encounter of the first target at $x = \lambda$:

1. Restarting from $x = \lambda + r_v + \Delta x$ and detecting the closest site at $x = \lambda + r_v$ (with probability $p_0/2$).
2. Restarting from $x = \lambda + r_v + \Delta x$ and detecting the distant site at $x = 2\lambda - r_v$ (with probability $p_\lambda/2$).
3. Restarting from $x = \lambda - r_v - \Delta x$ and detecting the closest site at $x = \lambda - r_v$ (with probability $p_0/2$).
4. Restarting from $x = \lambda - r_v - \Delta x$ and detecting the distant site at $x = r_v$ (with probability $p_\lambda/2$).

These events correspond to respective displacements $-\Delta x$, $(\lambda - 2r_v - \Delta x)$, Δx and $-(\lambda - 2r_v - \Delta x)$. In the limit $\lambda \gg r_v \gg \Delta x$, the generalization of this result for the possibilities that follow after the encounter of any target leads to a map of the search path onto a *distinct* random walk, which visits the sites $x = i\lambda$ with "steps" of approximate length $s = -\lambda$, 0 or λ, drawn from the pdf

$$\zeta(s) = p_0\delta(s) + \frac{p_\lambda}{2}\delta(s - \lambda) + \frac{p_\lambda}{2}\delta(s + \lambda). \tag{71}$$

Now, from the standard theory of random walks [39], with statistically independent steps taken from a pdf of finite first and second moments such as (71), we write the *actual* r.m.s. distance after $N_{\text{found}} \gg 1$ "steps" (i.e. $N_{\text{found}} \gg 1$ targets found) as

$$R = N_{\text{found}}^{1/2}[\langle s^2 \rangle - \langle s \rangle^2]^{1/2}, \qquad N_{\text{found}} \gg 1, \tag{72}$$

where, by using (71), we find $\langle s \rangle = 0$, reflecting the equiprobability to move left or right after each encounter, and $\langle s^2 \rangle = p_\lambda\lambda^2$, so that

$$R = \lambda p_\lambda^{1/2} N_{\text{found}}^{1/2}, \qquad N_{\text{found}} \gg 1. \tag{73}$$

Note the presence of Brownian dynamics (diffusion exponent $\nu = 1/2$) in the long-term regime, in agreement with the central limit theorem. In 2D or 3D, the rate of convergence to the Brownian diffusive regime may be slower than in 1D. This is so because higher spatial dimensions allow very large steps without encounter truncations. However, if infinite steps would be rigorously allowed, the possibility of non-convergence would exist, even in the long-run. Further analyses are needed to elucidate the robustness of the 1D analysis presented in this section at higher dimensional systems. Also important is to know up to which extent the 1D mapping between random walk steps and target encounters is valid at higher dimensions.

In Fig. 4 we compare the prediction of (73) with results from numerical simulation, with a nice agreement displayed. It is also worth to note that, even though the expected Brownian diffusion exponent, $\nu = 1/2$, arises for all values of μ considered, r.m.s. displacement values are larger for Lévy exponents μ closer to the ballistic ($\mu \to 1$) strategy.

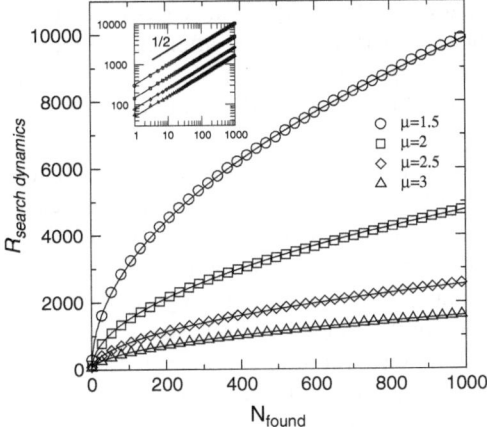

Fig. 4 R.m.s. distance related to the search dynamics diffusivity, R, defined in (52), versus the number of targets found, N_{found}, for asymmetric (non-destructive) initial condition. Notice the nice agreement between analytical (*solid lines*), (73), and numerical (*symbols*) results for all values of μ considered, with Brownian diffusion exponent, $\nu = 1/2$, as predicted by the central limit theorem (see inset). Simulation details: we let 10^4 searchers look for 10^3 targets each. The landscape was configured with 100 targets interspersed by a distance λ. The restarting distance to the last target found is fixed, $a = r_v + \Delta x$. Simulation parameters: $\Delta x = 0.2$, $r_v = \ell_0 = 1$ and $\lambda = 10^3$

One last comment regards the connection of the r.m.s. distance, (73), written as function of the number of targets found, with its time dependence. Indeed, as expected from standard theory of random walks [39], both dependences should be asymptotically described by the same diffusion exponent, as in (53). Indeed, this fact can be justified since, on average, the number of targets found grows linearly with time. Therefore, a Brownian search dynamics, $R \sim t^{1/2}$, also emerges in the long run.

7 Search in Heterogeneous Landscapes: Distributions of Starting Points

Up to this point, no heterogeneity in the targets landscape has been taken into account. In other words, in each search scenario considered (either asymetric or symmetric), the forager has always restarted the search from the same departing position $x_0 = a$. Though useful as limiting cases, these situations generally do not correspond to actual search processes, which usually take place in environments with distances to the last target found distributed according to some pdf. Therefore, we now consider (see [37]) a random search model in which diverse degrees of landscape heterogeneity are taken into account by introducing fluctuations in the starting distances to target sites in a 1D search space, with absorbing boundaries separated by the distance λ. The targets' positions remain fixed at the boundaries of the system. Fluctuations in the starting distances to the targets are introduced by sampling the searcher's departing position after each encounter from a pdf $\pi(x_0)$ of initial positions x_0. Importantly, $\pi(x_0)$ also implies a distribution of starting (a)symmetry conditions with respect to the relative distances between the searcher and the boundary targets.

This approach allows the typification of landscapes that, on average, depress or boost the presence of nearby targets in the search process. Diverse degrees of landscape heterogeneity can thus be achieved through suitable choices of $\pi(x_0)$.

For example, a pdf providing a distribution of nearly symmetric conditions can be assigned to a landscape with a high degree of homogeneity in the spatial arrangement of targets. In this sense, as discussed in Sect. 2, the destructive search represents the fully symmetric limiting situation, with the searcher's starting location always equidistant from all boundary targets. On the other hand, a distribution $\pi(x_0)$ which generates a set of asymmetric conditions is related to a patchy or aggregated landscape. Indeed, in a patchy landscape it is likely that a search process starts with an asymmetric situation in which the distances to the nearest and farthest targets are very dissimilar. Analogously, the non-destructive search corresponds to the highest asymmetric case, in which at every starting search the distance to the closest (farthest) target is minimum (maximum). Finally, a pdf $\pi(x_0)$ giving rise to an heterogeneous set of initial conditions (combining symmetric and asymmetric situations) can be associated with heterogeneous landscapes of structure in between the homogeneous and patchy cases.

More specifically, the limiting case corresponding to the destructive search can be described by the pdf with fully symmetric initial condition,

$$\pi(x_0) = \delta(x_0 - \lambda/2). \tag{74}$$

This means that every destructive search starts exactly at half distance from the boundary targets, just as considered in the previous sections. In this context, it is possible to introduce fluctuations in x_0 by considering, e.g., a Poisson-like pdf exponentially decaying with the distance to the point at the center of the search space, $x_0 = \lambda/2$:

$$\pi(x_0) = A \exp[-(\lambda/2 - x_0)/\alpha], \tag{75}$$

where $r_v < x_0 \leq \lambda/2$, A is the normalization constant and $\pi(x_0) = \pi(\lambda - x_0)$ due to the symmetry of the search space (see also below).

On the other hand, the highest asymmetric (non-destructive) limiting case is represented by

$$\pi(x_0) = \delta(x_0 - r_v - \Delta x), \tag{76}$$

so that every search starts from the point of minimum distance in which the nearest target is undetectable, $x_0 = r_v + \Delta x$, with $\Delta x \ll r_v$, as also previously discussed. Similarly, fluctuations in x_0 regarding this case can be introduced by considering a Poisson-like pdf decreasing with respect to the point $x_0 = r_v$:

$$\pi(x_0) = B \exp[-(x_0 - r_v)/\alpha], \tag{77}$$

where $r_v < x_0 \le \lambda/2$, B is the normalization constant, and $\pi(x_0) = \pi(\lambda - x_0)$. In (75) and (77), the parameter α controls the range and magnitude of the fluctuations. Actually, the smaller the value of α, the less disperse are the fluctuations around $x_0 = \lambda/2$ and $x_0 = r_v$, respectively.

In the case that, instead of always departing from the same location after an encounter, the searcher can restart from any initial position x_0 in the range $r_v < x_0 < \lambda - r_v$ chosen from a pdf $\pi(x_0)$, the fluctuating values of x_0 imply a distribution $w(\langle L \rangle)$ of $\langle L \rangle(x_0)$ values. Therefore, the average distance traversed between two successive target sites becomes

$$\overline{\langle L \rangle} = \int_{\langle L \rangle_{\min}}^{\langle L \rangle_{\max}} \langle L \rangle w(\langle L \rangle) d \langle L \rangle, \tag{78}$$

where $\langle L \rangle_{\min}$ and $\langle L \rangle_{\max}$ denote the minimum and maximum values of $\langle L \rangle(x_0)$, respectively. Notice that

$$w(\langle L \rangle) d \langle L \rangle = \int_{[\langle L \rangle, \langle L \rangle + d \langle L \rangle]} \pi(x_0') dx_0' = \int_{\langle L \rangle}^{\langle L \rangle + d \langle L \rangle} \pi(x_0') \left| \frac{d \langle L \rangle}{dx_0'} \right|^{-1} d \langle L \rangle. \tag{79}$$

Above, the lower symbol in the first integral means that the integrand only contributes to $w(\langle L \rangle)$ if x_0' is associated with a value in the range $[\langle L \rangle, \langle L \rangle + d \langle L \rangle)$. Since searches starting at x_0 are statistically indistinguishable from searches starting at $\lambda - x_0$ (in both cases the closest and farthest targets are at distances x_0 and $\lambda - x_0$ from the starting point), the symmetry of the search space with respect to the position $x = \lambda/2$ implies $\langle L \rangle(x_0) = \langle L \rangle(\lambda - x_0)$. As a consequence, any given value of $\langle L \rangle$ can be always obtained for two distinct starting positions x_0, one in each half of the search interval. By denoting these points as $x_0' = x_{0,A}$ and $x_0' = x_{0,B}$, with $r_v < x_{0,A} < \lambda/2$ and $\lambda/2 < x_{0,B} < \lambda - r_v$, where $x_{0,A} = \lambda - x_{0,B}$, we write

$$w(\langle L \rangle) d \langle L \rangle = \left[\left(\pi(x_0') \left| \frac{d \langle L \rangle}{dx_0'} \right|^{-1} \right)_{x_0' = x_{0,A}} + \left(\pi(x_0') \left| \frac{d \langle L \rangle}{dx_0'} \right|^{-1} \right)_{x_0' = x_{0,B}} \right] d \langle L \rangle, \tag{80}$$

which, when substituted into (78), leads to

$$\overline{\langle L \rangle} = \int_{r_v}^{\lambda/2} \langle L \rangle(x_{0,A}) \pi(x_{0,A}) dx_{0,A} + \int_{\lambda/2}^{\lambda - r_v} \langle L \rangle(x_{0,B}) \pi(x_{0,B}) dx_{0,B}. \tag{81}$$

Next, by dropping the subindices A and B, we obtain

$$\overline{\langle L \rangle} = \int_{r_v}^{\lambda - r_v} \langle L \rangle(x_0) \pi(x_0) dx_0 = 2 \int_{r_v}^{\lambda/2} \langle L \rangle(x_0) \pi(x_0) dx_0, \tag{82}$$

where we have used that $\pi(x_0) = \pi(\lambda - x_0)$, from which the average efficiency with a distribution of starting positions can be calculated:

$$\bar{\eta} = \left(\overline{\langle L \rangle}\right)^{-1}, \tag{83}$$

or explicitly [37],

$$\bar{\eta} = 1/\left(2\int_{r_v}^{\lambda/2} \langle L \rangle(x_0)\pi(x_0)dx_0\right). \tag{84}$$

In order to analyze the effect of fluctuations in the starting distances, the integral (84) must be evaluated. The detailed calculation of $\langle L \rangle(x_0)$ has been described in Sects. 3 and 4 for any pdf $p(\ell)$ of step lengths, and in Sect. 5 for Lévy searchers in particular. Nevertheless, no explicit analytic expression for $\langle L \rangle(x_0)$, (24), is known up to the present. This difficulty can be circumvented by successfully performing a multiple regression, so that

$$\langle L \rangle(x_0) = \sum_{i=0}^{N_x} \sum_{j=0}^{N_\mu} a_{ij} x_0^i \mu^j, \tag{85}$$

as indicated by the nice adjustment shown in Fig. 5c, obtained with $N_x = 10$ and $N_\mu = 8$. Thus, the integral (84) can be done using (75) or (77) and (85). Results are respectively displayed in Fig. 5a, b for several values of the parameter α.

By considering fluctuations in the starting distances to faraway targets through (75), we notice in Fig. 5a that the efficiency is qualitatively similar to that of the fully symmetric condition, (74). Indeed, in both cases the maximum efficiency is achieved as $\mu \to 1$. For $1 < \mu < 3$ the presence of fluctuations only slightly improves the efficiency. These results indicate that ballistic strategies remain robust to fluctuations in the distribution of faraway targets.

On the other hand, fluctuations in the starting distances to nearby targets, (77), are shown in Fig. 5b to decrease considerably the search efficiency, in comparison to the highest asymmetric case, (76). In this regime, since stronger fluctuations increase the weight of starting positions far from the target at $x = 0$, the compromising optimal Lévy strategy displays enhanced superdiffusion, observed in the location of the maximum efficiency in Fig. 5b, which shifts from $\mu_{opt} \approx 2$, for the delta pdf and (77) with small α, towards $\mu_{opt} \to 1$, for larger α (slower decaying $\pi(x_0)$). Indeed, both the pdf of (77) with a vanishing α and (76) are very acute at $x_0 = r_v + \Delta x$.

As even larger values of α are considered, fluctuations in the starting distances to the nearby target become non-local, and (77) approaches the $\alpha \to \infty$ limiting case of the uniform distribution, $\pi(x_0) = (\lambda - 2r_v)^{-1}$ (see Fig. 5b). In this situation, search paths departing from distinct x_0 are equally weighted in (84), so that the dominant contribution to the integral (and to the average efficiency $\bar{\eta}$ as well) comes from search walks starting at positions near $x_0 = \lambda/2$. Since for these walks the most efficient strategy is ballistic, a crossover from superdiffusive to ballistic optimal searches emerges, induced by such strong fluctuations. Consequently, the

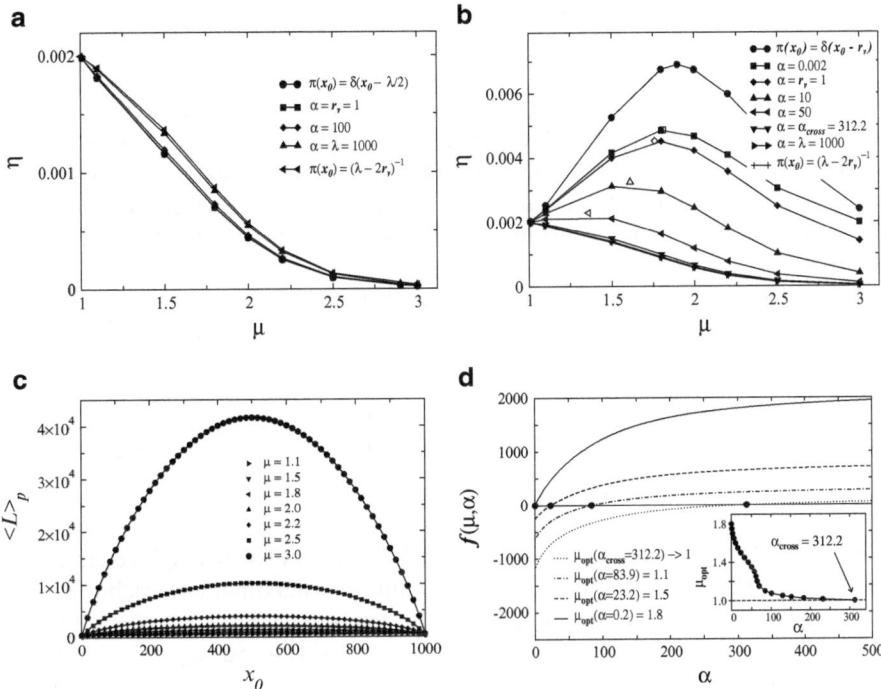

Fig. 5 (**a**) Robustness of the ballistic optimal search strategy with respect to fluctuations in the distances to faraway target sites. In the case of Lévy random searchers, the average search efficiency $\bar{\eta}$, (84), is always highest for $\mu \to 1$ (ballistic dynamics), for any value of the parameter α of the Poissonian fluctuations around the maximum allowed distance, $x_0 = \lambda/2$, (75). Cases with uniform and without any (δ-function) fluctuation are also shown. *Solid lines* are a visual guide. (**b**) Shift in the optimal search strategy towards an enhanced superdiffusive dynamical regime, as landscapes with distinct degrees of heterogeneity are considered. For Lévy random searchers (*solid symbols*), $\bar{\eta}$ is maximized for smaller $\mu_{\mathrm{opt}}(\alpha)$ (faster diffusivity) in the case of wider (larger-α) Poissonian fluctuations in the distances to nearby target sites, (77). *Solid lines* are a visual guide. *Empty symbols* locate the maximum $\bar{\eta}$ obtained from the condition $f(\mu = \mu_{\mathrm{opt}}, \alpha) = \partial \overline{\langle L \rangle} / \partial \mu |_{\mu = \mu_{\mathrm{opt}}} = 0$, with $f(\mu, \alpha)$ given by Eq. (86). (**c**) Nice adjustment of the average distance $\langle L \rangle$ traversed between consecutive findings by a Lévy random searcher starting at position x_0. Results were obtained by numerical discretization of (24) (*solid lines*) and multiple regression (*symbols*), (85). (**d**) Determination of the optimal search strategy of Lévy random searchers with Poissonian fluctuations in the distances to nearby targets, (77). The above mentioned condition provides the optimal Lévy exponent, μ_{opt}, associated with the strategy of maximum average efficiency. Inset: since strategies with $\mu \le 1$ are not allowed (non-normalizable pdf of step lengths), the highest efficiency is always obtained for $\mu \to 1$ as fluctuations with $\alpha > \alpha_{\mathrm{cross}} \approx 312.2$ are considered, marking the onset of a regime dominated by ballistic optimal search dynamics. We took $\lambda = 10^3$ and $r_v = 1$ in plots (**a**)–(**d**)

efficiency curves for very large α (Fig. 5b) are remarkably similar to that of the fully symmetric case (Fig. 5a).

We can quantify this crossover shift in μ_{opt} by defining a function $\mu_{opt}(\alpha)$ that identifies the location in the μ-axis of the maximum in the efficiency $\bar{\eta}$, for each curve in Fig. 5b with fixed α. We notice that eventually a compromising solution with $\mu_{opt}(\alpha) > 1$ cannot be achieved, and an efficiency function $\bar{\eta}$ monotonically decreasing with increasing μ arises for $\alpha > \alpha_{cross}$. In this sense, the value α_{cross} for which such crossover occurs marks the onset of a regime dominated by ballistic optimal search strategies.

The value of μ_{opt} for each α can be determined from the condition $f(\mu = \mu_{opt}, \alpha) = \partial \overline{\langle L \rangle} / \partial \mu |_{\mu = \mu_{opt}} = 0$, so that, by considering (77), (84) and (85),

$$f(\mu, \alpha) = 2A \sum_{i=0}^{N_x} \sum_{j=0}^{N_\mu} a_{ij} \, j \mu^{j-1} \left\{ \sum_{k=0}^{i} \left[\frac{i! \alpha^{k+1}}{(i-k)!} \left(e^{-\alpha r_v} r_v^{i-k} - e^{-\alpha \lambda/2} \left(\frac{\lambda}{2} \right)^{i-k} \right) \right] \right\},$$

$$(86)$$

with $A = \{2\alpha [\exp(-r_v/\alpha) - \exp(-\lambda/(2\alpha))]\}^{-1}$. Solutions are displayed in Fig. 5d and also in Fig. 5b as empty symbols, locating the maximum of each efficiency curve. In addition, the crossover value can be determined through $f(\mu \to 1^+, \alpha = \alpha_{cross}) = 0$. In the case of the pdf (77), we obtain (Fig. 5d) $\alpha_{cross} \approx 312.2$ for $\lambda = 103$ and $r_v = 1$ (regime $\lambda \gg r_v$).

We also note that the scale-dependent interplay between the target density and the range of fluctuations implies a value of α_{cross} which is a function of λ. For instance, a larger λ (i.e., a lower target density) leads to a larger α_{cross} and a broader regime in which superdiffusive Lévy searchers are optimal. In fact, the above qualitative picture holds as long as low target densities are considered.

Moreover, since ballistic strategies lose efficiency in higher dimensional spaces [40] (see Sect. 8), it might be possible that in 2D and 3D the crossover to ballistic dynamics becomes considerably limited. In spite of this, enhanced superdiffusive searches, with $1 < \mu_{opt} < 2$, should still conceivably emerge due to fluctuations in higher-dimensional heterogeneous landscapes.

From these results we conclude that, in the presence of Poissonian-distributed fluctuating starting distances with $\alpha \leq \alpha_{cross}$, Lévy search strategies with faster superdiffusive properties, i.e. $1 < \mu_{opt} \lesssim 2$, represent optimal compromising solutions. In this sense, as local fluctuations in nearby targets give rise to landscape heterogeneity, Lévy searches with enhanced superdiffusive dynamics actually maximize the search efficiency in aggregate and patchy environments. On the other hand, for strong enough fluctuations with $\alpha > \alpha_{cross}$, a crossover to the ballistic strategy emerges in order to access efficiently the faraway region where targets are distributed.

A recent numerical study [23], looking at how different starting positions in the target landscape modify optimal Lévy search strategies, found equivalent results. Our more detailed analysis provides further mechanistic explanation linking landscape features with optimal random search strategies.

8 Discussion

Optimal foraging is one of the most extensively studied optimization theory in ecology and evolutionary biology [25, 26, 28, 46, 48]. To fully develop a comprehensive theory, it is necessary to understand separately (but not isolatedly) the contribution and the evolutionary trade-offs of the different components of the foraging process (e.g., search, taxis, patch exploitation, prey handling, digestion times). Indeed, the foraging sequence [45, 46] can be divided in: pre-encounter (search and taxis), and post-encounter (pursuit, handling, digestion, and ingestion) events [8, 12, 45].

The framework developed here is focused exclusively on the search component of the foraging process. SOFT clearly illustrates that neither ballistic nor Lévy strategies should be considered as universal [22, 36], since realistic fluctuations in the targets distribution may induce switches between these two regimes. Most importantly, SOFT shows which basic elements need to be accounted for to perform efficient stochastic searches, and identifies why optimal solutions can change with the environment conditions. In particular, the theory demonstrates that the quantities $\langle L \rangle$ and $\langle |\ell| \rangle$ depend on the initial position of the searcher in the landscape (Sect. 3). Indeed, the initial searcher-to-target distances are essential to determine the best stochastic strategies (Sect. 7). When nearby and faraway targets exist, the trade-off between either scanning the closeby targets or exploring new territory for faraway ones needs to be efficiently solved. In such scenarios, stochastic laws governing *run and tumble* movement patterns come into play and have a clear impact on the search success, with Lévy-like features becoming beneficial [2, 5, 49, 51]. On the other hand, if such a trade-off does not exist, with all the targets being in average faraway, tumbling is not necessary at all, and ballistic strategies emerge as optimal.

In the overall search process, the diffusive properties of the searcher (Sect. 6) need to be considered both between targets (i.e. first passage times) and after encounter dynamics (i.e. flight truncation and reorientation due to encounter). Two regimes exist, a short-term superdiffusive optimal one, between encounters of successive targets, and a long-term Gaussian diffusive one, which describes the sum of the partial paths to find the targets. It is the balance between such diffusivities that sets the conditions for the emergence of particular optimal strategies.

Our main results are expected to hold in larger dimensions [6, 37, 49]. Although the random search problem in 1D is far from a mean field approximation, most of the results are qualitatively valid in higher-dimensions [6]. Indeed, in 2D and 3D the finding of targets does occur with much lower probability: the extra spatial directions yield a larger exploration space, thus smaller efficiencies. However, the straight lines, during single search steps, are *radial* 1D locomotions, hence, many of the properties observed in a 1D random search are kept at higher dimensions. Moreover, note that the proliferation of directions to go in high dimensional systems, can decrease the effect of fluctuations in target avaliability. In 2D and 3D systems, the influence of $\pi(x_0)$ on optimal search strategies is expected to be qualitatively similiar but weaker than in 1D systems. Being simple enough to allow a general and complete analysis, the 1D case is thus very useful in establishing maximum limits for the influence of landscape spatio-temporal heterogeneities in random search processes.

Further extensions of the theory should involve a better understanding of the landscape structure contribution to the search efficiency [37] and to the build up of the efficiency curves. In the latter case we can consider the encounter efficiency of nearby and faraway targets separately in order to identify the different contribution of each partial efficiency to the total efficiency. We believe that exciting theoretical advances will be achieved by adequately scaling foraging processes in time and space, seeking a fertile middle ground between the concepts derived from classic OFT and the ones coming from the here presented Stochastic Optimal Foraging Theory (SOFT).

Acknowledgements We thank CNPq, CAPES, FACEPE (Brazilian agencies), the Spanish Ministry of Science and Innovation (RyC-2009-04133 and BFU2010-22337) for financial support. FB specially thanks A. Oltra for helping in the edition of this chapter.

References

1. D.J. Anderson, Optimal foraging and the traveling salesman. Theor. Popul. Biol. **24**, 145–159 (1983)
2. F. Bartumeus, M.G.E. da Luz, G.M. Viswanathan, J. Catalan, Animal search strategies: a quantitative random-walk analysis. Ecology **86**, 3078–3087 (2005)
3. F. Bartumeus, Lévy processes in animal movement: an evolutionary hypothesis. Fractals **15**, 151–162 (2007)
4. F. Bartumeus, J. Catalan, Optimal search behavior and classic foraging theory. J. Phys. A: Math. Theor. **42**, 434002 (2009)
5. F. Bartumeus, S.A. Levin, Fractal intermittence in locomotion: linking animal behavior to statistical patterns of search. Proc. Natl. Acad. Sci. USA **105**, 19072–19077 (2008)
6. F. Bartumeus, P. Fernández, M.G.E. da Luz, J. Catalan, R.V. Solé, S.A. Levin, Superdiffusion and encounter rates in diluted, low dimensional worlds. Eur. Phys. J. Spec. Top. **157**, 157–166 (2008)
7. F. Bartumeus, J. Catalan, G.M. Viswanathan, E.P. Raposo, M.G.E. da Luz, The influence of turning angles on the success of non-oriented animal searches. J. Theor. Biol. **252**, 43–55 (2008)
8. W.J. Bell, *Searching Behaviour: The Behavioural Ecology of Finding Resources* (Cambridge University Press, Cambridge, 1991)
9. O. Bénichou, C. Loverdo, M. Moreau, R. Voituriez, Two-dimensional intermittent search processes: an alternative to Lévy flight strategies. Phys. Rev. E **75**, 020102(R) (2006)
10. S.V. Buldyrev, S. Havlin, A.Ya. Kazakov, M.G.E. da Luz, E.P. Raposo, H.E. Stanley, G.M. Viswanathan, Average time spent by Lévy flights and walks on an interval with absorbing boundaries. Phys. Rev. E **64**, 041108 (2001)
11. S.V. Buldyrev, M. Gitterman, S. Havlin, A.Ya. Kazakov, M.G.E. da Luz, E.P. Raposo, H.E. Stanley, G.M. Viswanathan, Properties of Lévy fights on an interval with absorbing boundaries. Physica A **302**, 148–161 (2001)
12. G.C. Collier, C.K. Rovee-Collier, A comparative analysis of optimal foraging behavior: laboratory simulations, in *Foraging Behavior: Ecological, Ethological, and Psychological Approaches*, ed. by A.C. Kamil, T.D. Sargent. Garland Series in Ethology (Garland STPM Press, New York, 1981)

13. I.C. Cuthill, P. Haccou, A. Kacelnick, J.R. Krebs, Y. Iwasa, Starlings exploiting patches: the effect of recent experience on foraging decisions. Anim. Behav. **40**, 625–640 (1990)
14. I.C. Cuthill, P. Haccou, A. Kacelnick, Starlings *Sturnus vulgaris* exploiting patches: response to long-term changes in travel time. Behav. Ecol. **5**, 81–90 (1994)
15. M.G.E. da Luz, S. Buldyrev, S. Havlin, E. Raposo, H. Stanley, G.M. Viswanathan, Improvements in the statistical approach to random Lévy flight searches. Physica A **295**, 89–92 (2001)
16. M.G.E. da Luz, A. Grosberg, E.P. Raposo, G.M. Viswanathan (eds.), The random search problem: trends and perspectives. J. Phys. A **42**(43), 430301 (2009)
17. A.S. Ferreira, E.P. Raposo, G.M. Viswanathan, M.G.E. da Luz, The influence of the Lévy random search efficiency: fractality and memory effects. Physica A **391**, 3234–3246 (2012)
18. T. Geisel, J. Nierwetberg, A. Zacherl, Accelerated diffusion in Josephson junctions and related chaotic systems. Phys. Rev. Lett. **54**, 616–619 (1985)
19. L.-A. Giraldeau, Solitary foraging strategies, in *Behavioural Ecology*, ed. by E. Danchin, L.-A. Giraldeau, F. Cézilly (Oxford University Press, Oxford, 2008)
20. R.F. Green, Bayesian birds: a simple example of Oaten's stochastic model of optimal foraging. Theor. Popul. Biol. **18**, 244–256 (1980)
21. Y. Iwasa, M. Higashi, N. Iamamura, Prey distribution as a factor determining the choice of optimal foraging strategy. Am. Nat. **117**, 710–723 (1981)
22. A. James, M.J. Planck, R. Brown, Optimizing the encounter rate in biological interactions: ballistic versus Lévy versus Brownian strategies. Phys. Rev. E **78**, 051128 (2008)
23. A. James, M.J. Planck, A.M. Edwards, Assessing Lévy walks as models of animal foraging. J. R. Soc. Interface **8**(62), 1233–1247 (2011). doi:10.1098/rsif.2011.0200
24. A. Kacelnick, I.A. Todd, Psychological mechanisms and the marginal value theorem: effect of variability in travel time on patch exploitation. Anim. behav. **43**, 313–322 (1992)
25. A.C. Kamil, T.D. Sargent (eds.), in *Foraging Behavior: Ecological, Ethological, and Psychological Approaches*. Garland Series in Ethology (Garland STPM Press, New York, 1981)
26. A.C. Kamil, J.R. Krebs, H.R. Pulliam (eds.), *Foraging Behavior* (Plenum Press, New York, 1987)
27. T. Koren, M.A. Lomholt, A.V. Chechkin, J. Klafter, R. Metzler, Leapover lengths and first passage time statistics for Lévy flights. Phys. Rev. Lett. **99**, 160602 (2007)
28. J.R. Krebs, N.B. Davies, *An Introduction to Behavioural Ecology*, 3rd edn. (Blackwell, Oxford, 1993)
29. M.A. Lomholt, T. Koren, R. Metzler, J. Klafter, Lévy strategies in intermittent search processes are advantageous. Proc. Natl. Acad. Sci. USA **105**, 1055–1059 (2008)
30. R.N. Mantegna, H.E. Stanley, Phys. Rev. Lett. **73**, 2946–2949 (1994)
31. J.M. McNamara, Optimal patch use in a stochastic environment. Theor. Popul. Biol. **21**, 269–288 (1982)
32. J.M. McNamara, Memory and the efficient use of information. J. Theor. Biol. **125**, 385–395 (1987)
33. R. Metzler, J. Klafter, The random walk's guide to anomalous diffusion: a fractional dynamics approach. Phys. Rep. **339**, 1–77 (2000)
34. A. Oaten, Optimal foraging in patches: a case for stochasticity. Theor. Popul. Biol. **12**, 263–285 (1977)
35. A. Okubo, S.A. Levin, *Diffusion and Ecological Problems: Modern Perspectives* (Springer, Berlin/Plenum Press, New York, 2001)
36. M.J. Plank, A. James, Optimal foraging: Lévy pattern or process? J. R. Soc. Interface **5**, 1077–1086 (2008)
37. E.P. Raposo, F. Bartumeus, M.G.E. da Luz, P.J. Ribeiro-Neto, T.A. Souza, G.M. Viswanathan, How landscape heterogeneity frames optimal diffusivity in searching processes. PLOS Comput. Biol. **7**(11), e10002233 (2011)
38. S. Redner, *A Guide to First-Passage Processes* (Cambridge Univeristy Press, Cambridge, 2001)
39. F. Reif, *Fundamentals of Statistical and Thermal Physics (Chap. 1)* (McGraw-Hill, Singapore, 1985)

40. A.M. Reynolds, F. Bartumeus, Optimising the success of random destructive searches: Lévy walks can outperform ballistic motions. J. Theor. Biol. **260**, 98–103 (2009)
41. A.M. Reynolds, A.D. Smith, R. Menzel, U. Greggers, D.R. Reynolds, J.R. Riley, Displaced honey bees perform optimal scale-free search flights. Ecology **88**, 1955–1961 (2007)
42. M.F. Shlesinger, J. Klafter, Comment on accelerated diffusion in Josephson junctions and related chaotic systems. Phys. Rev. Lett. **54**, 2551 (1985)
43. M.F. Shlesinger, J. Klafter, Lévy walks versus Lévy flights, in *On Growth and Form*, ed. by H.E. Stanley, N. Ostrowsky (Nijhoff, Dordrecht, 1986)
44. M.F. Shlesinger, G. Zaslavsky, U. Frisch (eds.), *Lévy Flights and Related Topics in Physics* (Springer, Berlin, 1995)
45. T.W. Shoener, Theory of feeding strategies. Ann. Rev. Ecol. Syst. **2**, 369–404 (1971)
46. T.W. Shoener, A brief history of optimal foraging ecology, in *Foraging Behavior*, ed. by A.C. Kamil, J.R. Krebs, H.R. Pulliam (Plenum Press, New York, 1987)
47. D.W. Stephens, J.R. Krebs, *Foraging Theory* (Princeton University Press, Princeton, 1986)
48. D.W. Stephens, J.S. Brown, R.C. Ydenberg, *Foraging: Behavior and Ecology* (The University Chicago Press, Chicago, 2007)
49. G.M. Viswanathan, S. Buldyrev, S. Havlin, M.G.E. da Luz, E. Raposo, H. Stanley, Optimizing the success of random searches. Nature **401**, 911–914 (1999)
50. G.M. Viswanathan, E.P. Raposo, F. Bartumeus, J. Catalan, M.G.E. da Luz, Necessary criterion for distinguishing true superdiffusion from correlated random walk processes. Phys. Rev. E **72**, 011111(6) (2005)
51. G.M. Viswanathan, M.G.E. da Luz, E.P. Raposo, H.E. Stanley, *The Physics of Foraging: An Introduction to Random Searches and Biological Encounters* (Cambridge University Press, Cambridge, 2011)

Lévy or Not? Analysing Positional Data from Animal Movement Paths

Michael J. Plank, Marie Auger-Méthé, and Edward A. Codling

Abstract The Lévy walk hypothesis asserts that the optimal search strategy for a forager under specific conditions is to make successive movement steps that have uniformly random directions and lengths drawn from a probability distribution that is heavy-tailed. This idea has generated a huge amount of interest, with numerous studies providing empirical evidence in support of the hypothesis and others criticising some of the methods employed in these. The most common method for identifying Lévy walk behaviour in movement data is to fit a set of candidate distributions to the observed step lengths using maximum likelihood methods. Commonly used candidate distributions are the exponential distribution and the power-law (Pareto) distribution, both on an infinite and a finite (truncated) range. Data sets for which the relative fit of a power-law distribution is better than that of an exponential are typically classified as Lévy walks. However, the movement pattern of the Lévy walk is similar to that of an animal that switches between two behavioural modes in a composite correlated random walk (CCRW) movement process. Recent studies have shown that standard approaches can misidentify the CCRW process as a Levy walk. This misidentification can be due to the methods used to sample and process the data, a failure to assess the absolute fit of the candidate distributions, or the lack of a strong alternative model. In this chapter, we simulate a CCRW process and show that including a composite exponential distribution in the set of candidate

M.J. Plank
Department of Mathematics and Statistics, University of Canterbury, Christchurch, New Zealand
e-mail: michael.plank@canterbury.ac.nz

M. Auger-Méthé
Department of Biological Sciences, University of Alberta, Edmonton, Canada
e-mail: marie.auger-methe@ualberta.ca

E.A. Codling (✉)
Departments of Mathematical Sciences and Biological Sciences, University of Essex, Colchester, UK
e-mail: ecodling@essex.ac.uk

M.A. Lewis et al. (eds.), *Dispersal, Individual Movement and Spatial Ecology*, Lecture Notes in Mathematics 2071, DOI 10.1007/978-3-642-35497-7_2, © Springer-Verlag Berlin Heidelberg 2013 distributions can alleviate the problem of misidentification. However, in some cases sampling and processing of the CCRW data can cause a power-law distribution to have a better fit than a composite exponential. In such cases, the absolute goodness-of-fit of the power-law distribution is typically poor, indicating that none of the candidate distributions are a good model for the data. We discuss the relevance of these results for the analysis of empirical movement data.

1   Introduction

1.1   Lévy Walks

The Lévy walk hypothesis, originally posed by Klafter et al. and Cole [19, 34] and subsequently by Viswanathan et al. [64], asserts that the optimal search strategy for a forager with limited perceptive range and no prior knowledge of the distribution of food in the environment is to move according to a Lévy walk (LW). This means taking a series of steps of (uniformly) random direction and of length l drawn from a probability distribution that is heavy-tailed, meaning that it does not have finite variance [55]. The most commonly used such distribution is the Pareto distribution, with probability density function

$$p(l) = C l^{-\mu}, \qquad l > l_{\min}, \tag{1}$$

where $1 < \mu \leq 3$ and C is a normalization constant given by $C = (\mu - 1)l_{\min}^{\mu-1}$. Note that, in genral, there may also be steps of length $l < l_{\min}$, but the distribution of these step lengths is not important and it is the power-law tail described by (1) that characterises a Lévy walk.

In foraging models, steps are truncated at points where the forager finds a food item [64] and so the search strategy is sometimes referred to as a truncated Lévy walk (TLW). Note that a 'pure' Lévy walk where steps are not truncated after encounters is no more efficient as a search strategy than movement in a straight line, as demonstrated in [10]. Similarly, almost all theoretical LW search models rest on the assumption that each new search begins with a food item just outside the perceptive range of the forager [64]. This can be thought of as representing a highly patchy distribution of food. However, the advantage of any LW strategy is rapidly diminished when each search begins with the nearest food item significantly further away than the forager's perceptive range [31, 51].

The original Lévy hypothesis was motivated by the presence of heavy-tailed power-law distributions in empirical movement data from plankton [34], fruit flies [19] and albatrosses [64]. The Lévy walk hypothesis has since generated a huge amount of interest, with numerous studies providing empirical evidence in support of the hypothesis [1, 3, 6, 15, 20, 30, 40, 48, 53, 57], and others criticising some of the methods employed in these [22–24, 56]. In tandem with the empirical evidence, several studies have investigated the efficiency of Lévy walks in various theoretical search scenarios. For recent reviews of Lévy walks as models of animal movement, see [31] and [67].

A key property of a non-truncated Lévy walk is that it is scale-free, meaning that the sampling scale used by the observer should not affect the observed properties [52]. In particular, a Lévy walk (LW) is known to be superdiffusive at all scales [65, 66]. However, a truncated Lévy walk cannot be truly scale-free at all spatial scales given the truncation inherent in the process when food items are encountered. Similarly, when considering an environment of finite size or the upper limit to the speed an animal can achieve, it is not possible to have arbitrarily large step lengths as could (theoretically) be generated in both the pure LW and the TLW. Hence the scale-free nature of Lévy walks is perhaps over-emphasised and looking for scale-free characteristics in observed movement data may not be a reliable way of detecting Lévy walk behaviour.

1.2   Correlated Random Walks and Composite Strategies

A more classical approach for modelling movement behaviour is the correlated random walk (CRW), in which there is some directional persistence from one step to the next [18, 42]. In a CRW, step lengths are drawn from a distribution with finite mean and variance, such as an exponential distribution. An important property of such distributions is that they satisfy the conditions of the central limit theorem, which implies that the random walk is diffusive in the long-term. Changes in direction between successive steps are not uniformly distributed, but are drawn from some circular distribution that is typically peaked about 0, for example the von Mises distribution [39]. The more concentrated this turning angle distribution is about 0, the more directional persistence the CRW will have in the short term.

In contrast to a Lévy walk, a simple CRW is not scale-free and the sampling rate used by the observer is known to have a significant effect on the apparent properties of the movement pattern in a CRW [14, 16, 28]. Although CRWs are always diffusive over sufficiently long timescale, they can appear superdiffusive over short timescales, depending on the level of persistence in the movement [5, 63, 65]. In this context, it should be noted that a TLW will also appear diffusive at large timescales due to the truncation of long steps.

The basic CRW essentially assumes that movement is modelled as a stationary process, meaning that the parameters governing the persistence in movement do not change with time or space. However, many animals have been observed to display intermittent behaviour, where the forager's movements consist of a mixture of movement strategies (possibly different types of CRWs) [7, 32, 35, 37, 41]. One approach to modelling this behaviour is to use a composite random walk. This is a random walk consisting of more than one distinct behavioural phase, e.g. an extensive phase, in which the forager covers large distances with relatively little turning, and an intensive phase, in which the forager searches a smaller area with a more tortuous path. (Intensive searching is sometimes termed an area-restricted search [36].) A composite random walk model was first proposed as an alternative to the Lévy walk strategy by Bénichou et al. [12, 13]. Benhamou [10] proposed a composite two-phase random walk model for a forager searching for food in a patchy food environment based on memory of encounters. In the first phase, termed the intensive phase, the forager moves according to a Brownian random walk. If after a predetermined amount of time (called the giving-up time) the forager has not located a food item, it switches to a ballistic (straight line movement) strategy until it finds a food item. It then reverts to the intensive strategy to begin the next search. Benhamou [10] showed that this simple composite strategy can be more efficient (i.e. the mean distance travelled between food items is lower) than a truncated Lévy walk.

Reynolds [49, 50] subsequently showed that in certain contexts an even higher efficiency could be obtained by switching to a TLW in the second phase of the composite process, rather than to ballistic motion (which can be viewed as a special case of a LW with $\mu \to 1$). Bartumeus and Levin [5] considered a "Lévy modulated" correlated random walk (CRW), where random reorientations, which break the short-term directional persistence of the CRW, occur after periods of time drawn from a power-law distribution. It was shown that this can increase the efficiency of the search strategy in certain contexts.

1.3 Determining Movement Processes from Observational Data

Given the current interest and the potential implications of Lévy walks being observed in real animal movement data, it is important to be able to determine robustly that: (i) the observed data set is well represented by a heavy-tailed distribution, and (ii) that the movement mechanism giving rise to this observed pattern is actually a LW process and not some other mechanism. It is the interpretation and validity of these two points across a range of studies that had caused much of the current controversy and discussion in the recent movement ecology literature. For example, with respect to (i), [22, 24] and [23] have demonstrated that many (but not all) of the recent studies that have reported LW behaviour in animal movement data may have been flawed or have wrongly interpreted the data. Similarly, with respect to (ii), a number of recent studies have shown that there are variety of movement mechanisms far removed from a LW that can give rise to heavy-tailed or scale-free characteristics in empirical observations of the movement process [17, 27, 31, 44, 45, 49]. Hence, although the the Lévy walk may be a suitable phenomenological description for a wide variety of movement processes, it may have limited relevance as an underlying mechanistic process in all but the simplest of biological scenarios.

2   Sampling and Processing of Movement Path Data

Standard techniques for analysing movement data are usually based on an arbitrary (spatial or temporal) discretization of the observed movement path [8, 9, 14, 16, 33, 59]. This discretization may be due to experimental constraints (as discussed in the next sections) or may be deliberate in order to determine how particular path properties change at different scales [e.g. 16, 28]. By considering features such as the distributions of turning angles and step lengths across the movement path, it is possible to determine the most likely underlying behavioural process(es) that generate the observed pattern, e.g. distinguishing taxis from kinesis [11, 18]. More sophisticated statistical techniques such as hidden Markov models (HMMs) and state-space models (SSMs), as reviewed in [43], have recently been developed to directly infer underlying behavioural processes and parameters from movement data sets. In contrast, many of the recent studies that look for LW characteristics in observed movement data are often based only on a simple analysis of the observed step lengths where a power-law and exponential distribution are the only candidate models considered in a maximum likelihood test [2, 22, 23].

2.1   Discrete Time Sampling

Animal foraging paths are often observed by recording the forager's position at equally spaced time intervals [e.g. 3, 40, 48]. The distances between successive positional fixes are then used to provide a sample of observed step lengths. Much of the empirical evidence for the Lévy walk hypothesis stems from fitting probability distributions to step length data obtained in this way and testing whether a power-law distribution provides a better fit than other candidate distributions. However, it is important to understand how the sampling rate imposed by the observer may affect the data that is subsequently generated. In most empirical studies the aim is to collect data with as high a resolution as possible and this sampling rate is, therefore, not imposed by choice but is due to experimental or technological limitations, e.g. a limited number of signals per day from a GPS tracker.

In such studies, the forager's path is sampled at discrete time points and this imposes a sampling scale on the random walk. Although true LW are scale-invariant, the question of whether truncated LW, composite CRW, or indeed other random search models are invariant to the sampling scale used by the observer has received relatively little attention. Reynolds [49] looked at the effect of subsampling a LW and analysing the rediscretised data. However, this study only examined the difference in the value of the exponent between the original step length distribution and the distribution fitted to the rediscretised data. No comparison was made between the power-law distribution and any other candidate distribution, nor was the dependence on the sampling rate or the exponent of the underlying LW investigated.

Studies on the scale-invariant properties of movement paths, in particular the scaling of mean squared displacement with respect to time, have compared LW to simple CRW observed over different time scales [5, 65], but have not looked at composite CRW.

Plank and Codling [45] considered a composite CRW model in which the forager has two behavioural phases: an extensive phase characterised by a high speed and low tortuosity; and an intensive phase characterised by low speed and high tortuosity. The forager's position was recorded using a range of different sampling frequencies. A power-law distribution and an exponential distribution were fitted to the resulting data and the relative goodness-of-fit of these two distributions compared. It was shown that the sampling scale can have a dramatic effect on the observed data and that this standard fitting method can produce potentially misleading results. At certain sampling scales, the composite CRW model (where the original step length distribution is not heavy-tailed) can produce data for which a power-law distribution fits better than an exponential distribution. This occurs more frequently when there are significant differences between the movement characteristics of the two phases. Plank and Codling [45] also simulated truncated Lévy walks and found that, whilst less sensitive to sampling scale, these can produce step length data for which an exponential distribution fits better than a power law. Similarly, Codling and Plank [17] demonstrated how the use of different sampling scales can cause the step length distribution in different types of movement data to fit a power-law better than an exponential distribution. This applies particularly to data from a set of CRWs with different levels of persistence or from a three-dimensional CRW viewed in one dimension.

2.2 Identification of Turning Points

A further issue with movement path sampling is that the sampling points do not necessarily correspond to actual turning or decision-making points in the underlying movement process. The discretisation of the movement path and the subsequent position of the turning points within the observed path is essentially an arbitrary choice imposed by the observer [8, 14] and hence it is often difficult to interpret true biological meaning from any subsequent analysis of the discretised path. To overcome this issue, attempts have been made to identify turning events as points where the forager undergoes a significant change of direction. There are many different ways in which this can be done. The most common is based on the method of [60] of splitting an observed path into a series of straight-line moves. The direction is monitored at each sampling point and a turning event is registered if the current direction deviates from the direction at the previous turning event by more than a specified threshold angle. This is a "non-local" or "cumulative" identification method as it allows a gradual accumulation of change in direction to be eventually registered as a turning event (Fig. 1a). An alternative is to use a "local" identification method, which only considers the change in direction

a   **b**

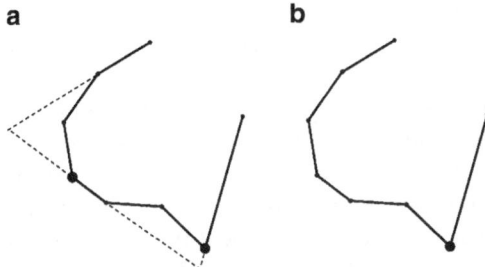

Fig. 1 Diagram illustrating: (**a**) non-local turn identification; (**b**) local turn identification. Identified turning points are indicated by *large solid circles*. With non-local turn designation, turns are identified when the cumulative change of direction from the previous turning event exceeds some threshold angle θ_0. In contrast, local turning identifies turns when the angle between two successive observed random walk steps exceeds some threshold angle θ_0. In this example a threshold angle of $90°$ is used. A gradual change in movement direction is not identified as a turning event using the local turn designation. Note that with both processing methods, small turns are removed from the data set, while the number of observed steps decreases but the lengths of these observed steps typically increase

between successive observations [17,53] (Fig. 1b). However, this method is sensitive to the sampling scale imposed by the observer. For example, at a high-resolution sampling scale, a local method may fail to identify large changes in direction if they are spread over several observations. For this reason, we will focus primarily on the non-local turn identification method in the rest of this chapter. The identification of turning events can be thought of as a post-processing step on sub-sampled data of the type considered by Turchin [58]. The straight-line distances between the identified turning points become the new "step lengths" and these data can be analysed using standard statistical methods. Reynolds and Frye [53] used a local and a non-local turn identification method to infer movement mechanisms from tracking data for honey bees. An alternative method for identifying turning points was used by de Jager et al. [20]: the autocorrelation between movement directions was monitored over a number of successive steps and once this autocorrelation reduced below some pre-defined threshold level, a turning point was identified. Results produced using this method are qualitatively similar to the non-local turn identification described above.

Codling and Plank [17] considered the effect of turn identification on data from a composite CRW model. The threshold angle used to identify turning events is essentially arbitrary and the sensitivity of results to this parameter, as well as to the sampling scale, was investigated. The results of [17] show that turn-identification can alter the results of a relative goodness-of-fit test of the power-law and exponential distributions. In some scenarios it was shown that the less sensitive the turn identification method used (i.e. the larger the threshold angle for registering a turn), the more likely the relative test is to favour a power-law distribution.

Note that processing of the movement data set to identify turning points using any of the methods described above will typically remove a large number of small turns from the data set; these small turns are usually assumed to be noise or minor directional corrections that do not correspond to a global reorientation event in the movement path [e.g. 6]. However, this sort of processing will clearly produce a non-uniform distribution of turning angles since small turning angles will have been removed. In contrast, a theoretical LW or TLW should have a uniform distribution of turning angles. However, as we show later, a true LW processed in the above manner would also produce a non-uniform distribution of turning angles. Hence, it remains unclear whether studies that identify a power-law distribution of step lengths should be classified as Lévy walks if the turning angles are non-uniform; there does not currently appear to be a robust method of using turning angle data to help identify LW patterns in movement data.

3 Analysing Data from a Composite Correlated Random Walk

In this section, we consider data generated using a composite CRW model and post-processed using a range of sampling scales and threshold angles for turn identification as described above. Motivated by the discussion in [2] and [46], our aim is to determine whether a composite exponential distribution will fit the data generated from a composite CRW better than a simple exponential or a power-law distribution, given that the data is sampled and processed in a similar way to [45] and [17]. In particular, we are interested to see if there are certain sampling and processing scenarios where the generated data fits a power-law distribution better (which may consequently be (mis)interpreted as the data having come from a Lévy walk process).

We also examine the distribution of observed turning angles, i.e. the changes in direction between successive sample points. This is motivated by the fact that LW have a uniform distribution of turning angles, whereas CRW (or any movement process incorporating directional persistence) have a turning angle distribution that is peaked around zero. In principle, this theoretical difference in the turning angle distribution could be used to distinguish LW from other movement processes and we investigate the efficacy of turning angle tests.

3.1 The Composite Correlated Random Walk Model

We use the composite CRW model of [45], which consists of two phases (note that other alternative models for a composite random walk are possible and we only consider a very simple case here). Phase 1 (the intensive phase) is characterised by small mean step length and a lack of directional persistence. Phase 2 (the extensive phase) is characterised by high mean step length and high directional persistence. At each step in phase i, the forager has a fixed probability $p_{\text{switch},i}$ of switching to

Table 1 Parameter values for composite CRW model

Parameter	Phase 1 value	Phase 2 value
Mean step length $\bar{l}$	1	10
Turning angle concentration κ	0	50
Probability of switching phases p_{switch}	0.01	0.01

Step lengths l are drawn from an exponential distribution with mean $\bar{l}$: $p(l) = \exp(-l/\bar{l})$. Turning angles ϕ are drawn from a zero-centred von Mises distribution with concentration parameter κ: $p(\phi) = Ce^{\kappa \cos \phi}$, where C is a normalization constant. All simulated random walks have a total of $N_{\text{RW}} = 1,000$ steps

the other phase. Thus the number of consecutive steps in phase i is geometrically distributed with mean $1/p_{\text{switch},i}$. The random walk is initialised in a statistically stationary state with respect to the two phases, i.e. the forager starts in phase i with probability $p_{\text{switch},i}/(p_{\text{switch},i} + p_{\text{switch},j})$. The parameter values and step length and turning angle distributions used in the simulations are given in Table 1.

In each case, N_{RW} steps of the composite CRW model were simulated, giving positional data (x_i, y_i) for $i = 0, \ldots, N_{\text{RW}}$. As in [17], these positional data were first sampled with a fixed sampling time step δ. This leads to a subsample of positions $(x_{j\delta}, y_{j\delta})$ for $j = 1, \ldots, N_{\text{RW}}/\delta$. This subsample was then subjected to a turn identification algorithm with threshold angle θ_0 [see 17, for full details], giving a sample of turning point locations. The cosine of the threshold angle θ_0 is denoted by c_0. Note that $\delta = 1$ corresponds to complete sampling (every step of the random walk is recorded) and $c_0 = 1$ (equivalent to $\theta_0 = 0$) corresponds to no turn identification (every recorded location is defined to be a "turning point"). The combination $\delta = 1$, $c_0 = 1$ is therefore a control case in which there is "perfect information". From the turning point location data, a sample of observed step lengths d_i and a sample of observed turning angles ϕ_i are constructed. An example realisation of the composite CRW model and illustration of the effects of sampling and turn identification are shown in Fig. 2.

3.2 Fitting a Composite Exponential Distribution

Auger-Méthé et al. [2] criticised the work of Plank and Codling [45] on the grounds that a simple exponential distribution should not be expected to provide a good fit to data from a process consisting of two distinct types of behaviour. Instead, it was suggested that the absolute goodness-of-fit of candidate distributions should be assessed, rather than simply conducting a relative test of two potentially poor models. This is certainly true and this motivates us to fit a composite exponential distribution to data generated from the composite CRW model of [45]. However, because of subsampling and turn identification, the observed data may not always be well described by a composite exponential distribution [46].

Fig. 2 An example
realisation of the composite
CRW model: (**a**) the actual
path (*solid* and *dashed lines*
indicate steps from the
extensive and intensive
phases respectively); (**b**) the
observed locations with a
sampling step of $\delta = 5$;
(**c**) the identified turning
points of the subsampled path
shown in (**b**), using a
threshold turning angle
cosine of $c_0 = 0.5$. Parameter
values for the composite
CRW are as given in Table 1
except for the switching
probabilities $p_{switch1} = 0.05$
and $p_{switch2} = 0.4$

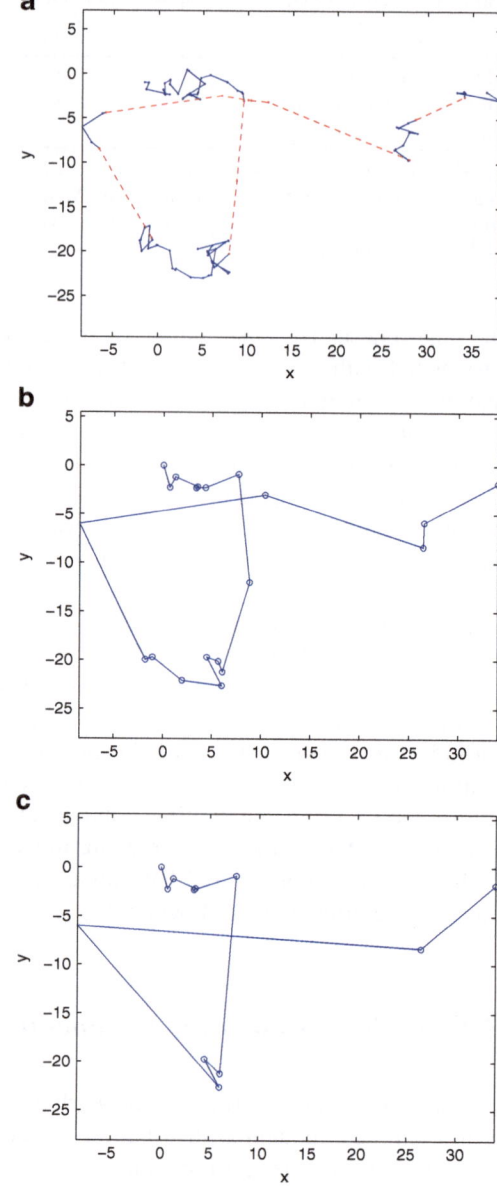

A composite exponential distribution is simply a combination of n exponential distributions, each with its own parameter λ_i and its own probability weighting P_i ($i = 1, \ldots, n$), such that

$$\sum_{i=1}^{n} P_i = 1. \tag{2}$$

This is an example of a mixture model. A random deviate from the composite distribution is simply a random deviate from the ith individual distribution with probability p_i. The composite distribution thus has probability density function (PDF)

$$p(d) = \sum_{i=1}^{n} P_i \lambda_i e^{-\lambda_i d}.$$

Since we are fitting data from a composite CRW model with two phases, we consider a double exponential distribution ($n = 2$). This distribution has three parameters: the parameter for each of the two exponential distributions, λ_1 and λ_2, and the proportion P_1 of step lengths that are drawn from the first exponential. (The parameter P_2 is then determined by the constraint (2).) Note that the simple exponential distribution and the power-law distribution each have just one fitted parameter.

In the "perfect information" case ($\delta = 1$ and $c_0 = 1$), the double exponential distribution provides an exact fit to the composite CRW model. The parameters λ_1 and λ_2 correspond to the mean step lengths in the two phases ($\lambda_i = 1/\bar{l}_i$) and the parameter P_1 corresponds to the proportion of steps that are in phase 1 ($P_1 = p_{\text{switch},1}/(p_{\text{switch},1} + p_{\text{switch},2})$). When there is imperfect sampling, the double exponential distribution will provide an imperfect fit and the fitted parameters will deviate from the underlying random walk parameters.

The log-likelihood of a sample $\{d_1, \ldots, d_N\}$ is

$$L = \sum_{i=1}^{N} \ln(p(d_i)). \tag{3}$$

The maximum likelihood estimates for the three model parameters $\mathbf{v} = \{\lambda_1, \lambda_2, P_1\}$ are found by solving the three simultaneous equations

$$\frac{\partial L}{\partial v_i} = 0, \qquad \text{for } i = 1, 2, 3. \tag{4}$$

These equations are highly nonlinear and must be solved numerically. This was achieved using Newton's method, taking care to ensure solutions satisfy $\lambda_1, \lambda_2 > 0$ and $0 \le P_1 \le 1$. Once the maximum likelihood values for the parameters have been calculated, these are substituted into (3) to calculate the log-likelihood. The Akaike information criterion (AIC) is then calculated according to

$$\text{AIC} = -2L + 2K,$$

where K is the number of fitted parameters ($K = 3$ for the double exponential distribution). The AIC therefore penalises the double exponential distribution relative to the single exponential and power-law distributions in accordance with its additional fitted parameters. The AIC was calculated for each of the three candidate models (power law, single exponential and double exponential) and these were used to calculate the Akaike weights w_{pow}, w_{exp1} and w_{exp2}. The Akaike weights sum to 1 and the weight for a given model measures the likelihood of that model being the best representation of the data out of the candidate models tested. In addition to the relative test based on AIC, we also carried out a G-test [see for example 21, ch. 11] to test the absolute goodness-of-fit of the preferred distribution, as advocated by Auger-Méthé et al. [2].

Note that although the composite exponential distribution is a more accurate representation of the composite CRW than the simple exponential distribution often used as an alternative to the Lévy walk, it is not the full representation of the composite CRW model used to simulate the data. While the behavioural phases represented in the composite exponential distribution are independent of one another, the behavioural phases of the composite CRW are related to one another through the Markov switching probability, p_{switch}. The full likelihood of the composite CRW would be a hidden Markov model (HMM) incorporating the Markovian dependencies between the behavioural phases. The composite exponential distribution, which is an independent mixture model, is the marginal distribution of the HMM [68] and thus can be an appropriate approximation of the full likelihood [62]. However, the AIC for the composite exponential distribution may differ from that for the full HMM. This potential discrepancy in AIC requires further investigation that is beyond the scope of this chapter. Here we will restrict ourselves to verifying that the AIC will select the composite exponential over the power-law or single exponential distributions for movement data produced by a composite CRW.

3.3 Testing Step Length Data

For each combination of sampling parameter values (δ and c_0), $M = 200$ replicate random walks, each with $N_{RW} = 1,000$ steps, were simulated. The Akaike weights presented in this section were obtained by averaging the Akaike weight across the M replicate simulations.

For all cases tested, the Akaike weight of the single exponential distribution w_{exp1} was always zero, indicating that this distribution is never the best fit of the three distributions tested. From here on, we only consider the Akaike weights for the double exponential and power-law distributions.

When there is no subsampling or turn identification ($\delta = 1$ and $c_0 = 1$), the Akaike weight for the double exponential distribution is always 1 (and the weight for the power-law distribution is 0). The absolute goodness-of-fit of the double exponential distribution is also good (G-test gives $P \gg 0.05$). These results are not surprising as, provided that the number of random walk steps is large enough so that the two phases of the composite random walk both occur within the movement path, the observed data will be drawn from a double exponential distribution. The underlying random walk parameters are almost exactly recovered by the maximum

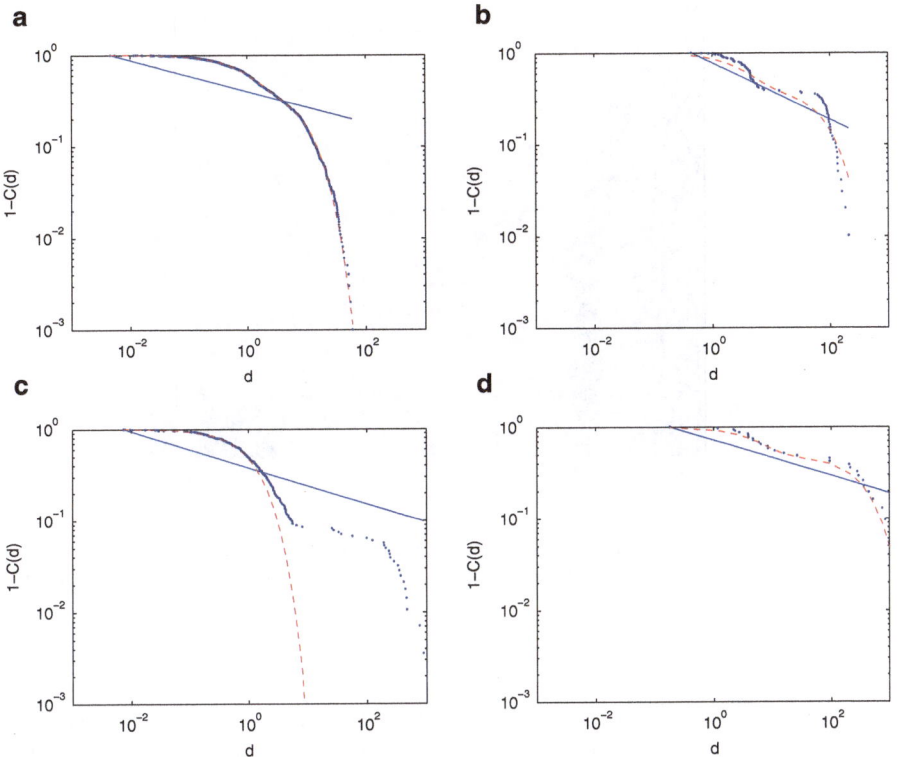

Fig. 3 Survival function $1 - C(d)$, where $C(d)$ is the cumulative distribution function, for observed step length data (*points*) together with the best-fit power-law distribution (*solid curve*) and double exponential distribution (*dashed curve*): (**a**) raw composite CRW data ($\delta = 1, c_0 = 1$); (**b**) subsampling but no turn identification ($\delta = 10$, $c_0 = 1$); (**c**) turn identification but no subsampling ($\delta = 1$, $c_0 = 0.5$); (**d**) subsampling and turn identification ($\delta = 10$, $c_0 = 0.5$). In (**a**), (**b**) and (**d**), the double exponential has the highest Akaike weight; in (**c**) the power-law has the highest Akaike weight ($\mu = 1.2$). Parameter values as in Table 1; $n = 1,000$ random walk steps

likelihood estimation (4) (i.e. $P_1 = 0.5$, $\lambda_1 = 1/\bar{l}_1$ and $\lambda_2 = 1/\bar{l}_2$). An example of the step lengths and fitted distributions for $\delta = 1$ and $c_0 = 1$ is shown in Fig. 3a.

Figure 3b–d show examples of the observed step lengths and fitted distributions for cases where there is either some subsampling of the forager's location or some turn identification (or both). In (b) and (d), where there is some subsampling, the double exponential distribution fits better than the power law. However, in (c), where there is turn identification but no subsampling, the power-law distribution provides the better fit. This is due mainly to the existence of a small number of very large step lengths (up to 10^5 times longer than the smallest observations), although it is clear in this case that neither model provides a good absolute fit to the data (G-test gives $P < 10^{-8}$ for both models).

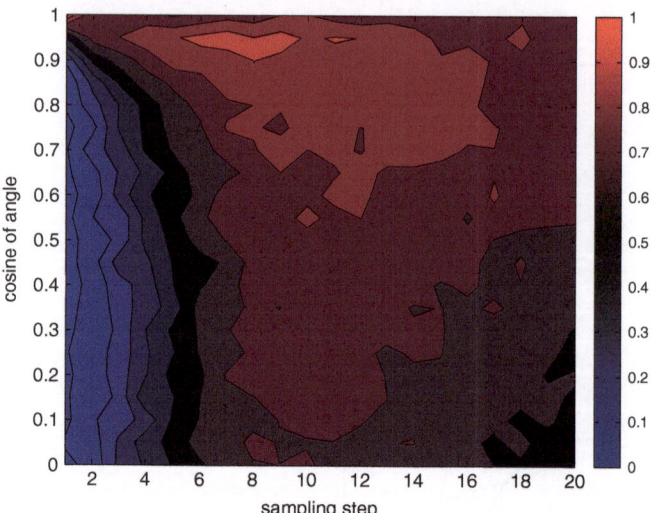

Fig. 4 Average Akaike weight for the double exponential distribution against sampling step size δ and cosine of threshold turn identification angle c_0. *Red areas* indicate cases where the double exponential distribution has the better fit ($w_{exp2} > 0.5$); *blue areas* indicate cases where the power-law distribution has the better fit ($w_{exp2} < 0.5$)

Figure 4 shows the outcome of the AIC test (Akaike weight w_{exp2} averaged over $m = 200$ realisations of the random walk model) for the double exponential distribution for a range of values of the sampling step size δ and cosine of threshold angle c_0. For most combinations of these two sampling parameters, the double exponential distribution is favoured over the power-law distribution ($w_{exp2} > 0.5$). However, as seen already in Fig. 3, when the sampling step δ is small (i.e. the sampling frequency is high) and some degree of turn identification is used ($c_0 < 1$), the power-law distribution is favoured ($w_{exp2} < 0.5$). This shows that, even when fitting candidate distributions that are a good representation of the underlying movement mechanism, the inclusion of a turn identification processing step can make the results of a relative goodness-of-fit test misleading.

The absolute fit of the candidate models was also assessed using a G-test. In cases where the double exponential distribution had a better fit than the power law, the double exponential distribution was not rejected at the 1 % level ($P > 0.01$). In all cases where the power-law has a better fit (blue areas of Fig. 4), the power-law was rejected at the 1 % level, indicating that neither distribution is a good model for the observed data. It should be noted, however, that the results of the G-test are sensitive to the sample size [46] and that if the sample size is sufficiently small, the power-law model is not rejected. Furthermore, because turn identification reduces the number of observed step lengths, this process tends to increase the likelihood of a given candidate distribution providing an acceptable fit to a given movement path.

Table 2 Results of the Rayleigh test of uniformity for a composite CRW and LW subsampled and processed to identify turning events

Random walk	(a) Control	(b) Subsampling	(c) Turn ID	(d) Subsampling and turn ID
Composite CRW	$R = 0.510$	$R = 0.438$	$R = 0.374$	$R = 0.279$
	$P = 0$	$P = 0$	$P = 0$	$P = 0$
Lévy walk	$R = 0.001$	$R = 0.006$	$R = 0.400$	$R = 0.406$
	$P = 0.738$	$P = 0.502$	$P = 0$	$P = 0$

For uniformity we expect $R \approx 0$; the P value gives the probability that the generated turning angle data comes from a uniform circular distribution. All values are given to three decimal places

3.4 Testing Uniformity of Turning Angles

To test for uniformity in the turning angles and to determine the effect that sampling and processing of the observed data may have, we complete a Rayleigh test for uniformity on the distribution of turning angles that is generated after sampling and processing (turn identification). The Rayleigh test is the simplest possible test of uniformity for circular data and is based on testing the mean resultant length, R, of the set of angles. A uniform distribution of angles should have $R \approx 0$ and hence the Rayleigh test considers the probability of a given R value being produced given the size of the data set and the assumption of uniformity. For further details see [39] and [61]; for our generated data we used the Rayleigh test that forms part of the CircStats package in R [47].

We considered three different sampling and turn processing scenarios and completed a Rayleigh test of uniformity for the generated turning angles from a composite CRW, with parameter values as in Table 1, and a LW with $\mu = 2.25$. In each case, the data from $M = 200$ replicate random walk simulations, each with $N_{RW} = 1,000$ random walk steps, were pooled into a single sample of $N_{RW}M = 200,000$ steps. The scenarios considered were: (a) no subsampling or turn processing (control); (b) subsampling only with $\delta = 10$; (c) non-local turn identification only with cosine threshold angle $c_0 = 0.5$; (d) both subsampling ($\delta = 10$) and turn identification ($c_0 = 0.5$). Although these choices of δ and c_0 are arbitrary, the results are not highly sensitive to variations in either of these parameters (Table 2).

The Rayleigh test demonstrates that the turning angles in a LW with or without subsampling, but with no turn processing, (scenarios (a) and (b)) would not be rejected as coming from a uniform distribution ($P > 0.5$ in both cases). However, any form of turn processing will clearly remove small turns from the data set and hence make the observed distribution of turn angles non-uniform. Hence, a LW that is processed in this manner no longer appears to have a uniform distribution of turning angles (scenarios c and d have $P \approx 0$). In all scenarios, the composite CRW has $P \approx 0$ and the assumption of uniformity of turning angles is rejected in all cases. These results illustrate how data on turning angles would only be useful to distinguish between a LW and a composite CRW if there is no turn identification mechanism in the analysis of the data.

4   Discussion

In this chapter, we have shown how the analysis of step-length data collected from regular sampling of an animal's movement path can be extended to include fitting a composite exponential distribution. In the case of a composite random walk movement process [e.g. 10,17,45], one would expect this to provide a better fit to the data than a simple exponential distribution, which is commonly used as an alternative candidate to the power-law distribution associated with a Lévy walk [23,31].

For the two-phase movement model considered here, the results show that a double exponential distribution always provides a better fit to observed step length data than a simple exponential. Furthermore, in most cases, the double exponential provides a better fit than a power-law distribution. However, this method is not foolproof and there are cases where the power-law fits better than the double exponential distribution, despite the fact that the movement process is not a Lévy walk. Of the scenarios considered here, the power-law has a better fit than the double exponential when using a relatively high-resolution sampling scale (small sampling time step) and performing turn identification. This is largely insensitive to the choice of threshold angle. Nevertheless, the absolute fit of the power-law distribution in these cases is poor, although if the sample size is small there may be insufficient evidence to reject the power-law.

A related issue is that the weight that AIC gives to simplicity decreases as sample size increases [25]. Therefore, the AIC comparison may be biased towards the composite exponential (with three fitted parameters) and away from the power-law (with one fitted parameter) for large sample sizes. Preliminary tests with data generated from a LW nevertheless indicate that the power-law distribution is correctly selected by the AIC test over the composite exponential. However, a thorough investigation of the dependency of this bias on sample size is still needed. The same applies to an investigation of the discrepancy between the AIC resulting from composite exponential likelihood function (4) and that of the full hidden Markov model of the composite CRW model.

Overall, the results presented in this chapter highlight the fact that commonly used sampling and data processing techniques can have a significant impact on the distribution of observed step lengths. The distribution that actually describes the underlying random walk step lengths (the double exponential distribution in the example considered here) may not provide a good fit to the observed data.

Not all movement data are collected by recording spatial locations at equally spaced time intervals. For example, smart position or temperature-transmitting tags (SPOT tags) function through radio transmissions and hence require the tag to have contact with air to send data [38]. These tags are often used to track marine mammals or marine predators [29,54] and hence spatial location is only recorded when the animal is at the surface. Consequently, it is likely that any data collected on the movements of these animals have been sampled at irregular intervals that depend on the frequency and distribution of times between surfacing for the particular species of interest. Techniques based on continuous-time random walks may aid in the analysis of this type of data set where the times between recordings are themselves a stochastic process. In this chapter, we have only looked at the effect of regular sampling and we have not considered irregular sampling. However, future studies with irregularly sampled positional data should consider this point in more detail.

We have also shown how analysing turning angle data can help distinguish between a Lévy walk, which has a uniform distribution of turning angles, and a random walk with directional persistence, which has a turning angle distribution peaked about zero. If an observed data set was closely approximated by a power-law distribution of step lengths and a uniform distribution of turning angles, it would be a convincing case for a Lévy walk. However, a turning angle test is not useful if a turn identification method, which removes small turning angles, has been applied to the data. In such a scenario, it would be possible for the original process to be a Lévy walk, but for the uniformity of the turning angle distribution to have been lost through turn identification.

A key feature of the Lévy walk hypothesis is that the animal is undergoing a purely random search (rather than interacting with the environment or relying on memory of encounters for example). In most biological scenarios this is unlikely to be true and hence the Lévy walk is arguably more useful as a descriptive tool for classifying particular types of movement process where both small and large steps (intensive and extensive phases) occur, rather than as a true mechanistic model for animal movement (although in very simple biological search scenarios it may be more appropriate). If foragers are performing a random Lévy search (or any other 'purely random' search), then one would expect bouts of intensive searching to occur at random in the environment (corresponding to a typical sequence of short steps in the LW). Conversely, if intensive searching behaviour is correlated with resource-rich areas of the environment, this would suggest that the forager is interacting with the environment. Unfortunately, most movement studies only record the movement data of the animal(s) of interest and do not collect information about the food distribution. However, if data on the occurrence of intensive movement periods can be combined with information about the distribution of resources then it may be possible to determine if the movement process really is a Lévy walk, or whether the path is generated by a composite movement process where the animal interacts with the environment [4, 5, 26].

Acknowledgements M.A.-M. thanks Drs. A. Derocher and M. A. Lewis and the Centre for Mathematical Biology for their support and Alberta Innovates-Technology Futures, Natural Sciences and Engineering Research Council of Canada, and the University of Alberta for graduate student scholarships.

References

1. R.P.D. Atkinson, C.J. Rhodes, D.W. Macdonald, R.M. Anderson, Scale-free dynamics in the movement patterns of jackals. OIKOS **98**, 134–140 (2002)
2. M. Auger-Méthé, C.C. St. Clair, M.A. Lewis, A.E. Derocher, Sampling rate and misidentification of Lévy and non-Lévy movement paths: comment. Ecology **92**, 1699–1701 (2011)

3. D. Austin, W.D. Bowen, J.I. McMillan, Intraspecific variation in movement patterns: modelling individual behaviour in a large marine predator. Oikos **105**, 15–30 (2004)
4. F. Bartumeus, Behavioral intermittence, Lévy patterns, and randomness in animal movement. Oikos **118**, 488–494 (2009)
5. F. Bartumeus, S.A. Levin, Fractal reorientation clocks: linking animal behavior to statistical patterns of search. Proc. Natl. Acad. Sci. **105**, 19072–19077 (2008)
6. F. Bartumeus, F. Peters, S. Pueyo, C. Marrasé, J. Catalan, Helical Lévy walks: adjusting searching statistics to resource availability in microzooplankton. Proc. Natl. Acad. Sci. USA **100**, 12771–12775 (2003)
7. S. Benhamou, Spatial memory and searching efficiency. Anim. Behav. **47**, 1423–1433 (1994)
8. S. Benhamou, How to reliably estimate the tortuosity of an animal's path: straightness, sinuosity, or fractal dimension? J. Theor. Biol. **229**, 209–220 (2004)
9. S. Benhamou, Detecting an orientation component in animal paths when the preferred direction is individual-dependent. Ecology **87**, 518–528 (2006)
10. S. Benhamou, How many animals really do the Lévy walk? Ecology **88**, 1962–1969 (2007)
11. S. Benhamou, P. Bovet, Distinguishing between elementary orientation mechanisms by means of path analysis. Anim. Behav. **43**, 371–377 (1992)
12. O. Bénichou, M. Coppey, P.-H. Suet, R. Voituriez, Optimal search strategies for hidden targets. Phys. Rev. Lett. **94**, 198101 (2005)
13. O. Bénichou, C. Loverdo, M. Moreau, R. Voituriez, Two-dimensional intermittent search processes: an alternative to Lévy flight strategies. Phys. Rev. E **74**, 020102 (2006)
14. P. Bovet, S. Benhamou, Spatial analysis of animals' movements using a correlated random walk model. J. Theor. Biol. **131**, 419–433 (1988)
15. D. Boyer, G. Ramos-Fernández, O. Miramontes, J.L. Mateos, G. Cocho, H. Larralde, H. Ramos, F. Rojas, Scale-free foraging by primates emerges from their interaction with a complex environment. Proc. R. Soc. Lond. B **273**, 1743–1750 (2006)
16. E.A. Codling, N.A. Hill, Sampling rate effects on measurements of correlated and biased random walks. J. Theor. Biol. **233**, 573–588 (2005)
17. E.A. Codling, M.J. Plank, Turn designation, sampling rate and the misidentification of power-laws in movement path data using maximum likelihood estimates. Theor. Ecol. **4**, 397–406 (2011)
18. E.A. Codling, M.J. Plank, S. Benhamou, Random walks in biology. J. R. Soc. Interface **5**, 813–834 (2008)
19. B.J. Cole, Fractal time in animal behaviour: the movement activity of *Drosophila*. Anim. Behav. **50**, 1317–1324 (1995)
20. M. de Jager, F.J. Weissing, P.M.J. Herman, B.A. Nolet, J. van de Koppel1, Lévy walks evolve through interaction between movement and environmental complexity. Science **332**, 1551–1553 (2011)
21. C. Dytham, *Choosing and Using Statistics: A Biologist's Guide*, 3rd edn. (Wiley, New York, 2011)
22. A.M. Edwards, Using likelihood to test for Lévy flight search patterns and for general power-law distributions in nature. J. Anim. Ecol. **77**, 1212–1222 (2008)
23. A.M. Edwards, Overturning conclusions of Lévy flight movement patterns by fishing boats and foraging animals. Ecology **92**(6), 1247–1257 (2011)
24. A.M. Edwards, R.A. Phillips, N.W. Watkins, M.P. Freeman, E.J. Murphy, V. Afanasyev, S.V. Buldyrev, M.G.E. da Luz, E.P. Raposo, H.E. Stanley, G.M. Viswanathan, Revisiting Lévy flight search patterns of wandering albatrosses, bumblebees and deer. Nature **449**, 1044–1048 (2007)
25. M.R. Forster, Key concepts in model selection: performance and generalizability. J. Math. Psychol. **44**, 205–231 (2000)
26. L. Giuggioli, F. Bartumeus, Animal movement, search strategies, and behavioural ecology: a cross-disciplinary way forward. J. Anim. Ecol. **79**, 906–909 (2009)
27. S. Hapca, J.W. Crawford, I.M. Young, Anomalous diffusion of heterogeneous populations characterized by normal diffusion at the individual level. J. R. Soc. Interface **6**, 111–122 (2009)

28. N.A. Hill, D.P. Häder, A biased random walk model for the trajectories of swimming micro-organisms. J. Theor. Biol. **186**, 503–526 (1997)
29. H.-H. Hsu, S.-J. Joung, Y.-Y. Liao, K.-M. Liu, Satellite tracking of juvenile whale sharks, *Rhincodon typus*, in the northwestern pacific. Fish. Res. **84**, 25–31 (2007)
30. N.E. Humphries, N. Queiroz, J.R.M. Dyer, N.G. Padel, M.K. Musy, K.M. Schaefer, D.W. Fuller, J.M. Brunnschweiler, T.K. Doyle, J.D.R. Houghton, G.C. Hays, C.S. Jones, L.R. Noble, V.J. Wearmouth, E.J. Southall, D.W. Sims, Environmental context explains Lévy and Brownian movement patterns of marine predators. Nature **465**, 1066–1069 (2010)
31. A. James, M.J. Plank, A.M. Edwards, Assessing lvy walks as models of animal foraging. J. R. Soc. Interface **8**, 1233–1247 (2011)
32. I.D. Jonsen, R.A. Myers, M.C. James, Identifying leatherback turtle foraging behaviour from satellite telemetry using a switching state-space model. Mar. Ecol. Prog. Ser. **337**, 255–264 (2007)
33. P.M. Kareiva, N. Shigesada, Analyzing insect movement as a correlated random walk. Oecologia **56**, 234–238 (1983)
34. J. Klafter, B.S. White, M. Levandowsky, Microzooplankton feeding behavior and the Lévy walk, in *Biological Motion*, ed. by G. Hoffmann, W. Alt (Springer, Berlin, 1989), pp. 281–296
35. R.H.G. Klassen, B.A. Nolet, D. Bankert, Movement of foraging tundra swans explained by spatial pattern in cryptic food densities. Ecology **87**, 2244–2254 (2006)
36. A.S. Knell, E.A. Codling, Classifying area-restricted search (ARS) using a partial sum approach. Theor. Ecol. **5**, 325–339 (2012)
37. D.L. Kramer, R.L. McLaughlin, The behavioural ecology of intermittent locomotion. Am. Zool. **41**, 137–153 (2001)
38. C.E. Kuhn, D.S. Johnson, R.R. Ream, T.S. Gelatt, Advances in the tracking of marine species: using GPS locations to evaluate satellite track data and a continuous-time movement model. Marine Ecol. Prog. Ser. **393**, 97–109 (2009)
39. K.V. Mardia, P.E. Jupp, *Directional Statitics* (Wiley, New York, 1999)
40. A. Mårell, J.P. Ball, A. Hofgaard, Foraging and movement paths of female reindeer: insights from fractal analysis, correlated random walks, and Lévy flights. Can. J. Zool. **80**, 854–865 (2002)
41. J.M. Morales, D.T. Haydon, J. Frair, K.E. Holsinger, J.M. Fryxell, Extracting more out of relocation data: building movement models as mixtures of random walks. Ecology **85**, 2436–2445 (2004)
42. C.S. Patlak, Random walk with persistence and external bias. Bull. Math. Biophys. **15**, 311–338 (1953)
43. T.A. Patterson, L. Thomas, C. Wilcox, O. Ovaskainen, J. Matthiopoulos, State-space models of individual animal movement. Trends Ecol. Evol. **23**, 87–94 (2008)
44. S. Petrovskii, A. Morozov, Dispersal in a statistically structured population. Am. Nat. **173**, 278–289 (2009)
45. M.J. Plank, E.A. Codling, Sampling scale and misidentification of Lévy and non-Lévy movement paths. Ecology **90**, 3546–3553 (2009)
46. M.J. Plank, E.A. Codling, Sampling rate and misidentification of Lévy and non-Lévy movement paths: reply. Ecology **92**, 1701–1702 (2011)
47. R Development Core Team, *R: A Language and Environment for Statistical Computing*. R Foundation for Statistical Computing, Vienna, Austria (2009). ISBN 3-900051-07-0
48. G. Ramos-Fernández, J.L. Mateos, O. Miramontes, G. Cocho, H. Larralde, B. Ayala-Orozco, Lévy walk patterns in the foraging movements of spider monkeys (*Ateles geoffroyi*). Behav. Ecol. Sociobiol. **55**, 223–230 (2004)
49. A.M. Reynolds, How many animals really do the Lévy walk? Comment. Ecology **89**, 2347–2351 (2008)
50. A.M. Reynolds, Adaptive Lévy walks can outperform composite Brownian walks in non-destructive random searching scenarios. Phys. A **388**, 561–564 (2009)
51. A.M. Reynolds, Balancing the competing demands of harvesting and safety from predation: Lévy walk searches outperform composite brownian walk searches but only when foraging under the risk of predation. Physica A **389**, 4740–4746 (2011)

52. A.M. Reynolds, M.A. Frye, Free-flight odor tracking in *Drosophila* is consistent with an optimal intermittent scale-free search. PLoS ONE **2**, e354 (2007)
53. A.M. Reynolds, A.D. Smith, R. Menzel, U. Greggers, D.R. Reynolds, J.R. Riley, Displaced honey bees perform optimal scale-free search flights. Ecology **88**, 1955–1961 (2007)
54. D. Rowat, M. Gore, Regional scale horizontal and local scale vertical movements of whale sharks in the Indian Ocean off Seychelles. Fish. Res. **84**, 32–40 (2007)
55. M.F. Shlesinger, B.J. West, J. Klafter, Lévy dynamics of enhanced diffusion: application to turbulence. Phys. Rev. Lett. **58**, 1100–1103 (1987)
56. D.W. Sims, D. Righton, J.W. Pitchford, Minimizing errors in identifying Lévy flight behaviour of organisms. J. Anim. Ecol. **76**, 222–229 (2007)
57. D.W. Sims, E.J. Southall, N.E. Humphries, G.C. Hays, C.J.A. Bradshaw, J.W. Pitchford, A. James, M.Z. Ahmed, A.S. Brierley, M.A. Hindell, D. Morritt, M.K. Musyl, D. Righton, E.L.C. Shepard, V.J. Wearmouth, R.P. Wilson, M.J. Witt, J.D. Metcalfe, Scaling laws of marine predator search behaviour. Nature **451**, 1098–1103 (2008)
58. P. Turchin, Translating foraging movements in heterogeneous environments into the spatial distribution of foragers. Ecology **72**, 1253–1266 (1991)
59. P. Turchin, *Quantitative Analysis of Movement: Measuring and Modeling Population Redistribution in Animals and Plants* (Sinauer Associates, Sunderland, 1998)
60. P. Turchin, F.J. Odendaal, M.D. Rausher, Quantifying insect movement in the field. Environ. Entomol. **20**, 955–963 (1991)
61. G.J.G. Upton, B. Fingleton, *Spatial Data Analysis by Example. Volume 2: Categorical and Directional Data* (Wiley, New York, 1989)
62. C. Varin, N. Reid, D. Firth, An overview of composite likelihood methods. Stat. Sinica **21**, 5–42 (2011)
63. A.W. Visser, T. Kiørboe, Plankton motility patterns and encounter rates. Oecologia **143**, 538–546 (2006)
64. G.M. Viswanathan, S.V. Buldyrev, S. Havlin, M.G.E. da Luz, E.P. Raposo, H.E. Stanley, Optimising the success of random searches. Nature **401**, 911–914 (1999)
65. G.M. Viswanathan, E.P. Raposo, F. Bartumeus, J. Catalan, M.G.E. da Luz, Necessary criterion for distinguishing true superdiffusion from correlated random walk processes. Phys. Rev. E **72**, 011111 (2005)
66. G.M. Viswanathan, E.P. Raposo, M.G.E. da Luz, Lévy flights and superdiffusion in the context of biological encounters and random searches. Phys. Life Rev. **5**, 133–150 (2008)
67. G.M. Viswanathan, M.G.E. da Luz, E.P. Raposo, H.E. Stanley, *The Physics of Foraging: An Introduction to Random Searches and Biological Encounters* (Cambridge University Press, London, 2011)
68. W. Zucchini, I.L. MacDonald, *Hidden Markov Models for Time Series: An Introduction Using R* (Chapman and Hall/CRC, London, 2009)

Beyond Optimal Searching: Recent Developments in the Modelling of Animal Movement Patterns as Lévy Walks

Andy Reynolds

Abstract Lévy walks first entered the biological literature when Shlesinger and Klafter (Growth and Form, Martinus Nijhof Publishers, Amsterdam, 1986, pp 279–283) proposed that they can be observed in the movement patterns of foraging ants. The fractal and superdiffusive properties of Lévy walks can be advantageous when searching for randomly and sparsely distributed resources, prompting the suggestion that Lévy walks represent an evolutionary optimal searching strategy. The suggestion is supported by a plethora of empirical studies which have revealed that many organisms (a diverse range of marine predator, honeybees, *Escherichia coli*) have movement patterns that can approximated by Lévy walks. Nonetheless, Lévy walks with their strange fractal geometry appear alien to biology and their relevance to biology has been hotly debated. Here I describe some of my own recent contributions to Lévy walk research. This research has sought to identify biologically plausible mechanisms by which organisms can execute Lévy walks and to demonstrate that these movement patterns have a utility beyond the understanding and prediction of optimal searching patterns. This work has made apparent that Lévy walks do not stand outside of the now well-established correlated random walk paradigm but are instead natural consequences of it. I also describe some recent advances in Lévy walk search theory.

> Where then but there see another.
> Bit by bit an old man and child.
> In the dim void bit by bit an old man and child.
> Any other would do as ill.
> Samuel Beckett

A. Reynolds (✉)
Rothamsted Research, Harpenden, Hertfordshire, AL5 2JQ, UK
e-mail: andy.reynolds@rothamsted.ac.uk

M.A. Lewis et al. (eds.), *Dispersal, Individual Movement and Spatial Ecology*,
Lecture Notes in Mathematics 2071, DOI 10.1007/978-3-642-35497-7_3,
© Springer-Verlag Berlin Heidelberg 2013

1 Introduction

In 1828 the Scottish botanist Robert Brown reported that minuscule pollen particles suspended in water have seemingly random movements. Einstein's subsequent 1905–6 [29, 30] mathematical description of these random "Brownian" movement patterns has been hugely successful and now lies at the heart of the "correlated random walk paradigm"—the dominant conceptual framework for modelling animal movement patterns [104]. Then just over two decades ago, physicists suggested that some animals have Lévy walk movement patterns. Lévy walks, named after the French mathematician Paul Lévy, arose in a purely mathematical context in the first half of the last century [54]. Lévy walks first entered the biological literature when [94] proposed that they can be observed in the movement patterns of foraging ants. Lévy walks comprise clusters of short step lengths with longer movements between them. This pattern is repeated across all scales with the resultant clusters creating fractal patterns that have no characteristic scale and such that the distribution of move lengths has an inverse power-law tail, $p_l(l) \sim l^{-\mu}$ where $1 < \mu < 3$. Over much iteration, a Lévy walk will be distributed much further from its starting position than a Brownian walk of the same length. The fractal and "superdiffusive" properties of Lévy walks can be advantageous when searching for randomly and sparsely distributed resources [108], prompting the suggestion that Lévy walks represent an evolutionary optimal searching strategy [9, 11]. Nonetheless, Lévy walks with their strange fractal geometry appear alien to biology and their relevance to biology has been hotly debated [18, 40, 101], Auger-Méthé et al. (2011). It seemed to some that physicists and mathematicians had lost touch with biology, and especially so after it became apparent that early empirical analyses of the flight patterns of the wandering albatross [107] (Fig. 1), which had provided the impetus for nearly two decades of research into Lévy walks, were flawed [27].

But the humble pollen has other tales to tell (which show that Lévy movements are pertinent even in the simplest of situations). Occasionally, one of Robert Brown's pollen grains would have come into contact with the bottom of the dish. It is readily seen that the distribution of straight-line distances between successive contact points has an inverse-square power-law tail. The contact points thus form a "Lévy flight" pattern with $\mu = 2$, a random jump process in which the distribution of jump lengths has an inverse-square power-law tail. The distribution of time intervals between consecutive contacts has an inverse power-law tail, $p(t) \propto t^{-3/2}$, by virtue of the Sparre Andersen Theorem [99, 100]. Net horizontal displacements made in a time interval, t, are Gaussian distributed with mean zero and variance $\sigma^2 = 2D't$ where D' is the bulk diffusivity. Taken together these two characteristics imply that the distribution, of distances $p_l(l) \propto \int_0^\infty \frac{e^{-l^2/4D't}}{4\pi D't} t^{-3/2} dt \propto l^{-2}$, between consecutive contact points has an inverse-square power-law tail. Observations of the pollen grains made at the bottom of the dish can therefore be modelled as Lévy flights with $\mu = 2$ (Fig. 2). Analogous behaviour has been predicted for bulk-mediated effective surface diffusion at liquid surfaces [20]. The Lévy flights have fractal dimension $D = \mu - 1$ [95]. The key ingredients of a Lévy walk movement pattern,

Fig. 1 Lévy walk research owes much to the wandering albatross. Early analyses of flight pattern data were flawed but fruitful and continue to provoke and inspire. Photograph courtesy of Corbis

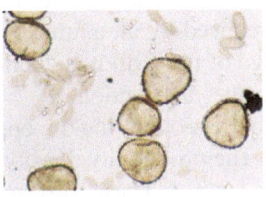 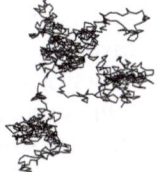 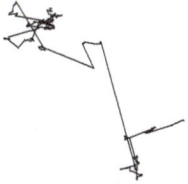

Fig. 2 The humble pollen does both the Brownian walk and the Lévy flight. Bracken pollen (*left*). An example of a simulated Brownian walk representative of pollen movements within a fluid (*middle*). An example of a Lévy flight representative of pollen movements across the bottom of the Petridish. Photograph by Jon S. West, Rothamsted Research

a power-law distribution of move lengths and fractal scaling, are thus lurking within Brownian walks and so are present within the correlated random walk paradigm, despite qualms about the biological plausibility of such properties [13, 103]. Lévy flights also abound once the pollen grains are liberated from watery confines and are at the mercy of the wind [4, 79, 93]. Although these airborne movements are clearly divorced from searching they are not without consequence as they result in a patchy, fractal-like, spatial population structures very different from the structure of a homogeneous front produced by Brownian movements [21, 53, 92, 112]. Here I take up this theme and describe some of my own recent contributions to Lévy walk research, made since my last review with Christopher Rhodes [85]. This research has sought to identify biologically plausible mechanisms by which organisms can execute Lévy walks and to demonstrate that these movement patterns have a utility beyond the understanding and prediction of optimal searching patterns. This work has made apparent that observations of Lévy walks do not stand outside of the correlated random walk paradigm but are instead natural consequences of it. I also describe some recent advances in Lévy walk search theory.

2 Underlying Mechanisms: The Key to Prediction and Understanding

Following the seminal work of [108], Lévy walk research has been mainly focused on establishing the conditions under which Lévy walks constitute an optimal searching strategy, and on establishing statistically reliably means of identifying such movement patterns in telemetry data [8, 9, 11, 13, 27, 28, 110]. Nonetheless, the key to prediction and understanding of movement patterns lies in the elucidation of mechanisms underlying the observed patterns [52]. "Without an understanding of mechanisms, one must evaluate each new stress on each new system *de novo*, without any scientific basis for extrapolation; with such an understanding, one has the foundation for understanding" [52].

One of the simplest candidate mechanisms could give rise to Lévy walk movement patterns in terrestrial ecotones such as riparian forests, dune systems or rocky shores where strong environmental gradients force animals to forage within a narrow strip [10]. This restriction would be realised by an animal with straight-line movements, if each time it arrives at an edge of the strip it is "deflected" back at a random angle, $0 < \theta < \pi$ drawn from the distribution $p_\theta = 1/\pi$. The horizontal distance travelled along the strip before encountering the opposing edge is $l = L \tan \theta$ where L is the width of the strip. The probability density function of l is determined by $p_l dl = p_\theta d\theta$ and so $p_l = \frac{1}{\pi} \frac{L}{l^2 + L^2}$. These movement patterns are a Lévy walk with $\mu = 2$. Random changes in direction at the edges of an ecotone could thus have adaptive value, as $\mu = 2$ Lévy walk movement patterns can be advantageous in random search scenarios [8, 86, 108]. Random scattering from locations on the perimeter of broad two-dimensional landscapes (that do not have strip-like geometries), on the other hand, does not result in Lévy walk movement patterns. Nonetheless, two-dimensional Lévy walk movement patterns with $\mu < 2$ would be produced if the random scattering occurred within the landscape at markers (e.g. vegetation patches) that have a patchy fractal distribution [38]; a scenario which pollen dispersal studies have made plausible [21, 53, 92, 112].

I next describe four other biologically plausible mechanisms that can give rise to two-dimensional observed Lévy walk movement patterns:

- Serial correlation
- Random reorientation at cues left by correlated random walkers
- By products of advantageous foraging behaviours
- Innate physiology

2.1 Serial Correlation

For many years the dominant conceptual framework for describing non-oriented animal movements has been the correlated random walk (CRW) model in which an individual's trajectory through space is regarded as being made up of a sequence of distinct, independent randomly-oriented "moves" [104]. It has long been recognized that the transformation of the animal's continuous movement path into a broken line is necessarily arbitrary and that probability distributions of move lengths and turning angles are model artefacts [104, and references therein]. Dunn and Brown [26], and Alt [1, 2] were perhaps the first to address the problem. They formulated "continuous-time" CRW models. In these models, velocities rather than positions evolve as a Markovian process and are exponentially autocorrelated. Integration of the velocity process gives the position process. The approach pioneered by Dunn and Brown [26], and by Alt [1, 2] has recently been developed by Johnson et al. [46] who demonstrated its utility in an analysis of telemetry data for harbor seals (*Phoca vitulina*) and northern fur seals (*Callorhinus ursinus*). Reynolds [75] showed that velocity autocorrelation inevitable leads to Lévy walk movement patterns on timescales less than the autocorrelation timescale.

Autocorrelation must be present in all movements but is not captured in discrete correlated random walk modelling. Autocorrelation has been quantified in cell motility studies [91, and references therein] but until recently it has received scant attention in the literature on the movement patterns in "higher" animals. A notable exception to this is the analysis by Johnson et al. [46] of seal telemetry data, where it was reported that autocorrelation timescales are several hours long. Lévy walks should be discernible over these timescales.

This advance has resonance with recent developments in the understanding of spontaneous movement of HaCaT and NHDF cells (cells of the epidermis) made in the absence of external guiding stimuli. These movements are well represented by generalizations of the Langevin equation [91]. This modelling is phenomenological as model components are *inspired* by fits to experimental data. Nonetheless, a slight re-parameterization and re-interpretation of the model components leads to the model of [59] which realises Lévy walks as Markovian stochastic processes [76]. This suggests that spontaneous cell movement patterns can be approximated by Lévy walks, as first proposed by Schuster and Levandowsky [89] and that Lévy walks could be lurking under the skin! These movement patterns could have adaptive value because cells of the epidermis form new tissue by locating and then attaching on to one another—a random search scenario.

2.2 Random Reorientation at Cues Left by Correlated Random Walkers

Traces of movement patterns in the form of odour trails can remain within the landscape for some time. In addition to these unintentional and perhaps unavoidable trails animals may also make deliberate scent marks. Mammalian scent marks might, for instance, act as: a deterrent or a substitute for aggression to warn conspecifics away from occupied territory; a sex attractant or stimulant; a system for labeling the habitat for an animal's own use in orientation or to maintain a sense of familiarity with an area; an indicator of individual identity; an alarm signal to conspecifics; and an indicator of population size [44].

Reynolds [77] noted that the odour trails left behind by correlated random walkers will be fractal with fractal dimension $D = 1.33$, illustrating once again that fractal scaling is a property of Brownian motion. By disrupting the movements of other animals these odours can result in reorientation. The locations at which these changes in the direction of travel occur will therefore be fractal. Odour-cued reorientation is therefore expected to give rise to movement patterns that can be approximated by Lévy walks since the turning points in a Lévy walk have fractal dimension $D = \mu - 1$. With the aid of numerical simulations [77] showed that animals which randomly reorient whenever they encounter the odour trail of a Brownian walker but otherwise move in straight-lines because of "directional persistence" will, indeed, have $\mu = 2.33$ Lévy walk movement patterns. These movement patterns are advantageous when searching for sparsely distributed resources without prior knowledge of resource locations or when information obtained during the search is difficult to process so that deterministic search rules cannot be used [8,86,108]. Consequently there will be strong selection pressures for the aforementioned reorientation process when resources are sparsely distributed within unpredictable environments inhabited by correlated random walkers. The presence of correlated random walkers may therefore drive the evolution of Lévy walks when the fitness advantage exceeds the environmental noise. Stochasticity in the form of random reorientations upon encountering odour cues could therefore have adaptive value when sensorial or cognitive animal skills do not override the need for randomness.

In this picture the emergence of Lévy walks from within the correlated random walk paradigm is dependent upon just three simple and well established ingredients: (1) landscapes are inhabited by animals that have CRW movement patterns and either intentionally or unintentionally leave behind odour trails; (2) some other animals also trace out near-straight line paths through the landscape; (3) but after encountering an odour trail abruptly change their direction of travel.

2.3 Lévy Walks as by Products of Advantageous Foraging Behaviours

The flight patterns of foraging bumblebees are of considerable interest because these bees, with a specialized worker caste, do nothing but forage uninterrupted for long periods of time (Fig. 3). They are not distracted by sex or territorial defence and have few predators [33]. They are therefore ideal for testing the clear-cut outcomes of predictive mathematical models of foraging. And this has led to a long running debate about whether bumblebees forage optimally within patches, and whether it matters [34, and references therein], This debate has been enriched by the possibility that bumblebees are executing optimal Lévy flight searching patterns when foraging within patches [27]; an analysis based on Heinrich's [33] classic observational study of bumblebee (*Bombus terricola*) movements (distances and turning angles) at clover (*Trifolium repens*) patches.

Bumblebees foraging within a flower patch tend to approach the nearest flower but then often depart without landing or probing it, if it has been visited previously; unvisited flowers are not rejected in this manner. Reynolds [73] replicated this foraging behaviour in numerical simulations. Lévy walk patterns with $\mu \approx 2$ were found to be an inconsequential emergent property of a bumblebees' foraging behaviour and, in this case, are not part of an innate, evolved optimal searching strategy.

Fig. 3 Bumblebees are ideal
for testing the clear-cut
outcomes of predictive
mathematical models of
foraging. Their foraging
flights are consistent with
Lévy searching theory but
they are not necessarily Lévy
searching. Photograph
courtesy of Corbis

The results thereby provide a vivid demonstration that the key to understanding the biological, ecological and evolutionary consequences of any movement pattern lies in the elucidation of underlying mechanisms [52]. The significance of a particular movement patterns can, in fact, vary markedly even across closely related species and perhaps even within the same organisms under different scenarios. Honeybees (*Apis mellifera*), for example, unlike foraging bumblebees, do seem to execute Lévy flights as part of an innate, evolved searching strategy, at least when searching for their hive and when searching after a known food resource has become depleted [83, 84, 87].

Lévy walk patterns with $\mu \approx 2$ are also known to be an emergent property of predators that use chemotaxis (odour gradient following) to locate randomly and sparsely distributed prey items [71]. Chemotaxis also provides good solutions to the "travelling salesman problem" often minimising the total distance travelled between prey items and so often minimising the energetic costs of foraging [80]. Taken together these findings suggest that $\mu \approx 2$ Lévy walk *patterns* are a frequent emergent property of advantageous searching when searchers have some information about target locations (i.e. when the position of the nearest potential food source is known or when predators can detect the presence of odours emanating from distant food sources). This intriguing possibility complements the widely held view that Lévy walk *processes* are symptomatic of advantageous searching when searchers have little or no knowledge of target locations, and provides a new perspective on the ongoing debate about whether Lévy walks are patterns or processes [13, 68]. Much of this debate is a rerun of earlier deliberations about what "randomness" actually means in the context of random walks [104]. Turchin [104] remarked "that even if animals were perfect automaton we might still choose to model such behaviour stochastically because we might not have perfect knowledge of the deterministic rules driving these animals or, if we did, because including them would require very accurate representation of all environmental "micro-cues". Randomness is thus a modelling convention which is useful when deterministic modelling is impractical or even unhelpful." The approach termed "behavioural minimalism" [55] is directly analogous to thermodynamic theory in which the essentially unpredictable motion of individual molecules is described probabilistically. The underlying philosophy is not that the finer detail does not exist, but that it is irrelevant for producing the observed patterns [52]. That is, the collective behaviour of large numbers of automaton may be indistinguishable (at the scale of the population) from that of random walkers.

2.4 Innate Physiology

Over recent years there has been an accumulation of evidence that many animal behaviours are characterised by common scale-invariant patterns of switching between two contrasting activities over a period of time. This is evidenced in mammalian wake–sleep patterns [15, 56, 57], in the intermittent stop–start locomotion of *Drosophila* fruit flies [60], and in even the nest building behaviours of Large White x Landrace gilts (a species of the wild boar *Sus scrofa*) [35]. Reynolds [81] showed that these dynamics can be modelled by a stochastic variant of Barabási's model [6] for bursts and heavy tails in human dynamics. The new model captures a tension between two competing and conflicting activities. The durations of one type of activity are distributed according to an inverse-square power-law, mirroring the ubiquity of inverse-square power-law scaling seen in empirical data. The durations of the second type of activity follow exponential distributions with characteristic time-scales that depend on species and metabolic rates. This again is a common feature of animal behaviour. In contrast to animal dynamics, "bursty" human dynamics, are characterised by power-law distributions with scaling exponents close to -1 and -3/2. The model may account for some occurrence of Lévy walk movement patterns where an animal is resolving a tension between two competing and conflicting actions: moving in a straight line and turning. And in this regard Lévy walks are no stranger than sleep–wake patterns, stop–start locomotion, and nesting building where construction competes with the need for vigilance.

3 Translating Observations Taken at Small Spatiotemporal Scales into Expected Patterns at Greater Scales

Translating observations taken at small spatial and temporal scales into expected patterns at greater scales is a major challenge in spatial ecology [48]. The ability to scale up from observational scales is of crucial importance when assessing the potential effects of landscape heterogeneity and changes in behaviour, and in applying traditional behavioural ecology to landscape-level ecological problems [55]. To scale from limited observations to the landscape, we must understand how to aggregate and simplify, retaining essential information without getting encumbered by unnecessary detail. In principle this can and has been achieved by associating different modes of movement with different parameterizations of a single CRW model [32, 45, 46, 61, 62, 66, 67]. Depending on the diffusivity (mode), K, a CRW model could, for instance, produce either long straight movements, random meanderings, or more circuitous movements. Difficulties arise when the available observational data are not sufficient to parameterize accurately the probability distribution function of the modes, $p(K)$. In these cases the principle of scientific objectivity dictates that we be maximally uncommitted about what we do not know concerning the distribution $p(K)$. The most conservative, non-committal $p(K)$ that is consistent with the data (e.g. with estimates for the mean value of K) is obtained by maximising Shannon's differential entropy [41–43]. Any other distribution would assume more information than is known from the data. In this context, Shannon's differential entropy, $H = -\int_K p(K) \log_e p(K) dK$, is a measure of the average surprise of seeing an animal in a particular movement mode, K, given a distribution of modes $p(K)$. A highly improbable outcome is very surprising. If there are two movement modes, K_1 and K_2, then the entropy is zero when there is no uncertainty, i.e. when $p(K_1) = 1$ and $p(K_2) = 0$ or when $p(K_1) = 0$ and $p(K_2) = 1$. It is maximized when $p(K_1) = p(K_2) = 1/2$ as there is less uncertainty when $p(K_1) \neq p(K_2)$ because then one or other of the modes is more likely to be seen.

Reynolds [82] showed that truncated $\mu = 2$ Lévy walk movement patterns are the most conservative, maximally non-committal model of movement patterns beyond the scale of data collection when (a) CRW models embody observed movement patterns and (b) minimal or partial information/assumptions about landscape and behavioural heterogeneity are in the form of reliable estimates for the lower order moments of diffusivity (e.g. when given estimates for the mean diffusivity, or the mean and variance of the diffusivity). Lévy walks therefore provide a robust, universal scaling-law which describes how movement patterns change across scale, and which has the potential to become a valuable modelling tool when scaling up from limited observational data in order to assess the likely effects of landscape heterogeneity and changes in behaviour. Reynolds' [82] result also indicates that with landscape and behavioural heterogeneity, the unusual thing is not truncated Lévy walk movement patterns but their absence. In fact, large-scale, Gaussian, diffusive movement patterns, if they arise at all, would be an emergent phenomenon, not a mathematically self-evident state from which any deviation is a worrisome anomaly. Standard methods in spatial ecology do, however, consider Gaussian statistics and diffusion as two basic ingredients of animal movement at the long-time limit [14, 65].

4   Enlarging the Framework of Lévy Walk Search Theory

The foregoing as illustrated that a diverse range of processes can give rise to Lévy walk movement patterns. Some of these processes are not selected for, thus illustrating that Lévy walk movement patterns may have utility beyond the understanding and prediction of optimal searching patterns. Other processes (e.g. random reorientation at cues left by correlated random walkers) will only operate if there are selection pressures for of Lévy walks.

The association of Lévy walks with optimal searching can be traced back to the theoretical and computational work of [108] which produced an idealised model of Lévy walk searching. In this model a searcher moves on a straight-line towards the nearest target if this target lies within the "direct perceptual range", r; otherwise it chooses a direction at random and a distance, l, drawn from a power-law distribution, $P(l) = (\mu - 1)r^{\mu-1}l^{-\mu}$ for $l \geq r$ and $P(l) = 0$ for $l < r$. The searcher then moves incrementally towards the new location whilst constantly seeking for targets within a radius, r. If no target is detected, it stops after traversing the distance l and chooses a new direction and a new distance; otherwise it proceeds to the target. Viswanathan et al. [108] showed that $\mu = 2$ Lévy walks are an optimal Lévy walk searching strategy for the location of randomly and sparsely distributed targets that can be repeatedly revisited because they are not depleted or rejected once visited. Lévy walks with $\mu < 2$ are nearly equally effective and outperform their $\mu > 2$ counterparts when searching "destructively" in either two- or three-dimensional arenas [8, 86]. From a mathematical perspective the difference between non-destructive and destructive searching lies in the specification of the initial conditions for the search. In a non-destructive search each new search begins close to a previously visited target but distant from many other targets. In a destructive search, each new search begins from a location that is distant from the surviving targets. Reynolds [78] and then James et al. [40] noticed that the optimal Lévy walk search strategy can be extremely sensitive to the initial conditions. The advantages that Lévy walks have over ballistic movements in random search scenarios are greatly reduced or removed if searches do not begin in the immediate vicinity of a target. James et al. [40] suggested that this sensitivity shows that the optimality of Lévy walk search is not as robust as previously thought thereby creating the impression that Lévy walk searches are optimal in just a few special circumstances. For two-dimensional searches this sensitivity stems, in part, from the use of point targets in numerical simulations and is less pronounced when targets are large compared with the perceptual range of the forager (Fig. 4), or are patchily distributed. Previously, it had been suggested wrongly that target size does not affect the optimality of searching patterns [39]. Nonetheless, this revised understanding leaves open the specification of biologically-realistic initial conditions for Lévy walk searches. In the next sections I show how the ambiguity in the specification of the initial conditions for a Lévy walk search can be resolved and argue that Lévy walk searches can be optimal when searching under the risk of predation [78]. I also show that Lévy walks searches are expected to be optimal when searching for prey that can occasionally evade capture [86], and when searching is intermittent such that bouts of active searching alternate with relocation bouts during which prey cannot be detected [70, 58]. This strand of research enlarges the framework of Lévy search theory, and may provide a new insights into the movement patterns of a diverse range of marine predator (basking shark *Cetorhinus maximus*, Atlantic cod *Gadus morhua*, bigeyed tuna *Thunnus obesus*, leatherback turtles *Dermochelys coriacea*, and Magellanic penguins *Spheniscus magellanicus*) and *Escherichia coli* which can be modelled as Lévy walks with $\mu \approx 2$ [37, 49, 96, 102]. It is, however, important to acknowledge from the outset that foragers may show plasticity and

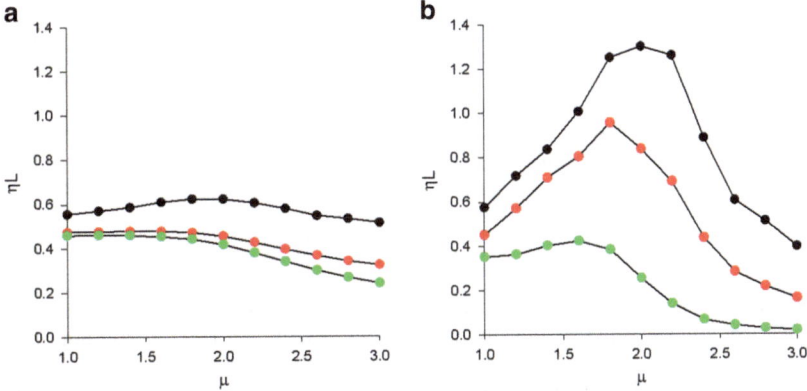

Fig. 4 Simulation data for the searching efficiencies, η, of non-destructive foragers with Lévy walks movement patterns as a function of the Lévy exponent μ. The searching efficiency is taken to be the reciprocal of the mean distance travelled before encountering a target. In other words, it is the mean number of targets located during a search divided by the total length of the search. The two-dimensional square search arena with side length $L = 1000$ arbitrary space units contains 50,000 stationary point-sized targets. The perceptual range of the searchers is $r = 0.1$ space unit. Data ensemble averaged over 5,000 realisations are shown for (**a**) random and uniformly distributed targets and for (**b**) for patchily distributed targets. Each patch contains 10,000 targets that are uniformly distributed within non-overlapping circles of diameter 100 arbitrary space units. Patches were randomly and uniformly distributed within the search arena. Data are shown for the cases when each new search begins $r(black)$, $5r(red)$ and $50r(green)$ from the last target to be located

change strategies depending on circumstances (as illustrated in Fig. 5) and that trade-offs might prevent a universal solution [12].

4.1 Balancing the Demands of Foraging and Safety from Predation

Benhamou [13] and then Plank and James [68] devised a composite Brownian walk model for the location of patchily distributed targets that once visited become temporally unavailable either because they have become depleted or because of the increased risk of predation. In this model searchers travel out from the origin of their search in a straight line until they encounter a target and then proceed to search destructively within the patch that contains this target using Brownian movements, i.e. using an area restricted search. If a target is not located within a prescribed time, the "giving up time" then the searcher switches back to the original straight-line motion. Benhamou [13] showed that his composite Brownian walk model is more efficient than any Lévy walk that is not responsive to conditions found in the search. Reynolds [71, 72] subsequently pointed out that the composite Brownian walk model can, in fact, be interpreted as being an "adaptive" or responsive Lévy

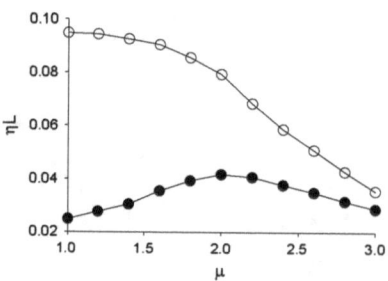

Fig. 5 Optimality hanging in the balance. The presence of a gentle breeze could be sufficient to switch the optimal searching strategy from a ballistic walk to a Lévy walk. Simulation data (unpublished) is shown for the efficiency, η, of Lévy walk searches as a function of the Lévy exponent μ. The searching efficiency is the mean number of prey items encountered per unit distance travelled. The search arena, a square with side length L, contains 10^5 prey items that are randomly and uniformly distributed. Prey are consumed once detected. The perceptual range of the predator $r = 10^{-4}L$ arbitrary space units. Data are shown for the cases when prey can be detected equally well when approached from any direction (*open circle*) and only when approached from an upwind location (*filled circle*) because unbeknown to the predator, prey flee from olfactory cues produced by predators

walk search. This correspondence arises because straight-line movements between targets correspond to truncated $\mu \to 1$ Lévy walks. Benhamou [13] and Plank and James [68] have therefore demonstrated that an adaptive Lévy walk is better than any non-adaptive Lévy walk when searching destructively in patchy environments. Moreover, predictions from the composite Brownian walk are entirely consistent with standard Lévy walk search theory; this predicts that straight-line movements are advantageous when searching destructively for sparsely distributed patches whilst Brownian movements are optimal for within-patch searching [108].

Reynolds [74] then developed a new class of adaptive Lévy walk searching model which encompassed composite Brownian models as a special case. In these models, bouts of Lévy walk searching alternate with bouts of more intensive Brownian walk searching. As with the composite Brownian model switching from extensive to intensive searching is prompted by the detection of a target and switching back to extensive searching arises if a target is not located after travelling a distance equal to the giving-up time. The model reconciles Lévy walk search theory with the ubiquity of two modes of searching by predators and with their switching searching model immediately after finding a prey [50]. This model reduces to the "composite Brownian walk" model when $\mu \to 1$. It should be noted that the model presupposes that the prey are patchily distributed and that the predator perhaps through past experience is aware of this. The models are thus fundamentally different from non-adaptive Lévy walk search models [108] where it is assumed that animals have *no* prior knowledge of the target distribution.

Prey capture does not always trigger an area restricted search [111]. This is probably because decisions to modify behaviour after prey capture are dependent on many parameters, including the presence of other predators, the state of the forager, the cost of catching the prey, the quality of the prey patch, or predation risks. Adaptive Lévy walk searching models have been used to examine the trade-offs between searching efficiency and safety from predation [78]. Only if the benefits of advantageous foraging outweigh these costs can there be strong selection pressures for Lévy walk movement patterns. In the absence of predation the giving-up time can be chosen to maximise foraging efficiency and in this case the searching efficiency of adaptive Lévy walks is no better than that of composite Brownian walks. But when foraging under the risk of predation this unconstrained optimal may not be realised because a forager must trade off food harvesting with safety [17]. When the realised giving-up distance is much shorter than the unconstrained optimal one, Lévy walks with $\mu \approx 2$ are advantageous. This finding has resonance with that of [109] who argued that convoluted movement patterns confer greater fitness than straight-line paths because they reduce the risk of predation. Straight-line paths present the most efficient means of searching for prey while also exposing the forager to maximum predation risk. Animals that manage to trade-off food and safety by vigilance to predators while feeding from a food patch can remain within the patch for long times and are not be expected to have Lévy walk movement patterns. Animals that use little vigilance and manage risk via time allocation by demanding a higher feeding rate to compensate for a higher risk of predation may have Lévy walk movement patterns. And so despite having fundamentally different properties, Lévy walks and composite Brownian walks can compete a priori as possible models of foraging movements. Lévy walks are expected in tritrophic systems and where intra-guild predation (a ubiquitous interaction, differing from competition or predation, defined as killing and eating among potential competitors) operates.

4.2   Red Queen Dynamics

The co-evolution of predators and their prey can lead to situations in which neither improves its fitness because both populations co-adapt to each other [25, 106]. In these evolutionary arms races, improvements in the ability of a predator to detect and capture prey (e.g., heightened sensitivity to chemical, mechanical or visual signals, stronger attack reactions) are matched by compensating improvements in the ability of prey to evade detection and capture (e.g. crypsis, feigning death, strong jumps, sudden increase of size, confounding signals). These "Red Queen" type of dynamics [105] preclude the possibility of a perfect searching/capture process. Reynolds and Bartumeus [86] showed that $\mu \sim 2$ Lévy walks can be optimal when searching destructively if targets occasionally evade detection and/or capture. Searches for escapees begin close to escaped prey but distant from other prey— a scenario mirroring "non-destructive" foraging which favours $\mu \sim 2$ Lévy walk searching. This suggests that accounting for the co-evolutionary arms races at the predator–prey detection/reaction scales can explain to some extent the $\mu \sim 2$ Lévy walks searching patterns at larger scales.

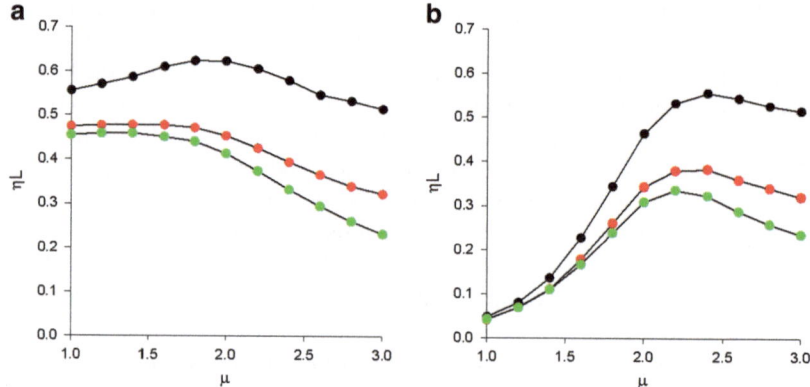

Fig. 6 Simulation data for the searching efficiencies, η, of non-destructive foragers with Lévy walks movement patterns as a function of the Lévy exponent μ. The searching efficiency is taken to be the reciprocal of the mean distance travelled before encountering a target. In other words, it is the mean number of targets located during a search divided by the total length of the search. Fifty thousand stationary targets were randomly and uniformly within a two-dimensional square search arena with side length $L = 1000$ arbitrary space units. The perceptual range of the searchers is $r = 0.1$ space unit. Data are shown for (**a**) non-intermittent (i.e. standard) Lévy walk searching and (**b**) intermittent Lévy walk searching where targets can be only detected using relatively short moves with length $l < 100r$. Data are shown for the cases when each new search begins $r(black)$, $5r(red)$ and $100r(green)$ from the last target to be located. Similar results (not shown) were obtained for patchily distributed targets

4.3 Intermittent Searches

The movements of many foragers (e.g. planktivorous fish, ground-foraging birds, and lizards) are intermittent with pauses or bouts of relatively slow movements lasting from milliseconds to minutes [51, 63]. This intermittency can have a variety of energetic benefits. Endurance can also be improved by partial recovery from fatigue. Perceptual benefits can arise because pauses increase the capacity of the sensory systems to detect relevant stimuli. Several processes, including velocity blur, relative motion detection, foveation, attention and interference between sensory systems could be involved [51]. Searching could therefore be salutatory such that "scanning" phases during which prey can be detected alternate with "relocation" phases during which prey cannot be detected. This trait can be incorporated into Lévy walk searching models by associating the short moves ($l < l_0$) with the scanning phases whilst longer moves are associated with the relocation phases. Intermittent Lévy walks with $\mu \approx 2$ are an optimal search strategy for both destructive and non-destructive foragers [58, 70]. In other words, this strategy is robustly optimally with respect to the initial conditions of the search, and so markedly different from non-intermittent Lévy walk searching which are extremely sensitive to initial conditions [78] (Fig. 6).

Here following [108] the searching efficiency is taken to be the reciprocal of the mean distance travelled before encountering a targets, i.e., it is the mean number of targets located during a search divided by the total length of the search. Foragers that minimize the average distance travelled between targets will therefore maximize their expected energy gain when energy expenditure increases linearly with distance travelled. The energy costs of intermittent locomotion are, however, more complex and the energy expended in accelerations and decelerations can be more than offset by a variety of energetic benefits and by recovery from fatigue [51, and references therein], Fish such as cod and Pollack can, for instance, save energy by "burst-coasting swimming" as the drag while coasting with the body straight is only about one third of the drag while swimming. The energetic costs of intermittent locomotion warrant further investigation because they could favour $\mu \neq 2$ Lévy walk movement patterns for some taxa.

4.4 Optimizing the Encounter Rate in Biological Interactions

Encounter rates set bounds on prey-consumption, the risk of predation, the likelihood of mate-location and the spread of infectious diseases and so play a crucial role in population dynamics. To date, however, there have not been any reported studies on the relative merits of Lévy walk search strategies for the location of mobile targets in two-search arenas.

James et al. [39] reported that ballistic movements outperform Lévy walks and Brownian walks when searching randomly and destructively for mobile prey in one-dimensional environments, thereby overturning the previous analysis of [7]. Data (previously unpublished but comparable to that reported on by [10]) from numerical simulations of destructive searching in two-dimensional arenas show that Lévy walks with $\mu \leq 2$ are equally effective and outperform Lévy walks with $\mu > 2$ when predators move with speeds that are faster than or comparable to that of their prey (Fig. 7a–c). Maximal encounter rates are then largely insensitive to the movement pattern of the prey. This is not surprising and entirely consistent with numerical simulations of destructive searching for immobile targets [8, 86]. It is evident from Fig. 7a–c that the prey cannot adapt their movement patterns so as to reduce the likelihood of predation. This suggests that prey movement patterns are determined by their foraging and mating-location requirements and not by the costs of predation. Predator movement patterns do, of course, become irrelevant when predators move much more slowly than their prey (Fig. 7d). A "sit-and-wait" strategy and a Brownian search are then just as effective as a Lévy walk search. It is thus possible for Brownian searches to have evolved naturally as one search strategy. Nonetheless, Lévy searches are more versatile and outperform Brownian walks when (if) searching for slowly moving prey (Fig. 7a–c) in addition to fast moving ones. This leads to the expectation that Lévy searches are predominant in generalist predators whilst Brownian and correlated random walk searching is likely in some specialist predators with a narrow prey range. Note also that ballistic movements are predictable, making the forager more vulnerable to predation (Fig. 8).

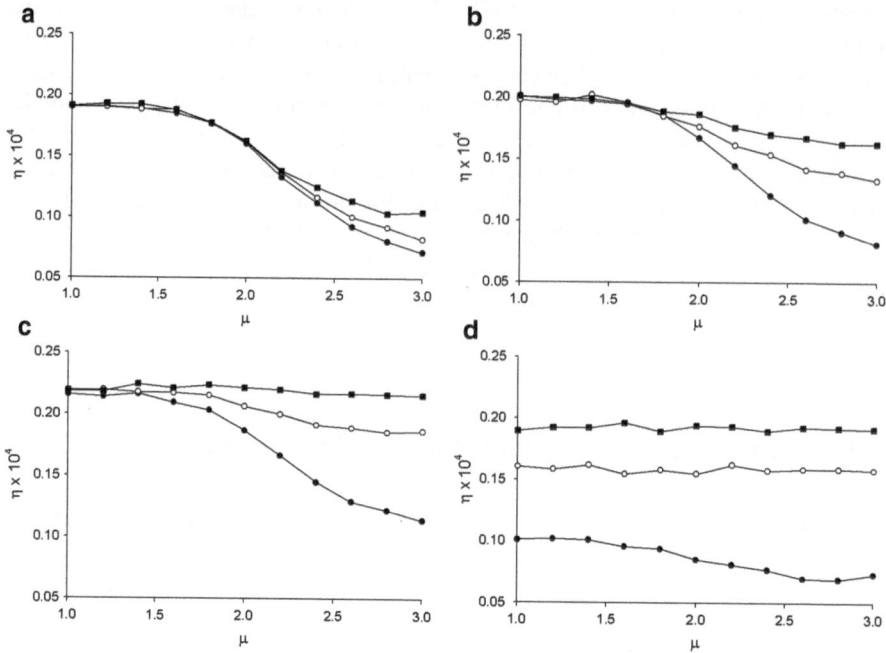

Fig. 7 Simulation data for the mean encounter rates, η, for predators with Lévy walks movement patterns as a function of the Lévy exponent, μ. Predators search within a two-dimensional square arena with size 1,000×1,000 arbitrary space units containing ten prey. The encounter rate is the average number of prey items encountered per unit distance travelled. Data are shown for predators that move ten times faster than their prey (1.0 and 0.1 space units in unit time) (**a**), for predators that move two times faster than their prey (1.0 and 0.5 space units in unit time) (**b**), for predators that move at the same speed as their prey (1.0 space units in unit time) (**c**) and for predators that move ten times slower than their prey (0.1 and 1.0 space units in unit time) (**d**). In all cases the perceptual range $r = 1$ space unit and predators travel for a time of 10^5 time units. Encounter rates for each case were obtained by ensemble averaging for 500 realizations of the initial prey distribution. Simulation data are shown for prey with Brownian walk ($\mu = 3$) (*filled circle*), $\mu = 2$ Lévy walk (*open circle*) and ballistic (*filled square*) movement patterns. Prey are deleted once encountered. To maintain a constant density of prey, each deleted prey is replaced by a new prey placed at a randomly selected location within the search arena. Analogous results (not shown) have been obtained for prey at lower densities (square arena with size $1,000 \times 1,000$ arbitrary space units containing 5, 2 and 1 prey) and for searching within three-dimensions (cube arena with size $100 \times 100 \times 100$ arbitrary space units containing ten prey)

5 Some Closing Remarks and Some Open Questions

5.1 Opening the Lévy Gates

The research reported on here has shown that Lévy walks do not stand outside of the correlated random walk paradigm [104] but rather are natural consequences of

Fig. 8 Still fishing for answers. There is strong evidence for Lévy walks in the swimming patterns of the Magellanic penguin (Spheniscus magellanicus) and other marine predators [96] and these appear to be associated with foraging. The idealised Lévy walk searching model of [108] suggest that these movement patterns are an optimal foraging strategy. Much subsequent work paints a more complicated picture. Photo courtesy of Corbis

it and that the utility of Lévy walk models extends well beyond the description of search behaviours.

The apparent strangeness of Lévy walks was shown to be innocuous. After all, a pollen grain does both the Brownian and the Lévy walk. The seemingly peculiar fractal properties of Lévy walks are also seen in Brownian walks [77]. Power-law scaling, the hallmark of Lévy walks, is necessarily present in continuous-time correlated random walks that take explicit account of serial correlations [75]. And when correlated random walks represent accurately observed movement patterns, Lévy walks are the most conservative model of movements at larger scales [82]. This strand of research is also bridging between the separate disciplines of animal movement patterns and plant disease epidemiology. This is generating new perspectives and questions at the interface between these two disciplines and thereby contributing to the emergence of a new synthesis that transcends traditional boundaries. Other work [31,98] is bridging the gap between the separate disciplines of animal and human movement patterns.

The research has also extended the reach of Lévy walk search theory to encompass the predator–prey co-evolutionary arms race [86], dynamic adaption to conditions found along the search [72, 74, 78], and physiological constraints [58, 70]. Taken together this research suggests that $\mu \sim 2$ Lévy walk searches represent an evolutionarily stable strategy in changing or dynamic environments [97]. This warrants further investigation because it would reveal the extent of selection pressures for $\mu \sim 2$ Lévy walk movement patterns.

5.2 Lévy Walks in Collective Motions: How the Blind Could Lead the Blind

Collective movement behaviour is seen in almost every taxa and is arousing considerable amongst behavioural ecologists as well as physicists and mathematicians [5, 16, 19, 22, 23]. On focus of attention has been group decision making. In a seminal study, using idealised mathematical models, Couzin et al. [23] showed how information about the location of a food source or a migration route can be transferred within groups both without signalling and when group members do not which individuals, if any, have pertinent information. This work has demonstrated how a few individuals (approximately 5%) within honeybee swarms can guide the group to a new nest site [90] and how relatively few informed individuals within fish schools can influence the foraging behaviours of the group [69].

This leaves open the question of whether effective leadership and decision-making can arise when no individual in the group has pertinent information about the location of resources, i.e. the question of whether Lévy walks movement patterns can arise in groups from social interactions. Lévy walks patterns of movement in groups can, as in individuals, be advantageous in random search scenarios [88].

Preliminary considerations in this direction have shown that Lévy walks could be an emergent property of collective movements ranging from "swarming" where there is no parallel alignment among members, as in often seen in insects, particular the *Diptera*, through to the high parallel movements displayed in some fish shoals. This investigation has drawn also out further connections between Brownian and Lévy walks.

Consider an idealised model of collective movements in which there is one "leader" and a "follower". The leader moves in a straight line, changing its direction of travel only when one of the followers comes within its immediate vicinity (collision avoidance). The follower keeps pace with the leader but has small random (Brownian) movements in the two directions orthogonal to the leaders' direction of travel. It can be shown (unpublished report) that the leader and so the pair are following a $\mu = 3/2$ Lévy walk. Truncated Lévy walks result when the follower cannot meander to arbitrarily long distances from the leader but instead remains within a "zone of attraction" that enforces group cohesion. These findings are broadly consistent with telemetry data for midges (*Anarete pritchardi Kim*) flying within swarms [64]. Okubo and Chiang [64] reported that midge flights may be classified into two distinct patterns; one is a "wide" pattern, the other is a "tight" pattern. In a wide pattern, the insect exhibits a relatively long, straight or slightly curved path that might be regarded as a free flight. After a straight path the insect shifts its motion from one direction to another. In the tight pattern, insects exhibit a relatively short, zigzag flight that might be regarded as random motion. How these different patterns are related to the behaviour of swarming midges is still unknown.

Similarly Lévy walk movement patterns with $\mu = 1 + N/2$ are predicted to arise in highly parallel groups consisting of one leader and N followers that keep pace with the leader whilst making one-dimensional random movements (traverse to the current direction of travel of the leader). The finding may explain the presence of Lévy walk movement patterns in some fish that forage in shoals [96]. The empirical observations are recovered if leaders are responsive to just two of their followers. Here it is worth noting that leaders and followers have been identified in shoals [69].

Intrinsic variability in the mobility of individuals within a group may therefore have adaptive value and Lévy walk movement patterns could be an overlooked benefit of group living. This warrants further investigation.

5.3 Mathematical Challenges

A more formidable challenge is to develop an analytical theory of Lévy walk searching that is applicable for two- and three-dimensions. This would serve to validate numerical codes and facilitate an examination of searching in regimes that are currently inaccessible to computation, e.g. at the threshold of starvation where targets are in very dilute concentration and so detected very infrequently. It remains to seen whether "mean field theories" of the kind developed by Viswanathan et al. [108] for one-dimensional searches can reproduce faithfully simulation data for two- and three-dimension searches.

The employment of entropy maximization in movement patterns also warrants further investigation because it offers new unexplored means for quantifying the information content of correlated random walk and Lévy walk models, and for establishing new connections between these models. The simplest of correlated random walk models can, for instance, be construed as being the most conservative, maximally non-committal models of animal movement patterns given only the arithmetic mean move length. This is simply because maximisation of Shannon entropy yields an exponential distribution of move lengths [24]. The arithmetic move length is a potentially meaningful characteristic of a movement pattern if the move lengths do not show a tendency to grow during the time course of a movement pattern. When move lengths do tend to grow then the geometric (or logarithmic) average move length can be useful. The geometric average of a set of

N moves with lengths $\{l_i\}$ is $\bar{l} = \left(\prod_{i=1}^{N} l_i \right)^{1/N} = \exp\left(\frac{1}{N} \sum_{i=1}^{N} \ln l_i \right) \equiv \exp\left(\langle \ln l \rangle \right)$

where $\langle \ln l \rangle$ is the logarithmic average. Maximisation of Shannon's entropy, subject to the condition that probabilities furnish the observed geometric average move length, gives a Pareto distribution of move lengths, $p(l) = (\mu - 1)a^{\mu-1}l^{-\mu}$ where $\mu = 1 - \frac{1}{\ln a - \langle \ln l \rangle}$ is the well known Hill's (maximum likelihood) estimate [36] for a power-law exponent [47]. Geometric constraints *per se* are not new [47] but until now had not appeared in movement ecology literature. Models utilizing move length distributions other than the Pareto or exponential distributions are less conservative if move lengths are characterized solely in terms of either the arithmetic or geometric average; a minimal requirement for any reasonable model of animal movement pattern. The Akaike information criterion which following [27] is now used widely to distinguish objectively between power-law and exponential distributions can, in this application, be interpreted as determining the relative appropriateness of the arithmetic and geometric averages as characterisations of the typical move length. This is because the Akaike weight for a power-law (i.e. weight of evidence in favour of a power-law) is determined by the logarithmic average whilst the Akaike weight for an exponential is determined by the arithmetic average. A bridge between the Lévy walk and correlated random walk models is formed if move lengths are simultaneously characterised in terms of arithmetic and geometric average move lengths (as would be the case if individuals occasionally switched between executing Lévy and correlated random walks or at the population level if Lévy walkers co-exist alongside correlated random walkers). In this case, maximisation of Shannon's entropy leads to a gamma distribution of length moves. This distribution has a power-law like core and an exponential tail, and was recently found to characterise accurately the flight patterns of the wandering albatross [27]. The wandering albatross may therefore bridge the apparent divide correlated random walks and Lévy walks. It seems that as with the humble pollen, Lévy walk research can still learn much from the wandering albatross.

Acknowledgements Rothamsted Research receives grant aided support from the Biotechnology and Biological Sciences Research Council. I am indebted to Don Reynolds and Chris Rhodes for their support and encouragement.

References

1. W. Alt, Modelling of motility in biological systems, in ed. by *ICIAM'87: Proceedings of the first international conference on industrial and applied mathematics,* J. McKenna, R. Temam. (SIAM, Philadelphia, 1988), pp 15–30
2. W. Alt, Correlation analysis of two-dimensional locomotion paths, in *'Biological Motion: Proceedings of a workshop held in Königswinter Germany'*, ed. by W. Alt, G. Hoffman. Lecture Notes in Biomathematics (Springer, Berlin, 1990), pp 254–268
3. M. Auger-Méthé, C. Cassady St. Clair, M.A. Lewis, A.E. Derocher, Sampling rate and misidentification of Lévy and non-Lévy movement patterns: comment. Ecology **92**, 1699–1701 (2011)
4. F.C.W. AusterlitzDick, C. Dutech, E.K. Klein, S. Oddou-Muratorio, P.E. Smouse, V.L. Sork, Using genetic markers to estimate the pollen dispersal curve. Mol. Ecol. **13**, 937–945 (2004)
5. M. Ballerini, N. Cabibbo, R. Candelier, A. Cavagna, E. Cisbani, I. Giardina, V. Lecomte, A. Orlandi, G. Parisi, A. Procaccini, M. Viale, V. Zdravkovic, Interaction ruling animal collective behavior depends on topological rather than metric distance: Evidence from a field study. Proc. Natl. Acad. Sci. **105**, 1232–1237 (2008)
6. A.L. Barabasi, The origin of bursts and heavy tails in human dynamics. Nature **435**, 207–211 (2005)
7. F. Bartumeus, J. Catalan, U.L. Fulco, M.L. Lyra, G.M. Viswanathan, Optimizing the encounter rate in biological interactions: Lévy versus Brownian strategies. Phys. Rev. Lett. **88**, article 097901 (2002)
8. F. Bartumeus, M.G.E. da Luz, G.M. Viswanathan, J. Catalan, Animal search strategies: a quantitative random search analysis. Ecology **86**, 3078–3087 (2005)
9. F. Bartumeus, Lévy processes in animal movement: an evolutionary hypothesis. Fractals **15**, 151–162 (2007)
10. F. Bartumeus, P. Fernández, M.G.E. da Luz, J. Catalan, R.V. Solé, S.A. Levin, Superdiffusion and encounter rates in diluted, low dimensional worlds. Eur. Phys. J. Spec. Top. **157**, 157–166 (2008)
11. F. Bartumeus, Behavioral intermittence, Lévy patterns, and randomness in animal movement. Oikos **118**, 488–494 (2009)
12. F. Bartumeus, J. Catalan, Optimal search behavior and classic foraging theory. J. Phys. A **42**(article), 434002 (2009)
13. S. Benhamou, How many animals really do the Lévy walk? Ecology **88**, 518–528 (2007)
14. H.C. Berg, *Random Walks in Biology* (Princeton University Press, Princeton, New Jersey, 1983).

15. M.S. Blumberg, A.M.H. Seelje, S.B. Lowen, A.E. Karlson, Dynamics of sleep-wake cyclicity in developing rats. Proc. Natl. Acad. Sci. **102**, 14680–14864 (2005)
16. N.W. Bode, D.W. Franks, A.J. Woods, Limited interactions in flocks: relating model simulations to empirical data. R. Soc. Interface (2010). doi:10.1098/rsif.2010.0397
17. J.S. Brown, Vigilance, patch use and habitat selection: Foraging under predation risk. Evol. Ecol. Res. **1**, 49–71 (1999)
18. M. Buchanan, The mathematical mirror to animal nature. Nature **453**, 714–716 (2008)
19. J. Buhl, D.J.R. Sumpter, I.D. Couzin, J.J. Hale, E. Despland, E.R. Miller, S.J. Simpson, From disorder to order in marching locusts. Science **312**, 1402–1406 (2006)
20. O.V. Bychuk, B. O'Shaughnessy, Anomalous surface diffusion: A numerical study. J. Chem. Phys. **101**, 772–780 (1994)
21. S.A. Cannas, D.E. Marco, M.A. Montemurro, Long range dispersal and spatial pattern formation in biological invasions. Math. Biosci. **203**, 155–170 (2006)
22. I.D. Couzin, J. Krause, R. James, G.D. Ruxton, N.R. Franks, Collective memory and spatial sorting in animal groups. J. Theor. Biol. **218**, 1–11 (2002)
23. I.D. Couzin, J. Krause, N.R. Franks, S.A. Levin, Effective leadership and decision-making in animal groups on the move. Nature **433**, 513–516 (2005)
24. T.M. Cover, J.A. Thomas, *Elements of Information Theory* (Wiley, New York, 2006)
25. R. Dawkins, J.R. Krebs, Arms races between and within species. Proc. R. Soc. Lond. B **205**, 489–511 (1979)
26. G.A. Dunn, A.F. Brown, A unified approach to analyzing cell motility. J. Cell Sci. Suppl. **8**, 81–102 (1987)
27. A.M. Edwards, R.A. Phillips, N.W. Watkins, M.P. Freeman, E.J. Murphy, V. Afanasyev, S.V. Buldyrev, M.G.E. da Luz, E.P. Raposo, H.E. Stanley, G.M. Viswanathan, Revisiting Lévy walk search patterns of wandering albatrosses, bumblebees and deer. Nature **449**, 1044–1048 (2007)
28. A.M. Edwards, Overturning conclusions of Lévy flight movement patterns by fishing boats and foraging animals. Ecology **926**, 1247–1257 (2011)
29. A. Einstein, Über die von der molekularkinetischen Theorie der Wärme geforderte Bewegung von in ruhenden Flüssigkeiten suspendierten Teilchen. Ann. Phys. **17**, 549–506 (1905)
30. A. Einstein, Zur theorie der Brownschen bewegung. Ann. Phys. **19**, 549–560 (1906)
31. M.C. González, C.A. Hidalgo, A.L. Barabási, Understanding individual human mobility patterns. Nature **453**, 779–782 (2008)
32. E. Gurarie, R.D. Andrews, K.L. Laidre, A novel method for identifying behavioural changes in animal movement data. Ecol. Lett. **12**, 395–408 (2009)
33. B. Heinrich, Resource heterogeneity and patterns of movement of foraging bumblebees. Oecologia **40**, 235–245 (1979)
34. B. Heinrich, Do Bumblebees Forage Optimally, and Does It Matter? Am. Zool. **23**, 273–281 (1983)
35. A. Harnos, G. Horváth, A.B. Lawrence, G. Vattay, Scaling and intermittency in animal behaviour. Phys. A **286**, 312–320 (2000)
36. B.M. Hill, A simple general approach to inference about the tail of a distribution. Ann. Stat. **3**, 1163–1174 (1975)
37. N.E. Humphries, N. Queiroz, J.R.M. Dyer, N.G. Pade, M.K. Musyl, K.M. Schaefer, D.W. Fuller, J.M. Brunnschweiler, T.K. Doyle, J.D.R. Houghton, G.C. Hays, C.S. Jones, L.R. Noble, V.J. Wearmouth, E.J. Southall, D.W. Sims, Environmental context explains Lévy and Brownian movement patterns of marine predators. Nature **465**, 1066–1069 (2010)
38. H. Isliker, I. Vlashos, Random walk through fractal environments. Phys. Rev. E **67**(article), 026413 (2003)
39. A. James, M.J. Planck, R. Brown, Optimizing the encounter rate in biological interactions: Ballistic versus Lévy versus Brownian strategies. Phys. Rev. E **78**(article), 051128 (2008)
40. A. James, M.J. Plank, A.M. Edwards, Assessing Lévy walks as models of animal foraging. J. R. Soc. Interface **8**, 1233–1247 (2011)
41. E.T. Jaynes, Information theory and statistical mechanics. Phys. Rev. **106**, 620–630 (1957a)

42. E.T. Jaynes, Information theory and statistical mechanics II. Phys. Rev. **108**, 171–190 (1957b)
43. E.T. Jaynes, in *Probability Theory: The Logic of Science*, ed. by Bretthorst, G.L. (Cambridge University Press, Cambridge, 2003)
44. R.P. Johnson, Scent marking in mammals. Anim. Behav. **21**, 521–535 (1973)
45. C.J. Johnson, K.L. Parker, D.C. Heard, M.P. Gillingham, Movement parameters of ungulates and scale-specific responses to the environment. J. Anim. Ecol. **71**, 225–235 (2002)
46. D.S. Johnson, J.M. London, M.A. Lea, J.W. Durban, The continuous-time correlated random walk model for animal telemetry data. Ecology **89**, 1208–1215 (2008)
47. J.N. Kapur, *Maximum-Entropy Models in Science and Engineering* (Wiley, New York, 1989)
48. P.M. Kareiva, Population dynamics in spatially complex environments: theory and data. Phil. Trans. R. Soc. Lond. B **330**, 175–190 (1990)
49. E. Korobkova, T. Emonet, J.M.G. Vilar, T.S. Shimizu, P. Cluzel, From molecular noise to behavioural variability in a single bacterium. Nature **428**, 574–578 (2004)
50. P. Knoppien, J. Reddingius, Predators with two modes of searching: a mathematical model. J. Theor. Biol. **114**, 273–301 (1985)
51. D.L. Kramer, R.L. McLaughlin, The behavioural ecology of intermittent locomotion. Am. Zool. **41**, 137–153 (2001)
52. S.A. Levin, The problem of pattern and scale in Ecology. Ecology **73**, 1943–1967 (1992)
53. C. Lett, H. Østergård, A stochastic model simulating the spatiotemporal dynamics of yellow rust on wheat. Acta Phytopathologica et Entomologica Hungarica **35**, 287–293 (2000)
54. P. Lévy, *Théorie de l'addition des Variables Aaléatoires* (Monographies des Probabilités, publiés sous la direction de E. Borel, no. 1.) (Guathier Villars, Paris, 1937)
55. S.L. Lima, P.A. Zollner, Towards a behavioral ecology of ecological landscapes. Trends Ecol. Evol. **11**, 131–135 (1996)
56. C.C. Lo, L.A. Nunes Amaral, S. Havlin, P.C.h. Ivanov, T. Penzel, J.H. Peterand, H.E. Stanley, Dynamics of sleep-wake transitions during sleep. Europhys. Lett. **57**, 625–631 (2002)
57. C.C. Lo, T. Chou, T. Penzel, T.E. Scammell, R.E. Strecker, H.E. Stanley, P.C.h. Ivanov, Common scale-invariant pattern of sleep-wake transitions across mammalian species. Proc. Natl. Acad. Sci. **101**, 17545–17548 (2004)
58. M.A. Lomholt, K. Tal, R. Metzler, J. Klafter, Lévy strategies in intermittent search processes are advantageous. Proc. Natl. Acad. Sci. **105**, 11055–11059 (2008)
59. I. Lubashevsky, R. Friedrich, A. Heuer, Realization of Levy flights as continuous processes. Phys. Rev. E **79**(article), 011110 (2009)
60. J.R. Martin, A portrait of locomotor behaviour in Drosophila determined by a video-tracking paradigm. Behav. Process. **67**, 207–219 (2004)
61. J.M. Morales, S.P. Ellner, Scaling up animal movements in heterogeneous landscapes: the importance of behaviour. Ecology **83**, 2240–2247 (2002)
62. J.M. Morales, D.T. Haydon, J. Frair, K.E. Holsinger, J.M. Fryxell, Extracting more out of relocation data: Building movement models as mixtures of random walks. Ecology **85**, 2436–2445 (2004)
63. W.J. O'Brien, H.I. Browman, B.I. Evans, Search strategies of foraging animals. Am. Sci. **78**, 152–160 (1990)
64. A. Okubo, H.C. Chiang, An analysis of the kinematics of swarming behaviour of Anarete pritchardi Kim (Diptera: Cecidomyiidae). Res. Popul. Ecol. **16**, 1–42 (1974)
65. A. Okubo, *Diffusion and Ecological Problems: Mathematical Models* (Springer, Berlin, 1980)
66. O. Ovaskainen, Habitat specific movement parameters estimated using mark-recapture data and diffusion model. Ecology **85**, 242–257 (2004)
67. O. Ovaskainen, H. Rekola, E. Meyke, E. Arjas, Bayesian methods for analyzing movements in heterogeneous landscapes form mark-recapture data. Ecology **89**, 542–554 (2008)
68. M.J. Plank, A. James, Optimal foraging: Lévy pattern or process? R. Soc. Interface **5**, 1077–1086 (2008)
69. S.G. Reebs, Can a minority of informed leaders determine the foraging movements of a fish shoal? Anim. Behav. **59**, 403–409 (2000)

70. A.M. Reynolds, On the intermittent behaviour of foraging animals. Europhys. Lett. **75**, 517–520 (2006)
71. A.M. Reynolds, Deterministic walks with inverse-square power-law scaling are an emergent property of predators that use chemotaxis to located randomly distributed prey. Phys. Rev. E **78**(article), 011906 (2008a)
72. A.M. Reynolds, How many animals really do the Lévy walk? Comment. Ecology **89**, 2347–2351 (2008b)
73. A.M. Reynolds, Lévy flight patterns are predicted to be an emergent property of a bumblebee's foraging strategy. Behav. Ecol. Sociobiol. **64**, 19–23 (2009a)
74. A.M. Reynolds, Adaptive Lévy walks can outperform composite Brownian walks in non-destructive random searching scenarios. Phys. A **388**, 561–564 (2009b)
75. A.M. Reynolds, Bridging the gulf between correlated random walks and Lévy walks: Autocorrelation as a source of Lévy walk movement patterns. R. Soc. Interface **7**, 1753–1758 (2010a)
76. A.M. Reynolds, Can spontaneous cell movements be modelled as Lévy walks? Phys. A **389**, 273–277 (2010b)
77. A.M. Reynolds, Animals that randomly reorient at cues left by correlated random walkers do the Lévy walk. Am. Nat. **175**, 607–613 (2010c)
78. A.M. Reynolds, Balancing the competing demands of harvesting and safety from predation: Lévy walk searches outperform composite Brownian walk searches but only when foraging under the risk of predation. Phys. A **389**, 4740–4746 (2010d)
79. A.M. Reynolds, Exponential and power-law contact distributions for windborne spores are representative of different atmospheric conditions. Phytopathology **101**, 1465–1470 (2011a)
80. A.M. Reynolds, Chemotaxis can provide biological organisms with good solutions to the travelling salesman problem. Phys. Rev. E **83**, 052901 (2011b)
81. A.M. Reynolds, On the origin of bursts and heavy tails in animal dynamics. Phys. A **390**, 245–249 (2011c)
82. A.M. Reynolds, Truncated Lévy walks are an expected outcome of correlated random walks models: Translating observations taken at small spatiotemporal scales into expected patterns at greater scales. J. R. Soc. Interface **9**, 528–534 (2012d)
83. A.M. Reynolds, A.D. Smith, R. Menzel, U. Greggers, D.R. Reynolds, J.R. Riley, Displaced honeybees perform optimal scale-free search flights. Ecology **88**, 1955–1961 (2007a)
84. A.M. Reynolds, A.D. Smith, D.R. Reynolds, N.L. Carreck, J.L. Osborne, Honeybees perform optimal scale-free searching flights when attempting to locate a food source. J. Exp. Biol. **210**, 3763–3770 (2007b)
85. A.M. Reynolds, C.J. Rhodes, The Lévy flight paradigm: random search patterns and mechanisms. Ecology **90**, 877–887 (2009)
86. A.M. Reynolds, F. Bartumeus, Optimising the success of random destructive searching: Lévy walks can outperform ballistic motions. J. Theor. Biol. **260**, 98–103 (2009)
87. A.M. Reynolds, J.L. Swain, A.D. Smith, A.P. Martin, J.L. Osborne, Honeybees use a Lévy flight search strategy and odour-mediated anemotaxis to relocate food sources. Behav. Sociobiol. **64**, 115–123 (2009)
88. M.C. Santos, E.P. Raposo, G.M. Viswanathan, M.G.E. da Luz, Can collective searches profit from Lévy walk strategies? J. Phys. A **42**(article), 4340174 (2009)
89. F.L. Schuster, M. Levandowsky, Chemosensory responses of *Acanthamoeba castellani*: Visual analysis of random movement and responses to chemical signals. J. Eukaryot. Microbiol. **43**, 150–158 (1996)
90. T.D. Seeley, *Honeybee Ecology: A Study of Adaptation in Social Life* (Princeton University Press, Princeton, 1985)
91. D. Selmeczi, S. Mosler, P.H. Hagedorn, N.B. Larsen, H. Flyvbjerg, Cell motility as persistent random motion: theories from experiments. Biophys. J **89**, 912–931 (2005)
92. M.W. Shaw, Simulation of population expansion and spatial pattern when individual dispersal distributions do not decline exponentially with distance. Proc. R. Soc. B **259**, 243–248 (1995)

93. M.W. Shaw, T.D. Harwood, M.J. Wilkinson, L. Elliott, Assembling spatially explicit land-scape models of pollen and spore dispersal by wind for risk assessment. Proc. R. Soc. B **273**, 1705–1713 (2006)

94. M.F. Shlesinger, J. Klafter, Lévy walks versus Lévy flights. In *Growth and Form*, ed. by H.E. Stanley, N. Ostrowski (Martinus Nijhof Publishers, Amsterdam, 1986), pp 279–283

95. Shlesinger, M.F., Zaslavsky, G., Frisch, U. (eds). *Lévy Flights and Related Topics in Physics* (Springer, Berlin, 1995)

96. D.W. Sims, E.J. Southall, N.E. Humphries, G.C. Hays, C.J.A. Bradshaw, J.W. Pitchford, A. James, M.Z. Ahmed, A.S. Brierley, M.A. Hindell, D. Moritt, M.K. Musyl, D. Righton, E.L.C. Shepard, V.J. Wearmouth, R.P. Wilson, M.J. Witt, J.D. Metcalfe, Scaling laws of marine predator search behaviour. Nature **451**, 1098–1102 (2008)

97. J.M. Smith, G.R. Price, The logic of animal conflict. Nature **246**, 15–18 (1973)

98. C. Song, T. Koren, P. Wang, A.L. Barabási, Modelling the scaling properties of human mobility. Nat. Phys. **6**, 818-823 (2010)

99. E. Sparre Andersen, On the fluctuations of sums of random variables. Math. Scand. **1**, 263–285 (1953)

100. E. Sparre Andersen, On the fluctuations of sums of random variables II. Math. Scand. **2**, 195–223 (1954)

101. J. Travis, Ecology: do wandering albatrosses care about math? Science **318**, 742–743 (2007)

102. Y. Tu, G. Grinstein, How white noise generates power-law switching in bacterial flagellar motors. Phys. Rev. Lett. **94**, 208101. (2005)

103. P. Turchin, Fractal analyses of animal movements: a critique. Ecology **77**, 2086–2090 (1996)

104. P. Turchin, *Quantitative Analysis of Movement: Measuring and Modelling Population Redistribution in Animals and Plants* (Sinauer Associates, Sunderland, Massachusetts, 1998)

105. L. Van Valen, A new evolutionary law. Evol. Theor **1**, 1–30 (1973)

106. G.I. Vermeij, *Evolution and Escalation: An Ecological History of Life* (Princeton University Press, Princeton, 1987)

107. G.M. Viswanathan, V. Afanasyev, S.V. Buldyrev, E.J. Murphy, P.A. Prince, H.E. Stanley, Lévy flight search patterns of wandering albatrosses. Nature **381**, 413–415 (1996)

108. G.M. Viswanathan, S.V. Buldyrev, S. Havlin, M.G.E. da Luz, E.P. Raposo, H.E. Stanley, Optimizing the success of random searches. Nature **401**, 911–914 (1999)

109. A.W. Visser, T. Kiørboe, Plankton motility patterns and encounter rates. Oecologia **148**, 538–546 (2006)

110. E.P. White, B.J. Enquist, J.L. Green, On estimating the exponent of power-law frequency distributions. Ecology **89**, 905–912 (2008)

111. H. Wiemerskirsch, D. Pinaud, F. Pawloswski, C.A. Bost, Does prey capture induce area-restricted search? A fine-scale study using GPS in a marine predator, the wandering albatross. Am. Nat. **170**, 734–743 (2007)

112. L.U. Wingen, J.K.M. Brown, M.W. Shaw, The population genetic structure of clonal organisms generated by exponentially bounded and fat-tailed dispersal. Genetics **177**, 435–448 (2007)

Part II
From Individuals to Populations

The Mathematical Analysis of Biological Aggregation and Dispersal: Progress, Problems and Perspectives

Hans G. Othmer and Chuan Xue

Abstract Motile organisms alter their movement in response to signals in their environment for a variety of reasons, such as to find food or mates or to escape danger. In populations of individuals this often this leads to large-scale pattern formation in the form of coherent movement or localized aggregates of individuals, and an important question is how the individual-level decisions are translated into population-level behavior. Mathematical models are frequently developed for a population-level description, and while these are often phenomenological, it is important to understand how individual-level properties can be correctly embedded in the population-level models. We discuss several classes of models that are used to describe individual movement and indicate how they can be translated into population-level models.

1 Introduction

The central topic of this chapter is the process of aggregation of biological organisms, which occurs in systems that range in scale from single-celled organisms such as the bacterium *E. coli*, to flocks of birds, schools of fish, and herds of ungulates. Aggregation is a broad term, which we use to mean a self-induced spatial localization of motile individuals that results from direct or indirect communication between them and produces a local density of individuals higher than would be observed under random motion. Depending on the organisms involved, more

H.G. Othmer (✉)
School of Mathematics and Digital Technology Center, University of Minnesota,
Minneapolis, MN 55455, USA
e-mail: othmer@math.umn.edu

C. Xue
Mathematical Biosciences Institute, Ohio State University, Columbus, OH 43210, USA
e-mail: cxue@mbi.osu.edu

M.A. Lewis et al. (eds.), *Dispersal, Individual Movement and Spatial Ecology*,
Lecture Notes in Mathematics 2071, DOI 10.1007/978-3-642-35497-7_4,
© Springer-Verlag Berlin Heidelberg 2013

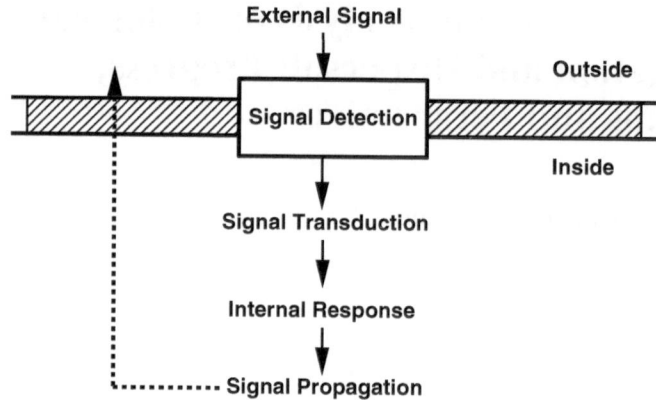

Fig. 1 The general steps involved in generating the response to an external signal

specific terms may be used: swarming in insects, flocking in birds, schooling in fishes and herding in mammals—but all refer to the same underlying process. In some aggregates there is large scale organization, such as alignment in fish schools, which undoubtedly involves at least nearest-neighbor interactions, whereas in other aggregates, such as the bacterial aggregates discussed later, there is no coherence to the motion even though there may be indirect interaction between individuals via the external medium. Whatever the scale or type of aggregation, locomotion—which we define to be self-induced movement that results from active forces generated by the individual—is an essential process in aggregation, but of course it also plays a role in numerous other contexts, including searching for food, mates or shelter. For example cell locomotion, either individually or collectively as tissues, is essential for early development, angiogenesis, tissue regeneration, the immune response, and wound healing in multicellular organisms, and plays a very deleterious role in cancer metastasis in humans. Directed locomotion, as opposed to random wandering, usually involves several steps (i) the detection and transduction of external signals, be they visual, chemical, mechanical, or of other types, (ii) integration of the signals into an internal signal, (iii) the control of the internal neural, biochemical and mechanical responses that lead to force generation and directed movement, and (iv) perhaps relay of the signal. A schematic of the sub-processes involved is shown in Fig. 1.

A detailed description of locomotion of higher organisms such as birds or fishes is extremely complex, and simpler descriptions are used for understanding aggregation. A starting point is to treat individuals as points and attempt to understand the collective behavior of an aggregate based on postulated interactions between individuals or between individuals and an external field, either imposed or generated by the population. In this framework the problem is mathematically similar to the study of interacting molecular species, and techniques established in that context can be carried over to biological problems. Because single cells are the simplest systems capable of self-locomotion, the description of cellular motion can be more complete in models of aggregation, but the principles that emerge from the analysis of cellular motion apply at higher levels as well. Thus several concrete examples of cell-level aggregation will be described in detail later.

Many single-celled organisms use flagella or cilia to swim, and the best studied example of this is *E. coli*. As we show later, much can be learned about 'run-and-tumble' organisms such as *E. coli* without a detailed description of the mechanical forces, but in eukaryotes forces play a more central role. There are two basic modes of movement used by eukaryotic cells that lack cilia or flagella—mesenchymal and amoeboid [10]. The former, which can be characterized as 'crawling' in fibroblasts or 'gliding' in keratocytes, involves the extension of finger-like filopodia or pseudopodia and/or broad flat lamellipodia, whose protrusion is driven by actin polymerization at the leading edge. This mode dominates in cells such as fibroblasts when moving on a 2D substrate. In the amoeboid mode, which does not rely on strong adhesion, cells are more rounded and employ shape changes to move—in effect 'jostling through the crowd' or 'swimming'. Recent experiments have shown that numerous eukaryotic cell types display enormous plasticity in locomotion in that they sense the mechanical properties of their environment and adjust the balance between the modes accordingly by altering the balance between parallel signal transduction pathways [85]. Thus pure crawling and pure swimming are the extremes on a continuum of locomotion strategies for eukaryotic cells, but many cells can sense their environment and use the most efficient strategy in a given context. Significant progress has been made in going beyond the point particle description in such systems (cf. [90] and references therein).

Some basic questions that arise in studying aggregation, either from the experimental or mathematical standpoint, are as follows.

- At what level of detail must individuals be described to explain the observed phenomena?
- What is the coarsest or highest-level description of the forces involved that suffices?
- What is the nature of the signal that is used to initiate aggregation? Is the signal externally-imposed, as for example, when bacteria move up the gradient of a desirable substance, is the signal relayed from individual to individual, and what is the range of the signal?
- What determines the size of an aggregate and how does it depend on the nature and range of the signal?
- When aggregates move coherently, by which we mean they locally adjust their speed and direction to those of their neighbors, the latter perhaps weighted in decreasing importance with distance, what is the time scale on which coherence is achieved beginning from an incoherent state, and how does the type of signal and its range affect this time.

There is a huge literature on the subject of aggregation, orientation and alignment, and other chapters in this volume will cover other aspects (see the chapters by Hillen and Painter and by Franz and Erban). Recent papers that discuss some of the topics treated herein are given in [6, 14, 21, 67, 94, 95]. Classic texts related to the topics herein include [9, 69]. We have two main objectives here: (i) to summarize some of the recent work on the derivation of macroscopic equations such as the Patlak-Keller-Segel chemotaxis equations from individual-based descriptions, and (ii) to illustrate the use of the macroscopic equations that result in cellular aggregation.

The classical taxis problem began with phenomenological equations in which a biased drift term was added to a diffusion equation to describe the movement of individuals in response to an imposed or self-generated signal [52], although a more fundamental approach along the lines described later was initiated earlier by Patlak [80], and the resulting taxis equation is called the PKS equation. To describe it more precisely, let $\Omega \subset R^n$ be a compact domain with smooth boundary, let n be the 'particle' density, and let S be the 'signal' density. The first of the following pair is the PKS equation, and the second describes the self-generated signal field, when applicable.

$$n_t = \nabla \cdot (\nabla n - n\nabla \Phi(S)) = \nabla \cdot (\nabla n - n\chi(S)\nabla S), \tag{1}$$

$$S_t = D\Delta S + f(n, S). \tag{2}$$

The first rigorous derivation of the coupled equations beginning with an interacting particle system is due to Stevens [89]. A review of the major developments from 1970 to about 2003 can be found in [45], and a 'user's guide' to these and other taxis equations can be found in [42]. The quantity $\chi \equiv \Phi_S(n, S, x, \dots)$ is called the chemotactic sensitivity, and $\mathbf{u}_c \equiv \chi(S)\nabla S$ is called the chemotactic velocity, and the fundamental problem we address is how knowledge of the internal dynamics governing signal transduction and response is reflected in these quantities. We develop the machinery for addressing this and describe some success for simple organisms such as $E.$ $coli$, and partial success for eukaryotic cells.

2   An Overview of Population-Level Descriptions

2.1   A Summary of the Levels of Description

We begin by summarizing classical approaches to the transition from equations of motion for individuals to population level distribution functions. The material in this section is standard and widely-discussed, but it is useful to remind the reader of some of the underlying assumptions. To understand the broad picture before delving into the details, we regard the particles or individuals as structureless, but we admit the possibility that they can exert forces and allow for external forces as well. We first consider point particles, and thus describe their motion by Newton's law. For later purposes we include an evolution equation for the internal state of the particles, but at present we do not include coupling of the latter to the movement. The general case of forcing on both position and velocity leads to the differential equations

$$\mathbf{dx}_i = \mathbf{v}_i \mathrm{dt} + \mathbf{dX}_i, \tag{3}$$

$$m_i \mathbf{dv}_i = \mathbf{F}_i \mathrm{dt} + \mathbf{dV}_i, \tag{4}$$

$$\frac{dy}{dt} = G(\mathbf{x}, \mathbf{v}, y, t). \tag{5}$$

Here $(\mathbf{x}, \mathbf{v}) \in R^n, n = 1, 2, 3$ are the positions and velocities, and $y \in R^s$ characterizes the internal state. If the imposed forces $\mathbf{X}$ and $\mathbf{V}$ are deterministic forces they can be written as $d\mathbf{X}_i = X_i dt$, and similarly for $d\mathbf{V}$, and (3) and (4) are the standard Newton equations for particles. When $\mathbf{X}$ and $\mathbf{V}$ are random forces these are stochastic differential equations, the integral forms of which are interpreted in the Ito sense [4, 13].

The two major types of random forcing processes that are widely used are Brownian motion and compound Poisson processes. Both Brownian motion and the Poisson process are examples of a more general class of random processes called Lévy processes [2, 86], which are stochastic processes that have independent, stationary increments, are stochastically continuous, i.e., for any $\epsilon > 0$, $\Pr\{|X_{t+s} - X_s| > \epsilon\} \to 0$ as $t \to 0$, and have sample paths that are right-continuous and have left limits. Brownian motion and Poisson processes differ in that the former have continuous sample paths whereas Poisson processes have discontinuities at the jump. Lévy processes with fat-tailed distributions will arise in Sect. 3.4 in the context of anomalous diffusion.

The formal differentials that appear in (3) and (4) are assumed to be white noise, which is a wide-sense stationary random process in which the component functions $d\mathbf{X}_i$ have zero mean and are uncorrelated, i.e.,

$$\langle \mathbf{dX}_i(t) \rangle = 0, \tag{6}$$

$$\langle \mathbf{dX}_i(t_1), \mathbf{dX}_i(t_2) \rangle = \sigma^2 \delta(t_1 - t_2). \tag{7}$$

Gaussian white noise is the generalized derivative of a single-variable Wiener process, i.e., of Brownian motion [4, 36].

As used here, a Poisson forcing function is a compound Poisson process, which can be thought of as a train of jumps distributed in time according to a Poisson law. Thus

$$\mathbf{X}_i(t) = \sum_{k=1}^{N(t)} Y_k H(t - t_k), \tag{8}$$

where the amplitudes Y_k are independent random variables, H is the step function, and $N(t)$ is a homogeneous Poisson counting process with parameter λ that counts the number of jumps in $[0, t]$, assuming that $N(0) = 0$ with certainty. A generalization of this allows coupling between the amplitudes of the impulses and their temporal occurrence, and can be defined by a random measure $M(dt, dY)$ that gives the number of jumps in $((t, t + dt) \times (y, y + dy))$. The derivative of the forcing, which is called Poisson white noise, is thus a train of impulses that arrive at the jump times of the underlying Poisson process.

$$dX_i(t) = \sum_{k=1}^{N(t)} Y_k \delta(t - t_k), \tag{9}$$

Later we allow the Poisson parameter to depend on external fields or on the internal state of individuals.

The simplest problem arises when there are no inter-particle interactions, and the forces stem from interactions with the environment. One example is the original Einstein model of a heavy particle in a bath that receives Gaussian-distributed momentum impulses from the surrounding bath [27]. In Einstein's formulation this leads to the diffusion equation for the position of the particle, and the probability to find a walker at $x \in R$, having started at the origin at $t = 0$, is

$$P(x,t) = \frac{1}{\sqrt{4\pi Dt}} e^{-x^2/4Dt}, \tag{10}$$

for $(x,t) \in R \times R^+$. In the next section we discuss descriptions that account for both velocity and position.

When there are impulsive forces, rather than Gaussian forces on the position in (3) we obtain the familiar random walk, in which there are instantaneous changes in position at random times. These are called space-jump processes [73], and later we show that the probability density for such a process satisfies the renewal equation

$$P(\mathbf{x}, t|0) = \hat{\Phi}(t)\delta(\mathbf{x}) + \int_0^t \int_{R^n} \phi(t - \tau)T(\mathbf{x}, \mathbf{y})P(\mathbf{y}, \tau|0)\, d\mathbf{y}\, d\tau. \tag{11}$$

Here $P(\mathbf{x}, t|0)$ is the conditional probability that a walker who begins at the origin at time zero is in the interval $(\mathbf{x}, \mathbf{x} + d\mathbf{x})$ at time t, $\phi(t)$ is the density for the waiting time distribution, $\hat{\Phi}(t)$ is the complementary cumulative distribution function associated with $\phi(t)$, and $T(\mathbf{x}, \mathbf{y})$ is the redistribution kernel for the jump process. In Sect. 3.2 we show that this also leads to diffusion equations in certain limits, which reflects the fact that under mild conditions on the distribution of jump sizes the compound Poisson process approaches Brownian motion in the limit $\lambda \to \infty$.

If we admit impulsive forces on the velocity in (4) then we arrive at the second major type of jump-driven movement, which is called a velocity jump process [73]. As described in detail later, the motion consists of a sequence of "runs" separated by re-orientations, during which a new velocity is chosen instantaneously. If we assume that the velocity changes are the result of a Poisson process of intensity λ, then in the absence of other forces we show later that we obtain the evolution equation

$$\frac{\partial p}{\partial t} + \nabla_{\mathbf{x}} \cdot \mathbf{v}p + \nabla_{\mathbf{v}} \cdot \mathbf{F}p = -\lambda p + \lambda \int T(\mathbf{v}, \mathbf{v}')p(\mathbf{x}, \mathbf{v}', t)\, d\mathbf{v}'. \tag{12}$$

A similar equation to describe the random movement of bacteria was first derived by Stroock [91].

2.2   The Fokker-Planck and Smoluchowski Equations

A generalization of the Einstein description of Brownian motion involves both velocity-dependent interaction of the particle with a fluid environment, and diffusion in velocity space. This is based on (3) and (4), in which we assume that the forcing on position is zero, the random forcing on velocity is Gaussian white noise, and we allow velocity-dependent frictional forces. In the standard notation of statistical physics, we write

$$dx_i = v_i dt, \tag{13}$$

$$m d\mathbf{v} = -m\zeta \mathbf{v} dt + \mathbf{F} dt + \sqrt{2\zeta m k_B T}\, dW(t), \tag{14}$$

where ζ is the friction coefficient, k_B is Boltzmann's constant and T is the temperature. This description is predicated on the assumption that the fluid particles are much lighter than the Brownian particle, and as a result, that the fluid motion relaxes on a much shorter time scale than the motion of the particle. Thus the hydrodynamic forces appear both via the deterministic friction force and the random forces, which are assumed to be Gaussian. If the assumption on the relaxation time of the fluid variables is not applicable the process is no longer Markovian, and a non-Markovian generalization of (14) has been derived [11].

The stochastic differential equations are equivalent, under the Gaussian assumption, to a partial differential equation for the conditional probability density $p(\mathbf{x}, \mathbf{v}, t | \mathbf{x}', \mathbf{v}', t')$, namely,

$$\frac{\partial p}{\partial t} + \nabla_{\mathbf{x}} \cdot \mathbf{v} p + \nabla_{\mathbf{v}} \cdot \left(\left(-\zeta \mathbf{v} + \frac{\mathbf{F}}{m} \right) p \right) = \frac{\zeta k_B T}{m} \nabla_{\mathbf{v}} \cdot \nabla_{\mathbf{v}} p. \tag{15}$$

This is commonly called the Fokker-Planck-Kramers-Klein equation [100], or simply the Fokker-Planck equation, although the latter is used for a much broader class of equations [18, 36, 50]. This is a mixed-type equation that describes drift-diffusion in the velocity component and drift in $\mathbf{x}$ due to the external force. If the latter vanishes it reduces to pure drift-diffusion in velocity space. The equation has also been formally generalized to describe the motion of multiple Brownian particles by incorporating an integral operator on the right-hand side to account for particle-particle interactions [63].

If the friction coefficient ζ is large, one may intuitively expect that the velocity relaxes on a time scale $\mathcal{O}(\zeta^{-1})$, and then (14) reduces to an algebraic equation that can be used to replace the velocity in (13). The result is the Smoluchowski equation

$$\frac{\partial n}{\partial t} = D \nabla_{\mathbf{x}} \cdot \left(\nabla_{\mathbf{x}} n - \frac{1}{k_B T} \mathbf{F} n \right), \tag{16}$$

for the number density $n(\mathbf{x}, t) = \int p \, d\mathbf{v}$, where the diffusion coefficient is defined by the Einstein relation

$$D = \frac{k_B T}{\zeta}.$$

Clearly this has the form of the PKS equation (1) for a suitable choice of the force. However, the reduction as described is formal, since the full equation is a singularly-perturbed hyperbolic equation, and (16) only describes the outer solution [100]. Smoluchowski equations have been widely used in the studies of aggregation, but the limitations are frequently not appreciated. Similar issues arise in the diffusion approximation of velocity-jump processes described in Sect. 4, and we will return to them there.

2.3   Interacting Particles, Liouville's Equation and Reduced Descriptions

Next we suppose that there are no external forces — only inter-particle forces. Newton's second law for the system reads

$$\frac{d\mathbf{x}}{dt} = \mathbf{v}$$

$$M\frac{d\mathbf{v}}{dt} = \mathbf{F}(\mathbf{x}) \tag{17}$$

where $\mathbf{x} = (\mathbf{x}_1, \mathbf{x}_2, \cdots, \mathbf{x}_N)$ is the vector of positions, $\mathbf{v} = (\mathbf{v}_1, \cdots, \mathbf{v}_N)$, $\mathbf{F} = (\mathbf{F}_1, \cdots, \mathbf{F}_N)$, and M is the diagonal matrix with $M_{ii} = m_i$. Note that we assume here that $\mathbf{F}_i$ does not depend on the velocity of any particle, nor does it depend explicitly on time. Velocity-dependence introduces dissipation and substantially changes the BBGKY hierarchy developed later. Thus there is no built-in friction-like force such as arises when an individual interacts with the background environment, nor is there a force for alignment, with the result that it may be difficult to obtain alignment of individuals for such models. This is in contrast to the force

$$\mathbf{F}_i(\mathbf{r}_{ij}) = \sum_j \phi(|\mathbf{r}_{ij}|)(\mathbf{v}_i - \mathbf{v}_j), \tag{18}$$

used in the Cucker-Smale model [22], where $\phi(s)$ is a monotone decreasing function. We assume hereafter that force between i and j depends only on their separation, i.e.,

$$\mathbf{F}_i(\mathbf{x}_j) = \mathbf{F}_i(\mathbf{r}_{ij}) \equiv \mathbf{F}(\mathbf{r}_{i1} \ldots, \mathbf{r}_{ii-1}, \mathbf{r}_{ii+1}, \ldots \mathbf{r}_{iN}),$$

where $\mathbf{r}_{ij} \equiv |\mathbf{x}_i - \mathbf{x}_j|$. Furthermore, we assume that the particles are identical, and that the forces are conservative. Then there is a potential Φ such that

$$\mathbf{F}_i(\mathbf{r}_{ij}) = -\nabla_{\mathbf{x}_i} \Phi,$$

and we assume that Φ can be written as the sum of pairwise interaction potentials

$$\Phi = \sum_{i<j}^{N} \varphi(\mathbf{r}_{ij}).$$

While (17) can be solved locally for N (in fact global solutions exist when Φ is the Newtonian potential and $N \geq 2$, as long as there are no collisions [97]), a less detailed description of the system for large N can be gotten by finding the joint probability $P_N(\mathbf{x}_1, \mathbf{x}_2, \ldots \mathbf{x}_N, \mathbf{v}_1, \ldots \mathbf{v}_N, t) = P_N(\mathbf{x}, \mathbf{v}, t)$ that particle i has position $\mathbf{x}_i$ and velocity $\mathbf{v}_i$. Denote the solution of (17) subject to $\mathbf{x}(0) = \mathbf{x}_o, \mathbf{v}(0) = \mathbf{v}_o$ as $(\mathbf{x}, \mathbf{v}) = (\chi(\mathbf{x}_o, \mathbf{v}_o, t), \mathcal{V}(\mathbf{x}_o, \mathbf{v}_o, t))$, which defines a unique curve in the 6N-dimension phase space for suitable $\mathbf{F}_i$, and implies that

$$P_N(\mathbf{x}, \mathbf{v}, t) = \Pi_{i=1}^{N} \, \delta \, (\mathbf{x}_i - \chi_i(\mathbf{x}_o, \mathbf{v}_o, t)) \, \Pi_{i=1}^{N} \, \delta \, (\mathbf{v}_i - V_i(\mathbf{x}_o, \mathbf{v}_o, t)) \, .$$

Thus if we specify an initial condition with certainty then the probability distribution at any later time is concentrated at one point. Now suppose that we run the 'experiment' many times or that we consider a large number of copies of the system. Given a distribution of initial conditions, P_N is no longer concentrated at a point or a finite number of points, but since there is no dissipation, the evolution of P_N follows from the Reynolds transport theorem [3]. Thus the N-particle distribution function evolves according to

$$\frac{\partial P_N}{\partial t} + \mathbf{v} \cdot \nabla_\mathbf{x} P_N + \frac{\mathbf{F}}{m} \cdot \nabla_\mathbf{v} P_N = 0, \tag{19}$$

which is called Liouville's equation. This is formally equivalent to Newton's equations and thus equally intractable for large numbers of particles, but one can derive equations for reduced or marginal distribution functions, defined as

$$P_l(\mathbf{x}_1, \ldots \mathbf{x}_l, \mathbf{v}_1, \ldots \mathbf{v}_l, t) \equiv \int P_N(\mathbf{x}, \mathbf{v}, t) d\mathbf{x}_{l+1} \ldots d\mathbf{x}_N d\mathbf{v}_{l+1} \ldots d\mathbf{v}_N.$$

Liouville's equation can be written

$$\frac{\partial P_N}{\partial t} + \sum_{k=1}^{N} \mathbf{v}_k \cdot \nabla_{\mathbf{x}_k} P_N - \sum_{\substack{j=1 \\ i, i<j}}^{N} \frac{1}{m} \left[\frac{\partial}{\partial \mathbf{x}_i} \varphi(\mathbf{r}_{ij}) \cdot \frac{\partial}{\partial \mathbf{v}_i} + \frac{\partial}{\partial \mathbf{x}_j} \varphi(\mathbf{r}_{ij}) \cdot \frac{\partial}{\partial \mathbf{v}_j} \right] P_N = 0 \tag{20}$$

and more compactly as

$$\frac{\partial P_N}{\partial t} + \mathscr{L}_N P_N = 0. \tag{21}$$

By integrating over $N - l$ particles one obtains the evolution equation for the l-particle distribution function [16]

$$\frac{\partial f_l}{\partial t} + \mathscr{L}_\ell f_l = -\sum_{i=1}^{l} \frac{\partial}{\partial \mathbf{v}_i} \cdot \int \frac{\mathscr{F}_{i,l+1}}{m} \ f_{l+1}(\mathbf{x}_1, \ldots \mathbf{x}_{l+1}, \mathbf{v}_1, \ldots \mathbf{v}_{l+1}, t) d\mathbf{x}_{l+1} d\mathbf{v}_{l+1}$$

(22)

where

$$\mathscr{F}_{ij} \equiv -\frac{\partial \varphi(\mathbf{r}_{ij})}{\partial \mathbf{x}_i}$$

is the force between particles i and j. These equations for $l = 1, \cdots, N$ are called the BBGKY hierarchy. Clearly the system is not closed unless $l = N$, because for any $l < N$ one must know P_{l+1} to solve (22). Thus we seem once again to have come full circle; the only self-contained equation is Liouville's equation and it is equivalent to Newton's equations.

Of particular use in this context are the evolution equations for the one- and two-particle number density functions.

$$\frac{\partial P_1}{\partial t} + \mathscr{L}_1 P_1 = -\frac{\partial}{\partial \mathbf{v}_1} \cdot \int \frac{\mathscr{F}_{1,2}}{m} \ P_2(\mathbf{x}_1, \mathbf{x}_2, \mathbf{v}_1, \mathbf{v}_2) d\mathbf{x}_2 d\mathbf{v}_2.$$

(23)

$$\frac{\partial P_2}{\partial t} + \mathscr{L}_2 P_2 = -\int \left[\frac{\partial}{\partial \mathbf{v}_1} \cdot \frac{\mathscr{F}_{1,3}}{m} + \frac{\partial}{\partial \mathbf{v}_2} \cdot \frac{\mathscr{F}_{2,3}}{m} \right] P_3(\cdots) d\mathbf{x}_3 d\mathbf{v}_3$$

(24)

As noted previously, when there are no collisions Liouville's equation has a smooth global solution for suitable potentials—it is the collisions that lead to Boltzmann's equation. When only binary interactions are involved, i.e., in the dilute limit, the two-particle distribution function factors and the equation for the single-particle distribution reduces to Boltzmann's equation [12]. Convergence of solutions of the BBGKY solution hierarchy to a smooth solution of a kinetic equation for a single particle distribution function is still an unresolved problem for general particle-particle interactions. A very accessible discussion of this, and in particular of the Boltzmann-Grad continuum limit $N \to \infty, \sigma^2 \to 0 \ N\sigma^2 = \text{constant}$, is given in Cercignani et al. [17]. More complete treatments of mathematical techniques for kinetic equations are given in [16, 17, 60, 83]. Application of the BBGKY hierarchy to derive reduced descriptions for flocking problems is widely-used [15, 38], but the use of idealized kinetic models frequently fails to capture some essential characteristics of animal movement [47].

3   Simple and Reinforced Random Walks in Space

3.1   The Pearson Random Walk

Consider a random jump process on R^n in which the walker executes a sequence of jumps of negligible duration, driven by Poisson forcing $\mathbf{x}$. This is called a random walk, and the earliest analyses of these processes apparently dates to Bachelier [5] around 1900, in the context of his analysis of financial time series. However

Fig. 2 Three steps in a Pearson walk of fixed step-length

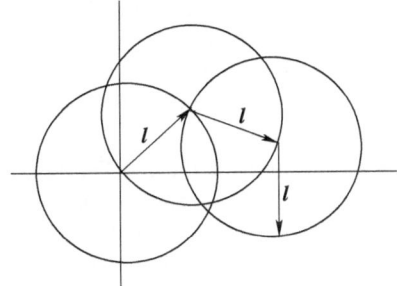

the term 'random walk' was apparently coined by Pearson [81], who proposed the following problem.

A man starts from the point O and walks ℓ yards in a straight line; he then turns through any angle whatever and walks another ℓ yards in a second straight line. He repeats this process n times. I require the probability that after n stretches he is at a distance between r and r + δr from his starting point O.

The solution to this problem had previously been obtained by Rayleigh [84] in a study of the superposition of sound waves. Later we will see that this walk fits into a more general framework that incorporates a waiting time distribution and a jump size distribution, but for now we treat the simple 2D walk shown in Fig. 2. Let $P_n(\mathbf{r})$ be the probability that a walker who begins at the origin is in the interval $(\mathbf{r}, \mathbf{r} + d\mathbf{r})$ at the nth step, and $T(\rho)$ be the probability of taking a step of length $|\rho|$ in the direction $\rho/|\rho|$. If the steps are uncorrelated then $P_n(\mathbf{r})$ satisfies the renewal equation

$$P_{n+1}(\mathbf{r}) = \int_{R^2} T(\rho) P_n(\mathbf{r} - \rho) d\rho. \tag{25}$$

In the Pearson walk the angular distribution is uniform on the circle of radius ℓ and thus

$$T(\rho) = \frac{\delta(|\rho| - \ell)}{2\pi\ell},$$

and for this kernel the probability at the $n + 1$st step is simply the average of the probabilities at the previous step over the circle of radius ℓ centered at $\mathbf{r}$. The solution of (25) is

$$P_n(\mathbf{r}) = \frac{1}{2\pi} \int_0^\infty J_0^n(k\ell) J_0(kr) k \, dk, \tag{26}$$

where $r = |\mathbf{r}|$ [7,55], and in the limit $n \to \infty$ this reduces to

$$P_n(\mathbf{r}) \sim \frac{1}{n\pi\,\ell^2} e^{-r^2/n\ell^2}. \tag{27}$$

The result sought by Pearson is just $2\pi r$ times this, i.e.,

$$P_n(r) \sim \frac{2r}{n\ell^2} e^{-r^2/n\ell^2}$$

which is Rayleigh's result. In an historical coincidence, Einstein's seminal paper [27] on Brownian motion also appeared in 1905, and the parallel between (10) and (27) for the discrete Pearson walk is evident. An isotropic diffusion equation is also derived from the Pearson walk in the chapter by Hillen and Painter, using different notation.

A variation of the 2D Pearson-Rayleigh random walk in which the steps are random vectors of exponential length and uniform orientation was considered in [33]. It is shown there that imposing a constraint of a fixed total length on a walk leads to a number of interesting results. For instance, by taking exactly three steps the probability distribution is uniform in the disc of radius l, while for fewer steps the distribution is concentrated near the boundary and for more it is concentrated near the origin.

3.2 The General Evolution Equation for Space-Jump or Kangaroo Processes

We generalize the simple random walk as follows. Suppose that the waiting times between successive jumps are independent and identically distributed. Let $\mathcal{T}$ be the waiting time between jumps and let $\phi(t)$ be the probability density function (PDF) for the waiting time distribution (WTD). If a jump has occurred at $t = 0$ then

$$\phi(t) = \Pr\{t < \mathcal{T} \leq t + dt\}.$$

The cumulative distribution function for the waiting times is $\Phi(t) = \int_0^t \phi(s)\,ds = \Pr\{\mathcal{T} \leq t\}$ and the complementary cumulative distribution function is $\hat{\Phi}(t) = \Pr\{\mathcal{T} \geq t\} = 1 - \Phi(t)$. If the jumps are exponentially distributed then $\Phi(t) = 1 - e^{-\lambda t}$, and $\phi(t) = \lambda e^{-\lambda t}$, and this is the only smooth distribution for which the jump process is Markovian ([31], p. 458).

In general the jumps in space may depend on the waiting time, and conversely, the WTD may depend on the size of the preceding jump, but to make the analysis tractable, we assume that the spatial redistribution that occurs at jumps is independent of the WTD. Let $T(\mathbf{x}, \mathbf{y})$ be the PDF for a jump from $\mathbf{y}$ to $\mathbf{x}$, i.e., given that a jump occurs at T_i,

$$T(\mathbf{x}, \mathbf{y})\,d\mathbf{x} = \Pr\{\mathbf{x} \leq X(T_i^+) \leq \mathbf{x} + d\mathbf{x} \mid X(T_i^-) = \mathbf{y}\}, \qquad (28)$$

where the superscripts $\pm$ denote limits from the right and left, respectively. If the underlying medium is spatially non-homogeneous and anisotropic, the transition probability depends on $\mathbf{x}$ and $\mathbf{y}$ separately, while in a homogeneous medium $T(\mathbf{x}, \mathbf{y}) = \tilde{T}(\mathbf{x} - \mathbf{y})$, where $\tilde{T}$ is the unconditional probability of a jump of length $|\mathbf{x} - \mathbf{y}|$. In either case, T is a probability kernel if and only if $\int_{R^n} T(\mathbf{x}, \mathbf{y})\,d\mathbf{x} = 1$. We further assume that T is a smooth function and that for any fixed $\mathbf{y}$ the first two

$\mathbf{x}$- moments of T are finite, though they depend on $\mathbf{y}$ unless the system is spatially homogeneous. Later we comment on the effect of infinite moments.

Let $P(\mathbf{x}, t|0)d\mathbf{x}$ be the probability that a jumper which begins at the origin at $t = 0$ is in the interval $(\mathbf{x}, \mathbf{x} + d\mathbf{x})$ at time t. It was shown in [73] that $P(\mathbf{x}, t|0)$ satisfies the renewal equation

$$P(\mathbf{x}, t|0) = \hat{\Phi}(t)\delta(\mathbf{x}) + \int_0^t \int_{R^n} \phi(t - \tau)T(\mathbf{x}, \mathbf{y})P(\mathbf{y}, \tau|0)\,d\mathbf{y}\,d\tau. \qquad (29)$$

Many of the standard jump processes can be recovered from this general result by particular choices of ϕ and T. For instance, if $\phi(t) = \delta(t - t_0)$ then $\Phi(t) = H(t_0 - t)$, where $H(\cdot)$ is the Heaviside function, and (29) reduces to

$$P(\mathbf{x}, t|0) = H(t_0 - t)\delta(\mathbf{x}) + [1 - H(t_0 - t)]\int_{R^n} T(\mathbf{x}, \mathbf{y})P(\mathbf{y}, t - t_0|0)\,d\mathbf{y}.$$

This is the governing equation for a discrete time, continuous space process in which jumps occur at intervals of t_0. If in addition the support of T is concentrated on the points of a lattice $Z^n \subset R^n$, then

$$P(\mathbf{x}_i, t|0) = H(t_0 - t)\delta_{i0} + [1 - H(t_0 - t)]\sum_j T_{ij} P(\mathbf{x}_j, t - t_0|0).$$

where δ_{i0} is the Kronecker delta, and $\mathbf{x}_i$ is a lattice point. This can be written in the more conventional Chapman-Kolmogorov form as follows.

$$P_{i0}(n + 1) = \sum_j T_{ij} P_{j0}(n) \quad n \geq 1$$

If the WTD is exponential, one obtains the continuous time random walk

$$\frac{\partial P}{\partial t}(\mathbf{x}, t|0) = -\lambda P(\mathbf{x}, t|0) + \lambda \int_{R^n} T(\mathbf{x}, \mathbf{y})P(\mathbf{y}, t|0)\,d\mathbf{y}. \qquad (30)$$

and if in addition the support of the kernel $T(\mathbf{x}, \mathbf{y})$ is a lattice then

$$\frac{\partial P}{\partial t}(\mathbf{x}_i, t|0) = -\lambda P(\mathbf{x}_i, t|0) + \lambda \sum_j T_{ij} P(\mathbf{x}_j, t|0). \qquad (31)$$

One can cast the latter into the form of a master equation for a countable state Markov process by applying the condition on T that guarantees conservation of walkers to obtain

$$\frac{\partial P}{\partial t}(\mathbf{x}_i, t|0) = -\lambda \sum_i T_{ij} P(\mathbf{x}_i, t|0) + \lambda \sum_j T_{ij} P(\mathbf{x}_j, t|0). \qquad (32)$$

A generalization of this that allows for other non-exponential WTDs takes the form

$$\frac{\partial P}{\partial t}(\mathbf{x}_i, t|0) = \int_0^t \Psi(t - \tau) \left[-\sum_i T_{ij} P(\mathbf{x}_i, t|0) + \sum_j T_{ij} P(\mathbf{x}_j, t|0) \right] d\tau, \quad (33)$$

and of course one can couple the jump probabilities with the WTD [53].

There is a large literature on the various special cases. For instance, the continuous-time random walk (CTRW) dates at least back to Irwin [48] and has been extensively developed for birth-death processes [40] and on lattices [54,66,98]. The general form (29) was first derived in [73].

3.3 The Evolution of Spatial Moments for General Kernels

To determine how the evolution of the spatial moments in time depends on the waiting time distribution, we assume that the medium is one-dimensional and spatially homogeneous—the generalization to n dimension is straightforward. Let

$$\langle x^n(t) \rangle = \int_{-\infty}^{+\infty} x^n P(x, t|0) \, dx$$

$$= \int_{-\infty}^{+\infty} \int_0^t \int_{-\infty}^{+\infty} x^n \tilde{T}(x - y)\phi(t - \tau)P(y, \tau|0) \, dy \, d\tau \, dx. \quad (34)$$

Denote by

$$m_k = \int_{-\infty}^{+\infty} x^k \tilde{T}(x) \, dx$$

the k-th moment about zero of the jump length distribution—then as shown in [73]

$$\langle x^n(t) \rangle = \int_0^t \sum_{k=0}^n \binom{n}{k} m_k \phi(t - \tau)\langle x^{n-k}(\tau) \rangle \, d\tau, \quad (35)$$

and thus the Laplace transform of the k-th moment is given by

$$X_n = \frac{\bar{\phi}(s)}{1 - \bar{\phi}(s)} \left[\sum_{k=1}^{n-1} \binom{n}{k} m_k X_{n-k} + \frac{m_n}{s} \right].$$

In particular the first two moments are

$$X_1(s) = \frac{m_1}{s} \frac{\bar{\phi}(s)}{1 - \bar{\phi}(s)}$$

$$X_2(s) = \left(2m_1 X_1(s) + \frac{m_2}{s} \right) \frac{\bar{\phi}(s)}{1 - \bar{\phi}(s)}. \quad (36)$$

The two most widely-used waiting time distributions are the exponential distribution and the gamma distribution. Suppose in either case that $m_1 = 0$, since a non-zero first moment simply adds a drift. Then for the exponential WTD one finds that $\bar{\phi}(s) = \lambda/(s + \lambda)$ and that $\langle x^2(t) \rangle = m_2 \lambda t$. If ϕ is a gamma WTD with parameters $(2, \lambda)$, then $\phi(t) = \lambda^2 t e^{-\lambda t}$, $\bar{\phi}(s) = \lambda^2/(s + \lambda)^2$, and

$$\langle x^2(t) \rangle = m_2 \int_0^t \mathcal{L}^{-1}\left(\frac{\lambda^2}{s(s + 2\lambda)}\right) d\tau = \frac{m_2 \lambda}{2}\left\{t - \frac{1}{2\lambda}(1 - e^{-2\lambda t})\right\}. \quad (37)$$

In general the asymptotic behavior of the moments can be gotten by applying limit theorems for Laplace transforms [99]. If we denote the kth moment of the WTD as M_k and suppose that $m_1 = 0$, then the leading terms in an asymptotic expansion of $X_2(s)$ are

$$X_2 = \frac{m_2}{M_1 s^2}\left[1 + \left(\frac{M_2 - 2M_1^2}{2M_1}\right)s + \mathcal{O}(s^2)\right].$$

Therefore, by (i) applying the limit result that $\lim_{s \to 0} f(s) = \lim_{t \to \infty} F(t)$, and (ii) using the fact that

$$\mathcal{L}(t^{\rho-1}) = \frac{\Gamma(\rho)}{s^\rho} \qquad \text{for} \quad \rho > 0,$$

one sees that if the mean waiting time M_1 is finite, then the mean-squared displacement for large t is given by

$$\langle x^2(t) \rangle \sim \frac{m_2}{M_1}t.$$

Thus so far as the mean-squared displacement is concerned, any jump process for which the jump distribution has a finite variance and the WTD has a finite mean behaves like a diffusion process with diffusion coefficient $D = m_2/(2M_1)$ for large t.

To make the connection with the PDE descriptions of motion more explicit, consider first the case of an exponential WTD, and suppose that the jump kernel is spatially homogeneous. If

$$\tilde{T}(\mathbf{x} - \mathbf{y}) = \frac{\delta(|\mathbf{x} - \mathbf{y}| - \ell)}{\omega_n \ell^{n-1}},$$

where $\omega_n = 2\pi^{\frac{n}{2}}/\Gamma(\frac{n}{2})$ is the surface measure of the unit sphere in R^n, one finds that

$$\frac{\partial P}{\partial t} = \lambda[\bar{P}(\mathbf{x}, \ell, t) - P(\mathbf{x}, t)],$$

where $\bar{P}$ is the average of P over the surface of a sphere of radius ℓ centered at $\mathbf{x}$. Expansion of P about $\mathbf{x}$ leads, in the diffusion limit $\lambda \to \infty, \ell \to 0, \lambda \ell^2/2n = D$, to

$$\frac{\partial P}{\partial t} = D\nabla^2 P, \tag{38}$$

provided that all higher-order derivatives are bounded. The Pearson walk described earlier falls into this class.

A similar conclusion holds for more general kernels, written in 1D for simplicity, of the form

$$\tilde{T}(x - y) = \frac{1}{\ell}T_0(\frac{|x - y|}{\ell}, \ell).$$

Then

$$\frac{\partial P}{\partial t} = \lambda \left(\ell \int_R T_0(r, \ell)r\,dr \right) \frac{\partial P}{\partial x} + \lambda \left(\frac{\ell^2}{2} \int_R T_0(r, \ell)r^2 dr \right) \frac{\partial^2 P}{\partial x^2} + \mathscr{O}(\ell^3). \tag{39}$$

Therefore if the first moment of T_0 is $\mathscr{O}(\ell)$ for $\ell \to 0$, if the second moment of T_0 tends to a constant, and if all higher moments are bounded, then in the diffusion limit we obtain a diffusion equation with drift. The diffusion coefficient is given by

$$D = \lambda \frac{\ell^2}{2} \lim_{\ell \to 0} \int_R T_0(r, \ell)r^2 dr \tag{40}$$

and the drift coefficient is given by

$$\beta = \lambda \frac{\ell^2}{2} \lim_{\ell \to 0} \int_R \frac{T_0(r, \ell)}{\ell}r\,dr. \tag{41}$$

The latter vanishes if the kernel is symmetric.

3.4 The Effects of Long Waits or Large Jumps

The fact that any jump process with a WTD that has a finite first moment and a jump distribution having a finite second moment evolves like a standard Brownian motion for large t is simply a reflection of the central limit theorem applied to the sum of the IID steps taken in the walk [51]. When the large-time limit of the mean-square displacement grows either sub- or super-linearly the process is said to exhibit anomalous diffusion. For example, if

$$\langle x^2(t) \rangle \sim \gamma t^\beta$$

for $\beta \neq 1$ and $t \rightarrow \infty$, it is called *subdiffusion* if $\beta < 1$ and *superdiffusion* if $\beta > 1$ [65]. Subdiffusion occurs when particles spread slowly, whether because they rest or are trapped for a long time, and in particular, if the mean waiting time between jumps is infinite. For example, if $m_1 = 0$, then from (36)

$$\langle x^2(t) \rangle = m_2 \mathscr{L}^{-1} \left(\frac{\bar{\phi}(s)}{s(1 - \bar{\phi}(s))} \right).$$

Therefore, if $\bar{\phi}(s) \sim 1/s^\rho$ for $\rho \in (0,1)$ and $s \rightarrow 0$, then $\langle x^2(t) \rangle \sim m_2 t^\rho$ for $t \rightarrow \infty$, i.e., movement is asymptotically subdiffusive. As another example, consider

$$\phi(t) = \frac{1}{(1+t)^2},$$

which is a well-defined distribution, but for which $M_k = \infty$ for all $k \geq 1$. The transform of ϕ is

$$\bar{\phi}(s) = \left(\frac{\pi}{2} - Si(s) \right) \cos s + Ci(s) \sin s$$

where Si and Ci are the sine and cosine integral functions [20]. From the asymptotic expansion of the integrals one finds that

$$\langle x^2(t) \rangle \sim \log t,$$

and thus the process is subdiffusive.

The superdiffusive case arises when the walk is highly persistent in time, for example, if the walker never changes direction, or for walks having a fat-tailed jump distribution. The simplest example of the first case arises when the walker never turns, which leads to a wave equation for which the mean square displacement scales as t^2. More generally this arises if $\bar{\phi}(s) \sim \Gamma(3)/(s^2 + \Gamma(3))$ for $s \rightarrow 0$. An application to bacteria that exhibit long runs is discussed in [64].

The latter case arises when the variance of the jump distribution diverges and the central limit theorem does not apply. The motion corresponds to a Lévy flight, which leads to alternate localized meandering punctuated by occasional long steps. A comparison of a Lévy flight for the jump distribution

$$\tilde{T}(\mathbf{x}) = A_\mu \frac{\sigma}{(\sigma|\mathbf{x}|)^{1+\mu}}$$

for $\mu = 1.5$ with Brownian motion is shown in Fig. 3. The applicability of Lévy flights as a description of animal movement is discussed in [26].

Fig. 3 An example of
Brownian motion (*lower left*)
in the X–Y plane, and a Lévy
walk (*upper right*)
(From [65])

3.5   *Biased Jumps Dependent on Gradients or Internal Dynamics*

Several generalizations of the preceding examples are possible. The WTD for the
jump process can depend on time or on the density of individuals, the redistribution
kernel may depend on the local density or a local average of the density, and of
course the WTD and jump distributions need not be independent. Examples of the
latter case include introduction of a resting phase in which the resting time depends
on the preceding jump length, or alternatively, the WTD distribution may depend
directly on the jump length. It is known that a resting phase with Poisson driven
entry and exits simply rescales the diffusion coefficient in simple random walks
[46, 98].

  If the waiting time distribution depends on the number density n and t, then

$$\phi(n,t) = \lambda(n(\mathbf{x},t),t)e^{-\int_0^t \lambda(n(\mathbf{x},s),s)\,ds}.$$

and the renewal equation for the number density is now the nonlinear equation

$$n(\mathbf{x},t) = e^{-\int_0^t \lambda(n(\mathbf{x},s),s)\,ds} F(\mathbf{x})$$
$$+ \int_0^t \int_R \lambda(n(\mathbf{x},t-\tau),t-\tau)e^{-\int_0^{t-\tau} \lambda(n(\mathbf{x},s),s)\,ds} T(\mathbf{x},\mathbf{y})n(\mathbf{y},\tau)\,d\mathbf{y}\,d\tau.$$

For suitable choices of the dependence on the density this can describe either
aggregation or dispersal. Dispersal at high densities would obtain if $\lambda(n,\cdot)$ is an
increasing function of n, in which case the mean waiting time between jumps is a
decreasing function n. On the other hand, density-dependent aggregation could be
modeled using a λ that decreases with n, in which case the waiting time between

jumps increases with the density. A different approach in which the parameters depend on internal state variables will be discussed later.

The kernel T may also depend on external fields such as the concentration of an attractant and on the internal state of the organism, and one expects this dependence to be reflected in the resulting limit equations. This will be discussed in greater detail in the context of velocity-jump processes, but here we briefly illustrate the issue for space-jump processes.

Let $\mathbf{x} = x\boldsymbol{\xi}$ and $\mathbf{y} = y\boldsymbol{\eta}$, where $\boldsymbol{\xi}$ and $\boldsymbol{\eta}$ are the directions of $\mathbf{x}$ of $\mathbf{y}$. For a fixed $\mathbf{y}$, the average $\mathbf{x}$ after a jump is defined as

$$\bar{\mathbf{x}} = \int T(\mathbf{x}, \mathbf{y}) \mathbf{x}\, d\mathbf{x} = \int T(\mathbf{x}, \mathbf{y}) \boldsymbol{\xi} x^n\, dx\, d\omega_n.$$

The angle between $\bar{\boldsymbol{\xi}}$ and $\boldsymbol{\eta}$ measures the tendency for the next jump to remain aligned with $\boldsymbol{\eta}$. Therefore we define an index of directional persistence as

$$\psi_d \equiv \langle \bar{\boldsymbol{\xi}}, \boldsymbol{\eta} \rangle, \tag{42}$$

and clearly $\psi_d \in [-1, +1]$. If the step lengths are fixed at Δ, as in the Pearson walk, and if the turning probability depends only on the cone angle

$$\theta(\mathbf{x}, \mathbf{y}) \equiv \cos^{-1}\left(\langle \boldsymbol{\xi}, \boldsymbol{\eta} \rangle\right)$$

between $\mathbf{y}$ and $\mathbf{x}$, then $T(\mathbf{x}, \mathbf{y})$ has the form

$$T(\mathbf{x}, \mathbf{y}) = \frac{\delta(|\mathbf{x} - \mathbf{y}| - \Delta)}{\Delta^{n-1}} h\left(\theta(\mathbf{x}, \mathbf{y})\right)$$

for any $n \geq 2$ and a normalized distribution h.

To illustrate how external fields can be incorporated we write

$$T(\mathbf{x}, \mathbf{y}) = \tilde{T}_0(x - y) + T_1(\mathbf{x}, \mathbf{y}),$$

and we suppose that the drift in $\tilde{T}_0$ vanishes, that the bias kernel T_1 has compact support and vanishing first moment, and that

$$\int_{R^n} T_1(\mathbf{x}, \mathbf{y}) P(\mathbf{y}) d\mathbf{y} = \int_{B_\delta(\mathbf{x})} (\mathbf{y} - \mathbf{x}) \cdot \mathbf{F}(S(\mathbf{y})) P(\mathbf{y}) d\mathbf{y}.$$

Here S is a specified field, $\mathbf{F}$ is a vector-valued function of S, and $B_\delta(\mathbf{x})$ is a ball of radius δ, the sensing radius, centered at $\mathbf{x}$. For example, let $\mathbf{F} = -\chi \nabla S$, define $\mathbf{y} - \mathbf{x} = \boldsymbol{\rho}$, and expand around $\mathbf{x}$; then one finds that

$$\int_{R^n} T_1(\mathbf{x}, \mathbf{y}) P(\mathbf{y}) d\mathbf{y} = -\gamma \left[\nabla S(\mathbf{x}) \nabla P(\mathbf{x}) + P(\mathbf{x}) \nabla \nabla S(\mathbf{x}) \right] : \int_{B_\delta(\mathbf{x})} \boldsymbol{\rho} \boldsymbol{\rho} d\rho \quad (43)$$

$$= -\gamma V_n \delta^3 \left[\nabla S(\mathbf{x}) \nabla P(\mathbf{x}) + P(\mathbf{x}) \nabla \nabla S(\mathbf{x}) \right] : \boldsymbol{\delta} \quad (44)$$

where

$$V_n = \pi^{\frac{n}{2}} / \Gamma(\frac{n}{2} + 1)$$

is the volume of B_1 in n dimensions, and $\boldsymbol{\delta}$ is the unit second rank isotropic tensor [71]. Thus the n-dimensional extension of the drift-free version of (39) to include the bias given above reads

$$\frac{\partial P}{\partial t} = D \Delta P - \chi \left(\nabla S \cdot \nabla P + P \nabla \cdot \nabla S \right), \quad (45)$$

which is a form of the chemotaxis equation discussed later.

3.6 Aggregation in Reinforced Random Walks

The rigorous analysis of random walks is more complicated when particle inter-actions, either direct or indirect, are taken into account (cf. [68, 88, 89]). As will be discussed later, *E. coli* releases a diffusible attractant, whereas myxobacteria gliding on a slime trail react to their own contribution to these trails and to the contributions of the other bacteria [101]. There is a growing mathematical literature on what are called reinforced random walks that began with the work of Davis [24]; a recent review can be found in [82]. Here we sketch the approach developed in [72], where the particle motion is governed by a jump process and the walkers modify the transition probabilities on intervals for subsequent transitions of an interval.

Davis [24] considered a reinforced random walk for a single particle in one dimension. Initially there is a weight w_n^i on each interval $(i, i + 1)$, $i \in \mathbb{Z}$ which is equal to w_n^0.[1] If at time n an interval has been crossed by the particle exactly k times, its weight will be

$$w_n^i = w_n^0 + \sum_{j=1}^{k} a_j,$$

where $a_j \geq 0$, $j = 1, \ldots, k$. Furthermore, the transition probabilities are given by

$$P(x_{i+1} = n + 1 | x_i = n) = \frac{w_n^i}{w_n^i + w_{n-1}^i}.$$

---

[1]In this section the weight w may be equivalent to the signal S used earlier, or some function of it.

Davis' main theorem asserts that localization of the particle will occur if the weight on the intervals grows rapidly enough with each crossing, as summarized in the following. Let x_i be the particle position at the ith step, let $X \equiv \{x_i, i \geq 0\}$, and let

$$\text{and} \quad \phi(a) \equiv \sum_{n=1}^{\infty} \left(1 + \sum_{i=1}^{n} a_i \right)^{-1}.$$

Theorem 1. *Suppose that* $w_n^0 = 1$. *Then*

(i) If $\phi(a) = \infty$ *then X is recurrent.*

(ii) If $\phi(a) < \infty$ *then X has finite range and there are random integers n and I such that* $x_i \in (n, n+1)$ *if* $i > I$.

Here recurrent means that every integer is visited infinitely often a.s., i.e., the walker does not become trapped. From this it follows that if $a_j = $ constant, for instance, which corresponds to linear growth of the weight, then X is recurrent almost surely, whereas if the growth is superlinear then the particle oscillates between two random integers almost surely after some random elapsed time. Since the result deals with a single particle it does not directly address the aggregation problem, but it does suggest that if the particles interact only through the modification of the transition probability there may be aggregation if this modification is strong enough.

This theorem motivated the following development, in which we begin with a master equation for a continuous-time, discrete-space random walk. and postulate a generalized form of (31) in which the transition rates depend on the density of a control or modulator species that modulates the transition rates [72]. We restrict attention to one-step jumps, although it is easy, using the framework given earlier, to apply this to general graphs, but one may not obtain diffusion equations in the continuum limit.

Suppose that the conditional probability $p_n(t)$ that a walker is at $n \in \mathbb{Z}$ at time t, conditioned on the fact that it begins at $n = 0$ at $t = 0$, evolves according to the continuous time master equation

$$\frac{\partial p_n}{\partial t} = \hat{\mathcal{T}}_{n-1}^+(W) \, p_{n-1} + \hat{\mathcal{T}}_{n+1}^-(W) \, p_{n+1} - (\hat{\mathcal{T}}_n^+(W) + \hat{\mathcal{T}}_n^-(W)) \, p_n. \quad (46)$$

Here $\hat{\mathcal{T}}_n^{\pm}(\cdot)$ are the transition probabilities per unit time for a one-step jump to $n \pm 1$, and $(\hat{\mathcal{T}}_n^+(W) + \hat{\mathcal{T}}_n^-(W))^{-1}$ is the mean waiting time at the nth site. We assume throughout that these are nonnegative and suitably smooth functions of their arguments. The vector W is given by

$$W = (\cdots, w_{-n-1/2}, w_{-n}, w_{-n+1/2}, \cdots, w_0, w_{1/2}, \cdots). \quad (47)$$

Note that the density of the control species w is defined on the embedded lattice of half the step size. The evolution of w will be considered later; for now we assume that the distribution of w is given. Clearly a time- and p-independent spatial distribution of w can model a heterogeneous environment, but this static situation is not treated here.

As (46) is written, the transition probabilities can depend on the entire state and on the entire distribution of the control species. Since there is no explicit dependence on the previous state the jump process may appear to be Markovian, but if the evolution of w_n depends on p_n, then there is an implicit history dependence, and the space jump process by itself is not Markovian. However, if one enlarges the state space by appending w one obtains a Markov process in this new state space.

Three distinct types of models are developed and analyzed in [72], which differ in the dependence of the transition rates on w; (i) strictly local models, (ii) barrier models, and (iii) gradient models. In the first of these the transition rates are based on local information, so that $\hat{\mathscr{T}}_n^\pm = \hat{\mathscr{T}}(w_n)$, and to simplify the analysis we assume that the jumps are symmetric, i.e., that $\hat{\mathscr{T}}^+ = \hat{\mathscr{T}}^- \equiv \hat{\mathscr{T}}$. In this case (46) reduces to

$$\frac{\partial p_n}{\partial t} = \hat{\mathscr{T}}(p_{n-1}, w_{n-1})p_{n-1} + \hat{\mathscr{T}}(p_{n+1}, w_{n+1})p_{n+1} - 2\hat{\mathscr{T}}(p_n, w_n)p_n.$$

If we assume that there is a scaling of the transition rates such that $\hat{\mathscr{T}} = \lambda \mathscr{T}$, and that the formal diffusion limit

$$\lim_{\substack{h \to 0 \\ \lambda \to \infty}} \lambda h^2 = \text{constant} \equiv D$$

exists, we obtain the nonlinear diffusion equation

$$\frac{\partial p}{\partial t} = D \frac{\partial^2}{\partial x^2}(\mathscr{T}(w)p). \tag{48}$$

The second type is one called a barrier model, for which there are two sub-cases, depending on whether or not the transition rates are re-normalized. In the first case one assumes that $\hat{\mathscr{T}}_n^\pm(W) = \hat{\mathscr{T}}(w_{n\pm1/2})$, which leads to the equation

$$\frac{\partial p}{\partial t} = D\nabla \cdot (\mathscr{T}\nabla p).$$

If one re-normalizes the transition rates so that

$$\lambda(\hat{\mathscr{T}}_n^+(W) + \hat{\mathscr{T}}_n^-(W)) = \text{constant} \equiv \lambda,$$

then after some analysis one finds that in the diffusion limit this leads to

$$\frac{\partial p}{\partial t} = D\frac{\partial}{\partial x}\left(p\frac{\partial}{\partial x}\left(\ln\frac{p}{\mathscr{T}}\right)\right). \tag{49}$$

Table 1 Dependence of the response on the sensing mechanism

	Type of sensing	Taxis velocity	Chemotactic sensitivity	Type of taxis
1.	Local	$-D\nabla\mathscr{T}$	$-D\mathscr{T}'(w)$	Negative if $\mathscr{T}'(w) > 0$
2.	Barrier without re-normalization	0	0	None
3.	Barrier with re-normalization	$D\nabla ln\mathscr{T}$	$D\,(ln\mathscr{T}(w))'$	Positive if $\mathscr{T}'(w) > 0$
4.	Nearest neighbor with re-normalization	$2D\nabla ln\mathscr{T}$	$2D\,(ln\mathscr{T}(w))'$	Positive if $\mathscr{T}'(w) > 0$
5.	Gradient without re-normalization	$2D\beta\nabla\tau$	$2D\beta\tau'(w)$	Positive if $\beta\tau'(w) > 0$
6.	Gradient with re-normalization	$D\dfrac{\beta}{\alpha}\nabla\tau$	$D\dfrac{\beta}{\alpha}\tau'(w)$	Positive if $\beta\tau'(w) > 0$

For later comparison with velocity jump processes we define the chemotactic velocity and sensitivity as

$$\chi = D\,(ln\mathscr{T})_w \qquad u = -D\frac{\partial}{\partial x}\ln p + D\,(ln\mathscr{T}(w))'\frac{\partial w}{\partial x}. \tag{50}$$

Thus the taxis is positive if $\mathscr{T}'(w) > 0$. The simplest form of w-dependence is to assume that $\mathscr{T}(w) = \alpha + \beta w$, and we use this form later in examples.

The last type of sensing leads to the gradient-based, or look-ahead model, for which $T_{n-1}^+ = \alpha + \beta(\tau(w_n) - \tau(w_{n-1}))$ and $T_{n+1}^- = \alpha + \beta(\tau(w_n) - \tau(w_{n+1}))$, $\alpha \geq 0$, and again there are two cases, depending on whether or not the rates are re-normalized. The chemotactic velocities and sensitivities for these and the preceding cases are summarized in Table 1.

Of course we also have to specify the local dynamics for the evolution of w, and here we use the general form

$$\frac{\partial w}{\partial t} = \frac{pw}{1 + \lambda w} + \gamma_r\frac{p}{K + p} - \mu w \equiv R(p, w) \tag{51}$$

in the examples shown in Fig. 4. For all cases we set $D = 0.36$, and in the first panel we show the solution of (49) and (51) for $\alpha = \gamma_r = \mu = 0$ and $\beta = 1, \lambda = 10^{-5}$. The second panel is as in the first, but with $\lambda = 0$, and in the third panel a more complicated transition rate is used (cf. [72]). One sees in that figure that both the dependence of the transition rates on the local modulator w, and the dynamics of w itself, play an important role in the dynamics of the system. In the first panel the solution stabilizes at some smooth distribution, in the second panel the solution blows up in finite time (around t = 9.3—this assertion is supported by analysis of

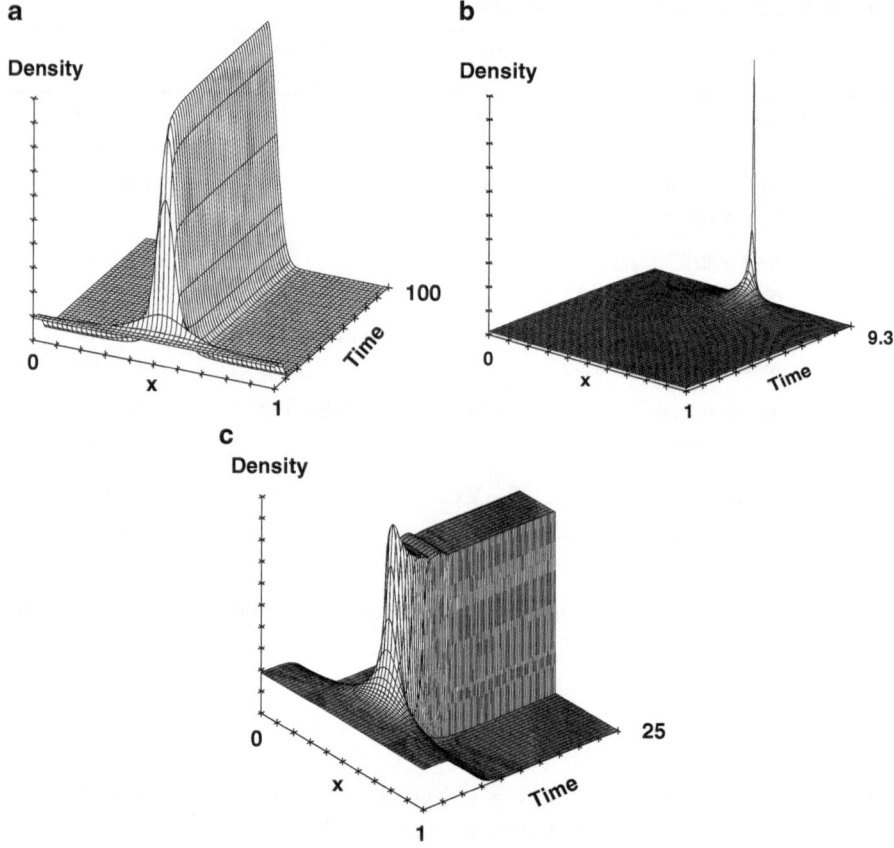

Fig. 4 The density profiles from three examples of the local dynamics. Reproduced from [72], copyright 1997 Society for Industrial and Applied Mathematics

the Fourier components—see [72]) and in the third panel the solution ultimately collapses, in a very interesting step-wise fashion that is not understood at present.

The analysis of reinforced random walks presented in [72] can be generalized in many directions. For example, consider the re-normalized transition rates

$$\hat{\mathscr{T}}_n^\pm(w) = \frac{w_{n\pm1/2}}{w_{n+1/2} + w_{n-1/2}}. \tag{52}$$

These can be regarded as the discrete version of the continuous forms

$$\hat{\mathscr{T}}^+(w) = \frac{\frac{1}{h}\int_x^{x+h} w(s)ds}{\frac{1}{h}\left[\int_{x-h}^{x+h} w(s)ds\right]}$$

$$\hat{\mathscr{T}}^-(w) = \frac{\frac{1}{h}\int_{x-h}^x w(s)ds}{\frac{1}{h}\int_{x-h}^{x+h} w(s)ds}.$$

The continuous version implies that the walker averages over the interval $(x, x + h)$ or $(x - h, x)$ to determine the transition rate. Of course one can incorporate a more general kernel. For example, one might use

$$\hat{\mathcal{T}}^{\pm}(w(x)) = \pm \frac{\int_x^\infty \lambda^2(y - x)^2 e^{-\lambda^2(y-x)^2} w(y)\, dy}{\int_{-\infty}^\infty \lambda^2(y - x)^2 e^{-\lambda^2(y-x)^2} w(y)\, dy}$$

which assigns the maximum weight to $x \pm 1/\lambda$. More generally, we may simply assume that

$$\hat{\mathcal{T}}^+(w(x)) = \frac{\int_x^\infty K(y - x, h)\, w(y)\, dy}{\int_x^\infty K(y - x, h)\, w(y)\, dy + \int_{-\infty}^x K(x - y, h)w(y)\, dy}$$

$$\hat{\mathcal{T}}^-(w(x)) = \frac{\int_{-\infty}^x K(x - y, h)\, w(y)\, dy}{\int_x^\infty K(y - x, h)w(y)\, dy + \int_{-\infty}^x K(x - y, h)w(y)\, dy}$$

for a suitable kernel K. To recover (52) we choose

$$K(y - x, h) = \delta(y - x - \frac{h}{2}).$$

4 Velocity Jump Processes and Taxis Equations

As described in Sect. 2, the velocity-jump (VJ) process is predicated on the assumption that particles make instantaneous jumps in velocity space, rather than in physical space [73]. By comparing the underlying basis of the FPKK equation with that of the Smoluchowski equation, one should expect that the VJ process gives rise to evolution equations that depend jointly on physical- and velocity-space operators. Just as the FPKK equation leads to the Smoluchowski equation in certain regimes, it is known that the long-time asymptotics of VJ processes lead to diffusion processes in space under suitable scalings of space and time [1,41,77]. In this section we define the general VJ process and summarize results on diffusion limits of this process. In the last subsections we describe the application of this process to two classes of biological organisms—swimming bacteria and crawling cells.

4.1 The General Velocity-Jump Process

We shall work directly with the differential equation form of the conservation equation for a phase space density function that depends only on the position, velocity, time and some intracellular variables. In essence the resulting equation is the analog of the Liouville equation (19) with an additional term to account for the gain or loss of particles at a point in phase space due to the underlying jump process. Throughout we focus on the evolution of a smooth density function, and do not address the question of how to connect this to limiting forms of the empirical density for an N-particle system.

Let $p(\mathbf{x}, \mathbf{v}, \mathbf{y}, t)$ be the density function for individuals in a $2n + m$-dimensional phase space with coordinates $(\mathbf{x}, \mathbf{v}, \mathbf{y})$, where $\mathbf{x} \in \mathbb{R}^n$ is the position of an individual, $\mathbf{v} \in \mathbb{R}^n$ is its velocity, and $\mathbf{y}$ is the set of intracellular state variables involved in cell movement. The evolution of p is governed by the equation

$$\frac{\partial p}{\partial t} + \nabla_{\mathbf{x}} \cdot (\mathbf{v}p) + \nabla_{\mathbf{v}} \cdot (\mathbf{F}p) + \nabla_{\mathbf{y}} \cdot (\mathbf{f}p) = \mathscr{R}, \tag{53}$$

where $\mathbf{F}$ denotes an external, velocity-independent force acting on the individuals, $\mathbf{f}$ is the rate of change of the internal variable $\mathbf{y}$, and $\mathscr{R}$ is the rate of change of p due to birth/death processes, a jump process that generates random changes of velocity, etc. Normally cell proliferation is independent of the velocity, and the rate of proliferation can be approximated by $r(n)p$, where $r(n)$ is the density-dependent growth rate, but here we only include random velocity changes. In addition we assume that cells are sufficiently separated and neglect cell-cell mechanical interactions.

The jump process for velocity changes is the direct analog of the stochastic process underlying space jumps. Initially we suppose that the waiting time between jumps and the changes in velocity are independent, and that the WTD is exponential. As a result, the turning can be described by two quantities, the turning rate λ, and the turning kernel $T(\mathbf{v}, \mathbf{v}')$, which defines the probability of a change in velocity from $\mathbf{v}'$ to $\mathbf{v}$, given that a reorientation occurs. $T(\mathbf{v}, \mathbf{v}')$ is non-negative and normalized so that $\int T(\mathbf{v}, \mathbf{v}') \, d\mathbf{v} = 1$, and at present we assume that it is independent of time and space. In light of the foregoing assumptions, (53) becomes

$$\frac{\partial p}{\partial t} + \nabla_{\mathbf{x}} \cdot (\mathbf{v}p) + \nabla_{\mathbf{v}} \cdot (\mathbf{F}p) + \nabla_{\mathbf{y}} \cdot (\mathbf{f}p) = -\lambda p + \lambda \int T(\mathbf{v}, \mathbf{v}') p(\mathbf{x}, \mathbf{v}', t) \, d\mathbf{v}', \tag{54}$$

and the underlying stochastic process is called a *velocity jump process*. For most purposes one does not need the distribution p, but only its first few velocity moments. The first three are the observable density of individuals $n(\mathbf{x}, t)$, the momentum, and the momentum flux.

$$n(\mathbf{x}, t) = \int p(\mathbf{x}, \mathbf{v}, \mathbf{y}, t) d\mathbf{v}d\mathbf{y}, \quad \mathbf{j}(\mathbf{x}, t) = \int p(\mathbf{x}, \mathbf{v}, \mathbf{y}, t) \mathbf{v} \, d\mathbf{v}d\mathbf{y}$$

$$\mathbf{P} = \int p(\mathbf{x}, \mathbf{v}, \mathbf{y}, t) \mathbf{v}\mathbf{v} \, d\mathbf{v}d\mathbf{y}.$$

The momentum $\mathbf{j}$ defines the average velocity $\mathbf{u} = \mathbf{j}/n$. Integration of (54) over $(\mathbf{v}, \mathbf{y})$ leads to

$$\frac{\partial n}{\partial t} + \nabla_{\mathbf{x}} \cdot n\mathbf{u} = 0. \tag{55}$$

When λ is independent of $\mathbf{y}$, multiplication of (54) by $\mathbf{v}$ and integration over $(\mathbf{v}, \mathbf{y})$ yields

$$\frac{\partial(n\mathbf{u})}{\partial t} + \nabla \cdot \mathbf{P} - \mathbf{F}n = -\lambda n\mathbf{u} + \lambda \int T(\mathbf{v}, \mathbf{v}')\mathbf{v}p(\mathbf{x}, \mathbf{v}', \mathbf{y}, t)\, d\mathbf{v}'d\mathbf{v}d\mathbf{y}. \tag{56}$$

These are not closed, except in a special case noted later, due to the presence of the momentum flux tensor $\mathbf{P}$ and the integral term on the right. Until stated otherwise, we assume that $\mathbf{F} = \mathbf{0}$.

It is observed experimentally that the movement of cells often exhibits directional persistence, and as a result, the turning kernel depends on the angle θ between the previous velocity $\mathbf{v}'$ and the new direction $\mathbf{v}$ [8, 39, 59, 62]. Let s denote the cell speed, and $\mathbf{e_v}$ denote the direction of the velocity, then, $\mathbf{v} = s\mathbf{e_v}$. For a fixed $\mathbf{v}'$, the average velocity $\mathbf{v}$ after reorientation is defined as

$$\bar{\mathbf{v}} = \int T(\mathbf{v}, \mathbf{v}')\mathbf{v}\, d\mathbf{v} = \int T(\mathbf{v}, \mathbf{v}')s^n\mathbf{e_v}\, ds\, d\omega_n$$

and the average speed is

$$\bar{s} \equiv \int T(\mathbf{v}, \mathbf{v}')\, \|\,\mathbf{v}\,\|\, d\mathbf{v} = \int T(\mathbf{v}, \mathbf{v}')s^n\, ds\, d\omega_n.$$

As in the space-jump framework, we characterize persistence via an index of directional persistence, defined as

$$\psi_d \equiv \frac{\bar{\mathbf{v}} \cdot \mathbf{v}'}{\bar{s}s'} \in [-1, +1], \tag{57}$$

which measures the tendency of the motion to persist in a given direction $\mathbf{e_{v'}}$. Of particular interest is the case in which the speed does not change with reorientation and the turning probability depends only on θ. Then $T(\mathbf{v}, \mathbf{v}')$ has the form

$$T(\mathbf{v}, \mathbf{v}') = h\left(\theta(\mathbf{v}, \mathbf{v}')\right) \tag{58}$$

for any $n \geq 2$. For such T, ψ_d is independent of $\mathbf{v}'$ and

$$\bar{\mathbf{v}} = \psi_d\mathbf{v}', \tag{59}$$

where

$$\psi_d = \begin{cases} 2\int_0^\pi h(\theta)\cos\theta\, d\theta & \text{for } n = 2 \\ 2\pi\int_0^\pi h(\theta)\cos\theta\sin\theta\, d\theta & \text{for } n = 3. \end{cases} \tag{60}$$

Observations of the movement of *Dictyostelium discoideum* (Dd) amoeba yield $\psi_d \approx 0.7$ [39], whereas the three-dimensional bacterial random walk data in [8] show $\psi_d \approx 0.33$.

External signals enter either through a direct effect on the turning rate λ and the turning kernel T, or indirectly via internal variables $\mathbf{y}$ that reflect the external signal and in turn influence λ and/or T. The first case arises when experimental results are used to directly estimate parameters in the equation [32], but the latter approach is more fundamental. The reduction of (54) to the macroscopic chemotaxis equations for the first case is done in [41, 70], and second case is done in [28–30, 104, 105]. In [104], external forces are also included. We summarize some of the important aspects of the reduction in the following sections.

4.2   The Telegraph Process

A simple example will illustrate both the reduction of the jump process to a diffusion process, and how the parameters of the jump process have to be controlled so as to produce aggregation. Suppose that the walkers are confined to the real line $\mathbb{R}$, that the speeds $s^{\pm}$ to the right and left may depend on position, and that direction is reversed at random instants governed by Poisson processes of intensity $\lambda^{\pm}$. Let $p^{\pm}$ denote the density of walkers moving to the right and left, respectively. Then the conservation equations for these densities are[2]

$$\frac{\partial p^+}{\partial t} + \frac{\partial (s^+ p^+)}{\partial x} = -\lambda^+ p^+ + \lambda^- p^-,$$

$$\frac{\partial p^-}{\partial t} - \frac{\partial (s^- p^-)}{\partial x} = \lambda^+ p^+ - \lambda^- p^-. \tag{61}$$

Let $n \equiv p^+ + p^-$ be the macroscopic density and note that the flux j is $(s^+ p^+ - s^- p^-)$; then (61) can be written in the alternative form

$$\frac{\partial n}{\partial t} + \frac{\partial j}{\partial x} = 0,$$

$$\frac{\partial j}{\partial t} + s^+ \frac{\partial (s^+ p^+)}{\partial x} + s^- \frac{\partial (s^- p^-)}{\partial x} = (s^+ + s^-)(-\lambda^+ p^+ + \lambda^- p^-). \tag{62}$$

To illustrate the essence of aggregation and taxis in this simple context, we ask how the walkers should modify their behavior so as to produce a nonuniform distribution in space at steady-state, and we consider three cases.

---

[2]These equations are the restriction of (54) to one-space dimension only when the speeds $s^{\pm}$ are constant, and in that case the moment equations close at the second level for constant λ [73]. We consider the more general case for illustrative purposes.

Case I: Constant and equal turning rates and speeds, $\lambda^+ = \lambda^- = \lambda_0$ and $s^+(x) = s^-(x) = s_0$

By combining the two equations at (62) we obtain the classical telegrapher's equation

$$\frac{\partial^2 n}{\partial t^2} + 2\lambda_0 \frac{\partial n}{\partial t} = s^2 \frac{\partial^2 n}{\partial x^2}, \tag{63}$$

and by formally taking the limit $\lambda_0 \to \infty, s \to \infty$ with $s^2/\lambda_0 \equiv 2D$ constant in (63), one obtains the diffusion equation. However the limiting procedure can be made more precise by considering the exact solution of (63), which is

$$n(x,t) = \begin{cases} \dfrac{e^{-\lambda_0 t}}{2}\left(\delta(x-st) + \delta(x+st) + \dfrac{\lambda_0}{s}\left[I_0(\Lambda) + \dfrac{\lambda_0 t}{\Lambda}I_1(\Lambda)\right]\right) + n_0 & |x| \le st, \\ n_0 & |x| > st. \end{cases}$$

Here I_0 and I_1 are modified Bessel functions of the first kind. By applying the asymptotic expansions

$$I_0(z) = \frac{e^z}{\sqrt{2\pi z}} + \mathscr{O}\left(\frac{1}{z}\right), \qquad I_1(z) = \frac{e^z}{\sqrt{2\pi z}} + \mathscr{O}\left(\frac{1}{z}\right), \qquad \text{as } z \to \infty,$$

one finds that

$$n(x,t) = \frac{1}{\sqrt{4\pi D t}} e^{-\frac{x^2}{4Dt}} + n_0 + e^{-\lambda_0 t}\mathscr{O}(\xi^2), \qquad \xi^2 \equiv (x/st)^2.$$

From this one sees that the telegraph process reduces to a diffusion process on space scales that are small compared with the ballistic scale st. This fact was known to Einstein and this process has since been studied by many [34, 37, 49, 73, 92].

If we define $\tau = \epsilon^2 t$ and $\xi = \epsilon x$, where ϵ is a small parameter, then (63) reduces to

$$\epsilon^2 \frac{\partial^2 n}{\partial \tau^2} + 2\lambda_0 \frac{\partial n}{\partial \tau} = s^2 \frac{\partial^2 n}{\partial \xi^2}. \tag{64}$$

In these coordinates $x/(st) = \epsilon\xi/(s\tau)$ and the diffusion regime only requires that $\xi/(s\tau) \le \mathscr{O}(1)$. In the limit $\epsilon \to 0$ the exact solution can be used to show that (64) again reduces to the diffusion equation, both formally and rigorously (for t bounded away from zero). However this shows that the approximation of the telegraph process by a diffusion process hinges on the appropriate relation between the space and time scales, not necessarily on the limit of speed and turning rate tending to infinity. In any case, it is clear that the spatial distribution of n is asymptotically constant, and thus there is no localization of walkers. Imposing no-flux boundary conditions on a finite interval does not change this conclusion.

Case II: Constant and equal turning rate $\lambda^+ = \lambda^- = \lambda_0$, distinct speed $s^+(x) \neq s^-(x)$

By assuming that the flux j at infinity vanishes, and solving for the steady state solutions of (61), one finds that

$$p^+(x) = \left[\frac{s^+(0)p^+(0)}{s^+(x)} \right] e^{\lambda_0 \int_0^x \frac{s^+ - s^-}{s^+ s^-} d\xi},$$

$$p^-(x) = \left[\frac{s^+(0)p^+(0)}{s^-(x)} \right] e^{\lambda_0 \int_0^x \frac{s^+ - s^-}{s^+ s^-} d\xi},$$

where the constant $p^+(0)$ is the cell density moving to the right at $x = 0$, which is determined from the conservation of total particle number. From this we see that, (a) aggregation can occur when the speed of the cell depends on the spatial location, i.e., $s^\pm$ are not constants, (b) the distributions for the right-moving cells and the left-moving cells differ if $s^+(x) \neq s^-(x)$, and (c) if $s^+ = s^-$, both left- and right-moving cells aggregate at points of low speed. This is somewhat similar to the scenario of traffic flow—when the road becomes narrower, cars slow down, and traffic jams may form.

Case III: Distinct turning rates $\lambda^+(x) \neq \lambda^-(x)$, constant and equal speeds $s^+ = s^- = s_0$

We write

$$\lambda^\pm = \frac{\lambda^+ + \lambda^-}{2} \pm \frac{\lambda^+ - \lambda^-}{2} =: \lambda_0 \pm \lambda_1,$$

then the density-flux form (62) becomes

$$\frac{\partial n}{\partial t} + \frac{\partial j}{\partial x} = 0,$$

$$\frac{\partial j}{\partial t} + s^2 \frac{\partial n}{\partial x} = -2\lambda_0 j - 2s\lambda_1 n. \tag{65}$$

When λ_0 is constant this reduces to

$$\frac{\partial^2 n}{\partial t^2} + 2\lambda_0 \frac{\partial n}{\partial t} = s^2 \frac{\partial^2 n}{\partial x^2} - 2s \frac{\partial}{\partial x}(\lambda_1 n). \tag{66}$$

We call this a hyperbolic aggregation or taxis equation, and we will see later how this emerges in general. The difference of the turning rate produces a drift in the dynamical evolution equal to $u_c = s\lambda_1/\lambda_0$. This is similar to what is observed in a 1D space-jump process when the probability of right and left jumps differ.

The steady-state solution of (65) is

$$n(x) = n_0 \exp\left\{-\frac{2}{s}\int_0^x \lambda_1(\xi)d\xi\right\},$$

and again there may be a non-constant solution, which is a result of the difference in turning of cells.

We see from the simple 1D process that non-uniform cell distributions can arise when either the cell speeds are different or the turning rates are different, and these two cases correspond to what are called chemotaxis and chemokinesis, resp. In particular, in case 4.2 cells aggregate where their speed is lowest, which is the case when amoeboid cells reach the peak of a potential attractant, while in case 4.2 cells aggregate most strongly when the turning rate deviation λ_1 returns to zero, which happens when run-and-tumble cells adapt to the signal gradient.

4.3   Reduction of the VJ Process to a Diffusion Process

In general, in higher space dimensions equations (55) and (56) do not specify n and $\mathbf{u}$ as they stand, for they involve the second $\mathbf{v}$ moment of p and the as yet unspecified kernel $T(\mathbf{v}, \mathbf{v}')$. We call the process unbiased when the turning rate and kernel depend only on $\mathbf{v}$ and $\mathbf{v}'$, and biased when external fields or internal state variables are included. Note that an unbiased kernel does not mean that reorientation is isotropic. We assume hereafter that λ is independent of the velocity, and we write (54) for the unbiased process as

$$\frac{\partial}{\partial t}p(\mathbf{x}, \mathbf{v}, t) + \mathbf{v} \cdot \nabla p(\mathbf{x}, \mathbf{v}, t) = -\lambda p(\mathbf{x}, \mathbf{v}, t) + \lambda \int_V T_0(\mathbf{v}, \mathbf{v}')p(\mathbf{x}, \mathbf{v}', t)d\mathbf{v}' \equiv \mathscr{L}_0 p(\mathbf{x}, \mathbf{v}, t).$$
(67)

We consider the spatial domain $\Omega = \mathbb{R}^n$, and we suppose that the velocities lie in a compact set $V \subset \mathbb{R}^n$ that is symmetric with respect to the origin.

To state some of the results from [41], we let $\mathscr{K}$ denote the cone of nonnegative functions in $L^2(V)$, and for fixed $(\mathbf{x}, t)$ define an integral operator $\mathscr{T}$ and its adjoint $\mathscr{T}^*$ by

$$\mathscr{T}p = \int_V T(\mathbf{v}, \mathbf{v}')p(\mathbf{x}, \mathbf{v}', t)d\mathbf{v}', \qquad \mathscr{T}^*p = \int_V T(\mathbf{v}', \mathbf{v})p(\mathbf{x}, \mathbf{v}', t)d\mathbf{v}'. \quad (68)$$

We impose the following conditions on the kernel and the integral operator:

(T1) $T(\mathbf{v}, \mathbf{v}') \geq 0$, $\int_V T(\mathbf{v}, \mathbf{v}')d\mathbf{v} = 1$,   and $\int_V \int_V T^2(\mathbf{v}, \mathbf{v}')d\mathbf{v}'d\mathbf{v} < \infty$.

(T2)   *There are functions u_0, ϕ, and $\psi \in \mathscr{K}$ with $u_0 \not\equiv 0$ and $\phi, \psi \neq 0$ a.e. such that for all $(\mathbf{v}, \mathbf{v}') \in V \times V$*

$$u_0(\mathbf{v})\phi(\mathbf{v}') \leq T(\mathbf{v}', \mathbf{v}) \leq u_0(\mathbf{v})\psi(\mathbf{v}'). \quad (69)$$

(T3) $\|\mathscr{T}\|_{\langle 1 \rangle^{\perp}} < 1$, where $\langle 1 \rangle^{\perp}$ is the orthogonal complement in $L^2(V)$ of the span of 1.
(T4) $\int_V T(\mathbf{v}, \mathbf{v}')d\mathbf{v}' = 1$.

Then the turning operator $\mathscr{L}_0 p(\mathbf{x}, \mathbf{v}, t)$ acts in $L^2(V)$, and has the following spectral properties [41].

Theorem 2. *Assume (T1)–(T4); then the following hold.*

1. *0 is a simple eigenvalue of $\mathscr{L}_0$, and the corresponding eigenfunction is $\phi(\mathbf{v}) \equiv 1$.*
2. *There is a decomposition $L^2(V) = \langle 1 \rangle \oplus \langle 1 \rangle^{\perp}$, and, for all $\psi \in \langle 1 \rangle^{\perp}$,*

$$\int_V \psi \mathscr{L}_0 \psi d\mathbf{v} \leq -\mu_2 \|\psi\|^2_{L^2(V)}, \quad \text{where} \quad \mu_2 \equiv \lambda(1 - \|\mathscr{T}\|_{\langle 1 \rangle^{\perp}}). \quad (70)$$

3. *All nonzero eigenvalues μ satisfy $-2\lambda < \text{Re } \mu \leq -\mu_2 < 0$, and to within scalar multiples there is no other positive eigenfunction.*
4. *$\|\mathscr{L}_0\|_{\mathbf{L}(L^2(V), L^2(V))} \leq 2\lambda$.*
5. *$\mathscr{L}_0$ restricted to $\langle 1 \rangle^{\perp} \subset L^2(V)$ has an inverse $\mathscr{F}$ with norm*

$$\|\mathscr{F}\|_{\mathbf{L}(\langle 1 \rangle^{\perp}, \langle 1 \rangle^{\perp})} \leq \frac{1}{\mu_2}. \quad (71)$$

If for example the turning kernel $T(\mathbf{v}, \mathbf{v}')$ is symmetric, then the constant μ_2 given in (70) is the negative of the second eigenvalue of the turning operator $\mathscr{L}_0$. This defines a time scale for relaxation of the reorientation process, and in particular, if 1 is not a simple eigenvalue of $\mathscr{T}$, the streaming character of the transport process dominates, and we can no longer expect to obtain a diffusion limit.

Under the preceding assumptions the parabolic scaling $\tau = \epsilon^2 t$ and $\xi = \epsilon \mathbf{x}$, where ϵ is a small dimensionless parameter, leads to a diffusion approximation of the transport equation [41]. In these variables we have

$$\epsilon^2 \frac{\partial p}{\partial \tau} + \epsilon \mathbf{v} \cdot \nabla_{\xi} \, p = -\lambda p + \lambda \int_V T(\mathbf{v}, \mathbf{v}')p(\xi, \mathbf{v}', \tau)d\mathbf{v}'. \quad (72)$$

where the subscript on ∇, which we drop hereafter, indicates differentiation with respect to the scaled space variable. The right-hand side of (72) is $\mathscr{O}(1)$ compared with the left-hand side, whatever the magnitude of p, and this leads to a diffusion equation for the lowest order term p_0 of an outer expansion, which we write as

$$p(\xi, \mathbf{v}, \tau) = \sum_{i=0}^{k} p_i(\xi, \mathbf{v}, \tau)\epsilon^i + \epsilon^{k+1} p_{k+1}(\xi, \mathbf{v}, \tau). \quad (73)$$

An approximation result for any order in ϵ that provides a bound on the difference between the solution of the transport equation and an expansion derived from the solution of the associated parabolic diffusion equation has also been proven.

Theorem 3. *[41] Assume* (T1)–(T4) *and the Hilbert expansion (73), where p_0 solves the parabolic limit equation*

$$\frac{\partial p_0}{\partial \tau} - \nabla \cdot (D \nabla p_0) = 0, \qquad p_0(\boldsymbol{\xi}, 0) = \int_V p(\boldsymbol{\xi}, \mathbf{v}, 0) d\mathbf{v}, \qquad (74)$$

with diffusion tensor

$$D = -\frac{1}{\omega} \int_V \mathbf{v} \mathscr{F} \mathbf{v} d\mathbf{v}. \qquad (75)$$

In addition, the higher order corrections are given by

$$p_1 = \mathscr{F}(\mathbf{v} \cdot \nabla p_0), \qquad p_2 = \mathscr{F}(p_{0,\tau} + \mathbf{v} \cdot \nabla \mathscr{F} \mathbf{v} \cdot \nabla p_0),$$

where $\mathscr{F}$ is the pseudoinverse defined in Theorem 2 and $\omega = |V|$. Then, for each $\vartheta > 0$, there exists a constant $C > 0$ such that for each $\vartheta/\epsilon^2 < t < \infty$ and each $x \in \mathbb{R}^n$

$$\|p(x, ., t) - q_2(\epsilon x, ., \epsilon^2 t)\|_{L^2(V)} \leq C \epsilon^3, \text{[3]}$$

and the constant C depends on $\mu_2, V, D,$ and ϑ.

In general, the approximate solution depends only on the solution of the limiting parabolic equation, and, therefore, it cannot be uniformly valid in time (cf. [41]). When the speed is constant and the outgoing directions are uniformly distributed on S^{n-1}, $\mathscr{F} = -\lambda^{-1}$, and

$$D = \frac{1}{\omega} \int_V \frac{\mathbf{v}\mathbf{v}}{\lambda} d\mathbf{v} = \frac{s^2}{\lambda n} I.$$

One can prove in general that the diffusion tensor is positive definite, and one can also derive necessary and sufficient conditions for it to be a scalar multiple of I.

Since the reduction depends critically on the existence of the parabolic scaling, we give an example of how it is determined. Let L be a characteristic scale associated with the macroscopic evolution, for instance, the size of the domain on which an experiment is done. Define the dimensionless velocity, space and time variables

$$\mathbf{u} = \frac{\mathbf{v}}{s} \quad \boldsymbol{\xi} = \frac{\mathbf{x}}{L} \quad \tau = \frac{t}{\sigma},$$

where s is a characteristic speed and σ is as yet undetermined. Then

$$\frac{1}{\sigma} \frac{\partial p}{\partial \tau} + \left(\frac{s}{L}\right) \mathbf{u} \cdot \nabla^* p = -\lambda p + \lambda \int T(\mathbf{u}, \mathbf{u}') p(\boldsymbol{\xi}, \mathbf{u}', \tau) d\mathbf{u}',$$

---

[3]In [41] this estimate appears with the L^2-norm squared, but it is clear from the proof that there should be no square.

We estimate a diffusion coefficient as the product of the characteristic speed times the average distance traveled between velocity jumps, which gives $D \sim \mathcal{O}(s^2/\lambda)$, and a characteristic drift time

$$\tau_{DIFF} \sim \frac{L^2}{D} = \frac{L^2\lambda}{s^2}, \qquad \tau_{DRIFT} = \frac{L}{s},$$

A characteristic speed for bacteria such as *E. coli* is 10–20 μ/s, and $\lambda^{-1} \sim \mathcal{O}(1)$ s. On a length scale of 1 mm $\tau_{DRIFT} \sim 50$–100 s and $\tau_{DIFF} \sim 2{,}500 - 10^4$ s. Therefore, in this example we have $\tau_{RUN} \sim \mathcal{O}(1)$ on the dimensional scale, and

$$\tau_{DRIFT} \sim O(1/\epsilon), \qquad \tau_{DIFF} \sim O(1/\epsilon^2)$$

where $\epsilon \sim \mathcal{O}(10^{-2})$. Then

$$\tau_{RUN} \equiv \lambda^{-1} \ll \tau_{DRIFT} \ll \tau_{DIFF}$$

and the scaled equation results for $\sigma = \tau_{DIFF}$.

When biases are introduced their magnitude relative to the base turning rate is critical. We write the kernel with bias as $T(\mathbf{v}, \mathbf{v}', p(\cdot))$, and if, for example, we assume the bias is linear in a signal gradient, then

$$T(\mathbf{v}, \mathbf{v}', p(\cdot)) = T_0(\mathbf{v}, \mathbf{v}') + \kappa(\mathbf{v} \cdot \nabla p)(\mathbf{v}' \cdot \nabla p).$$

One finds that

$$D(\boldsymbol{\xi}, \tau) = \frac{s^2}{\lambda_0 n}\left(I + \frac{\omega s^2}{n}\kappa \nabla p \nabla p\left(I - \frac{\omega s^2}{n}\kappa \nabla p \nabla p\right)^{-1}\right),$$

and as expected, there is no drift or taxis in this case.

On the other hand, if the perturbation is $\mathcal{O}(\epsilon)$, and linear in the gradient, then one finds that

$$\frac{\partial p_0}{\partial \tau} = \nabla \cdot (D \nabla p_0 - u_c p_0),$$

where the drift or chemotactic velocity is given by

$$\mathbf{u}_c \equiv -\frac{\lambda_0}{\omega} \int\int v \mathscr{F}_0 T_1(\mathbf{v}, \mathbf{v}') d\mathbf{v}' d\mathbf{v}.$$

Here F_0 denotes the pseudo inverse defined by the kernel T_0. If Q_1 has the particular form

$$Q_1 = k_1(\mathbf{v}', S)\mathbf{v}$$

then

$$\chi(p) = k(p)\frac{s^2}{\omega n}.$$

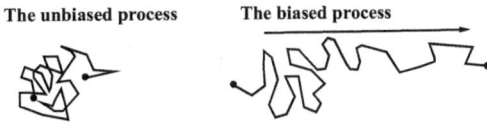

Fig. 5 The movement of a particle driven by a VJ process, in the absence (*left*) and presence (*right*) of an external bias

and the lowest-order approximation is the solution of

$$\frac{\partial p_0}{\partial \tau} = \nabla \cdot \left(\frac{s^2}{n} \nabla p_0 - p_0 \chi(p) \nabla p \right).$$

4.4   The Role of Internal Dynamics

The most widely-studied examples of organisms whose motion can be described as a velocity jump process are the flagellated bacteria, the most-studied of which is *E. coli*. *E. coli* generates the force needed for swimming by rotating flagella embedded in the cell membrane, and thus the swimming speed is fixed by the hydrodynamic loading, and can be taken to be essentially constant in a specified medium. To search for food or escape an unfavorable environment, *E. coli* alternates two basic behavioral modes, swimming in a more or less straight line called a run, and a highly erratic motion called tumbling, the purpose of which is to reorient the cell. Run times are typically much longer than the time spent tumbling, and when bacteria move in a favorable direction (i.e., either in the direction of foodstuffs or away from harmful substances) the run times are increased further. Conversely, when bacteria move in an unfavorable direction the run length decreases and the relative frequency of tumbling increases. The distribution of new directions is not uniform on the unit sphere, but has a bias in the direction of the preceding run. The effect of alternating these two modes of behavior, and in particular, of increasing the run length when moving in a favorable direction, is that a bacterium executes a three-dimensional random walk with drift in a favorable direction when observed on a sufficiently long time scale [9, 56] (cf. Fig. 5).

To illustrate the main points involved in the inclusion of internal dynamics in macroscopic equations, we begin with a simple example based on *E. coli*, and assume that there is no interaction between cells. This is a reasonable assumption, since typical bacterial densities are of the order of 10^8/ml and individual bacteria have a volume per cell of order $\pi \mu m^3$—thus the volume fraction is $\mathcal{O}(10^{-3})$. Therefore we can consider either the probability of a single walker being at a given position with a given velocity at time t, or the density of walkers, and we choose the latter here.

New technology has led to extensive experimental data at the cell and molecular level, and as a result, more complete descriptions of inter- and intracellular signal transduction are possible for use in population-level models of *E. coli*. Detailed models of the full signal transduction network exist [87, 102], but simplified cartoon models that capture the essential dynamics involved in aggregation and patterning have been used in recent studies [28, 29, 104]. By neglecting body forces and cell growth, the transport equation for the cell density becomes

$$\frac{\partial p}{\partial t} + \nabla_x \cdot (\mathbf{v}p) + \nabla_y \cdot (\mathbf{f}p) = -\lambda(\mathbf{y})p + \int_V \lambda(\mathbf{y})T(\mathbf{v},\mathbf{v}',\mathbf{y})p(\mathbf{x},\mathbf{v}',\mathbf{y},t)\,d\mathbf{v}', \quad (76)$$

where $\mathbf{y} = (y_1, y_2)^T$. The vector **y**s encodes the excitation and adaptation response of cells to external signals, and $\lambda(\mathbf{y})$ describes the motor response. The vector **y** evolves according to

$$\frac{dy_1}{dt} = \frac{G(S(\mathbf{x},t)) - (y_1 + y_2)}{t_e}, \quad (77)$$

$$\frac{dy_2}{dt} = \frac{G(S(\mathbf{x},t)) - y_2}{t_a}, \quad (78)$$

where $G(S)$ models signal detection via surface receptors and t_e and t_a specify the excitation and adaptation time scales, with $t_e \ll t_a$. A complete quantitative understanding of how different parameters at the cell level influence the population dynamics involves the incorporation the entire signal transduction of bacteria, but the cartoon description can predict biological aggregations and traveling bands of bacteria (cf. Fig. 6). Other intracellular variables, such as the metabolic state, can also be included in the transport equation, and this allows for a description of nutrient dependent cell growth. The existence of traveling wave solutions in the transport equations when coupled with the signal evolution equations was established in [106]. Further analysis on the comparison of the traveling waves obtained from the classical Keller-Segel equations are presented in the chapter by Frantz and Erban.

Macroscopic equations can be derived from the above multiscale models using perturbation methods and moment closure techniques, and this has been carried out successfully for the cartoon description above. The macroscopic equation

$$\frac{\partial n}{\partial t} = \nabla \cdot \left[\frac{s^2}{N\lambda_0} \nabla n - G'(S(x,t)) \frac{bs^2 t_a}{N\lambda_0(1 + \lambda_0 t_a)(1 + \lambda_0 t_e)} n \nabla S \right], \quad (79)$$

with $b = -\frac{\partial \lambda}{\partial y_1}|_{y_1=0}$ and N as the space dimension was derived in 1D first in [29], and extended to 3D in [28]. The major assumption used there and in earlier papers is that the signal gradient is shallow: $G'(S)\nabla S \cdot \mathbf{v} \sim \mathcal{O}(\epsilon) \sec^{-1}$ and $t_a \lambda_0 \sim \mathcal{O}(1)$, which results in a clear separation of the microscopic time scales from the macroscopic transport and diffusion time scales. Other assumptions include

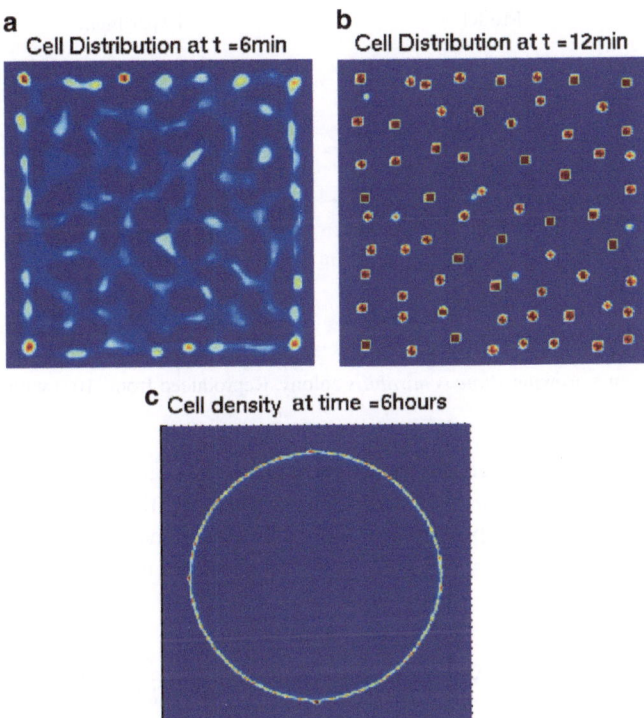

Fig. 6 Simulated *E. coli* patterns by a cell-based model. (**a**) Network formation from an uniform cell lawn; (**b**) Aggregate formation from the network; (**c**) Traveling wave formation from a single inoculum in the center. Adapted from [74] with permission

time-independent signals $S = S(\mathbf{x})$, a linear turning rate $\lambda = \lambda_0 - by_1$ and no directional persistence. From this equation one sees that if cells adapt instantly, i.e., if $t_a = 0$, then the taxis term vanishes and the population simply diffuses. In this case no aggregates will form, which is consistent with experimental observations.

New moment closure methods were developed in [104] to account for time dependent signals $S = S(\mathbf{x}, t)$ and nonlinear dependence of the turning rate on internal variables via $\lambda = \lambda_0 - by_1 + a_2 y_1^2 - \cdots$. In the general case considered there, the shallow gradient assumption becomes $\frac{b}{\lambda_0} G'(S)(\nabla S \cdot \mathbf{v} + \frac{\partial S}{\partial t}) \sim \mathcal{O}(\epsilon)\ \mathrm{s}^{-1}$. As before, the implication of this assumption is the separability of microscopic and macroscopic time scales. The same equation (79) results from the derivation, with the directional persistence appearing as a scaling of the turning rates by a factor of $1/(1-\psi_d)$. The method also applies for any finite system of internal dynamics $f(\mathbf{y})$ in polynomial form.

Chemotaxis equations in the presence of multiple signals and external forces were also derived in [104] in the context of bacterial chemotaxis. In general cells have multiple receptor types and thus can respond to many different signals. How a cell integrates these different signals and responds properly depends on the cell

Model   Experiment

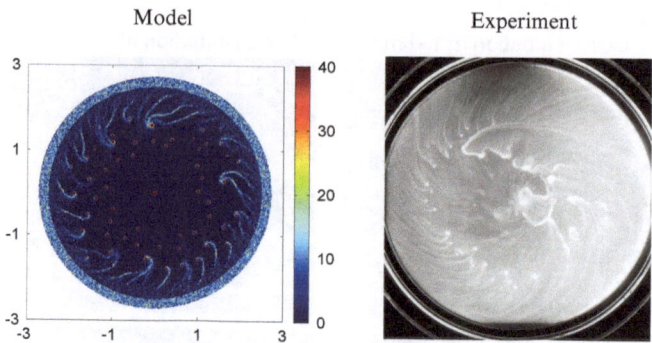

Fig. 7 Streams in a growing *Proteus mirabilis* colony. Reproduced from [107] with permission

type and is not known in general. However in bacteria different signaling pathways share the same network downstream of the receptors, and therefore different signals are integrated at the signal processing step. In this case, the function G is generally a function of all signals, $G = G(S_1, S_2, \cdots, S_m)$, and the macroscopic equation for cell density becomes

$$\frac{\partial n}{\partial t} = \nabla \cdot \left[D_n \nabla n - \chi_0 n \left(\frac{\partial G}{\partial S_1} \nabla S_1 + \cdots + \frac{\partial G}{\partial S_m} \nabla S_m \right) \right], \qquad (80)$$

where

$$D_n = \frac{s^2}{N\lambda_0(1 - \psi_d)},$$

and

$$\chi_0 = \frac{bs^2 t_a}{N\lambda_0(1 + \lambda_0(1 - \psi_d)t_a)(1 + \lambda_0(1 - \psi_d)t_e)}.$$

Other systems may involve separate transduction pathways for different signals, which will lead to different chemotactic sensitivities for different signals. Examples of how this affects pattern formation are given in [75].

When there are external forces that act on cells, then $\mathbf{F} \neq \mathbf{0}$ in 53, and additional terms appear in the chemotaxis equations. For example, when *E. coli* swims the flagella rotate counterclockwise when viewed from behind, and under typical conditions the Reynolds number is very small. As a result, the motion is both force and torque free, and thus the cell body must rotate clockwise. When cells swim close to a surface there is an imbalance in the viscous force between the top and bottom of the cell, which produces a clockwise swimming bias when viewed from above [25]. When this bias is incorporated into a cell-based model of aggregation, it leads to spiral density patterns as shown in Fig. 7 [107]. This was treated as a velocity-dependent force in a continuum description derived from the transport equations (54), and this led to the macroscopic chemotaxis equation

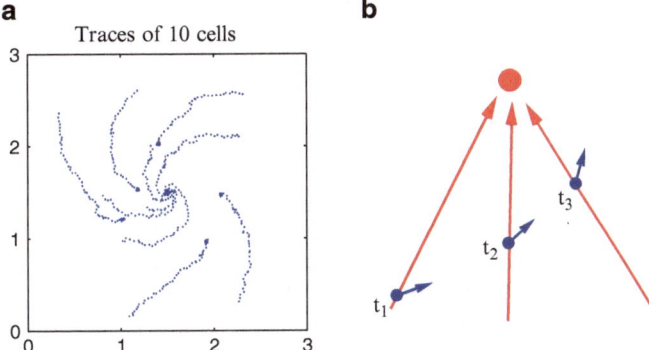

Fig. 8 (**a**) The positions of 10 randomly chosen cells from Fig. 7, each position recorded every 30 s by a *blue dot*. (**b**) the macroscopic drift given by (79) yields a qualitative explanation of the spirals. Reproduced from [107] with permission

$$\frac{\partial n}{\partial t} = \nabla \cdot \left(D_n \nabla n - G'(S)n \left(\chi_0 \nabla S + \beta_0 (\nabla S)^{\perp} \right) \right) \tag{81}$$

in two space dimensions [104]. Here $(\nabla S)^{\perp} = (\partial_{x_2} S, -\partial_{x_1} S)^T$ is a vector orthogonal to ∇S, and the diffusion coefficient and the chemotactic sensitivities, assuming fast excitation, are as follows:

$$D_n = \frac{s^2}{2\lambda_0(1 - \psi_d) + \frac{2\omega_0^2}{\lambda_0(1-\psi_d)}},$$

$$\chi_0 = \frac{b(1 - \psi_d)s^2[\lambda_0(1 - \psi_d)(\lambda_0(1 - \psi_d) + \frac{1}{t_a}) - \omega_0^2]}{2((\lambda_0(1 - \psi_d) + \frac{1}{t_a})^2 + \omega_0^2)(\lambda_0^2(1 - \psi_d)^2 + \omega_0^2)}, \tag{82}$$

$$\beta_0 = \frac{\omega_0 b(1 - \psi_d)s^2(2\lambda_0(1 - \psi_d) + \frac{1}{t_a})}{2((\lambda_0(1 - \psi_d) + \frac{1}{t_a})^2 + \omega_0^2)(\lambda_0^2(1 - \psi_d)^2 + \omega_0^2)}.$$

The parameter ω_0 measures the swimming bias, while ψ_d is the index of directional persistence. Notice that the swimming bias decreases the diffusion coefficient and the chemotactic sensitivity χ_0, and introduces a drift or a second taxis-like term in the direction orthogonal to the signal gradient. Since the force is not velocity-independent here, the moment analysis had to be modified accordingly. The method developed in [104] can be used to incorporate the effect of more general imposed forces as well.

Equation (81) leads to an heuristic explanation of the handedness of the spirals shown in Fig. 7. This is illustrated in Fig. 8, where the traces of 10 cells are shown in (a), and the path of an individual cell is shown in (b). At $t = t_1$ the blue cell detects a signal gradient (red arrow) roughly in the 1 o'clock direction, but according to (81) its average drift is in the direction of the blue arrow, due to the combined effects of the attractant gradient and the swimming bias. This balance is repeated at each successive time point, with the result that the cell approaches the signal source (the red dot) along an inward-spiraling, counterclockwise path as shown in (a).

As remarked earlier, (79) was derived for shallow signal gradients, i.e., $H \equiv \frac{b}{\lambda_0} G'(S)(\nabla S \cdot \mathbf{v} + \frac{\partial S}{\partial t}) \sim \mathscr{O}(\varepsilon)\,\mathrm{s}^{-1}$ with $\varepsilon = s/(L\lambda_0) \approx 10^{-2}$. It remains to be determined whether the chemotaxis equation or its variant forms gives an accurate representation of the population dynamics for bacterial chemotaxis under large spatial or temporal signal gradients. This hinges on how the macroscopic quantities relate to microscopic parameters, and, if the PKS equation fails under certain conditions, what macroscopic equation can be derived. For an ultra-small signal gradient, $H \leq \mathscr{O}(\varepsilon^2)\,\mathrm{s}^{-1}$, the chemotactic response of the population provides a small perturbation, via higher order terms, of the cell density, which evolves according to a diffusion process with $D_n = s^2/(N\lambda_0)$ [103]. For large signal gradients ($H \geq \mathscr{O}(1)\,\mathrm{s}^{-1}$) the macroscopic equation should include the nonlinear effects of the gradients in the macroscopic drift, for otherwise the linear approximation may predict a chemotactic velocity that exceeds the cell speed, which is unrealistic since there are no cell-cell or hydrodynamic interactions in the model. In this case the microscopic time scale and macroscopic time scales may overlap, and new techniques are needed to derive macroscopic equations.

In any case, the dependence of the diffusion coefficient D and the chemotactic velocity $\mathbf{u}_c$ on H can be obtained by stochastic simulations. Given different levels of H, 10^4 stochastic simulations are performed with the same initial conditions. The turning rate is given by $\lambda = \lambda_0(1 - \frac{y_1}{\gamma + |y_1|})$. The positions of the cell are recorded every half minute, and the data was analyzed to obtain the diffusion rate and the macroscopic drift. Figure 9 compares the diffusion rate and the macroscopic drift inferred from the stochastic simulations of the 2D cell-based model with the predictions from the macroscopic equation (79). It is shown that the macroscopic description gives a good approximation for $H \sim \mathscr{O}(\epsilon)\mathrm{s}^{-1}$, but the nonlinearity in the cell-based model for $H \sim \mathscr{O}(1)\mathrm{s}^{-1}$ can not be captured by the macroscopic equation with its linear dependence of H. More specificly, from the stochastic simulations, we see that the cell-based model reveals saturation in the macroscopic velocity, and gradient-dependent diffusion rates.

4.5 Macroscopic Descriptions of Eukaryotic Cell Movement

Many single-celled organisms such as E. coli use flagella or cilia to swim, but eukaryotic cells that lack such structures use one of two basic modes of movement—mesenchymal and amoeboid [10]. The former can be characterized as 'crawling' and involves the extension of structures whose protrusion is driven by actin polymerization at the leading edge. This mode dominates in cells such as fibroblasts when moving on a 2D substrate. In the amoeboid mode cells are more rounded and employ shape changes to move—in effect 'jostling through the crowd' or

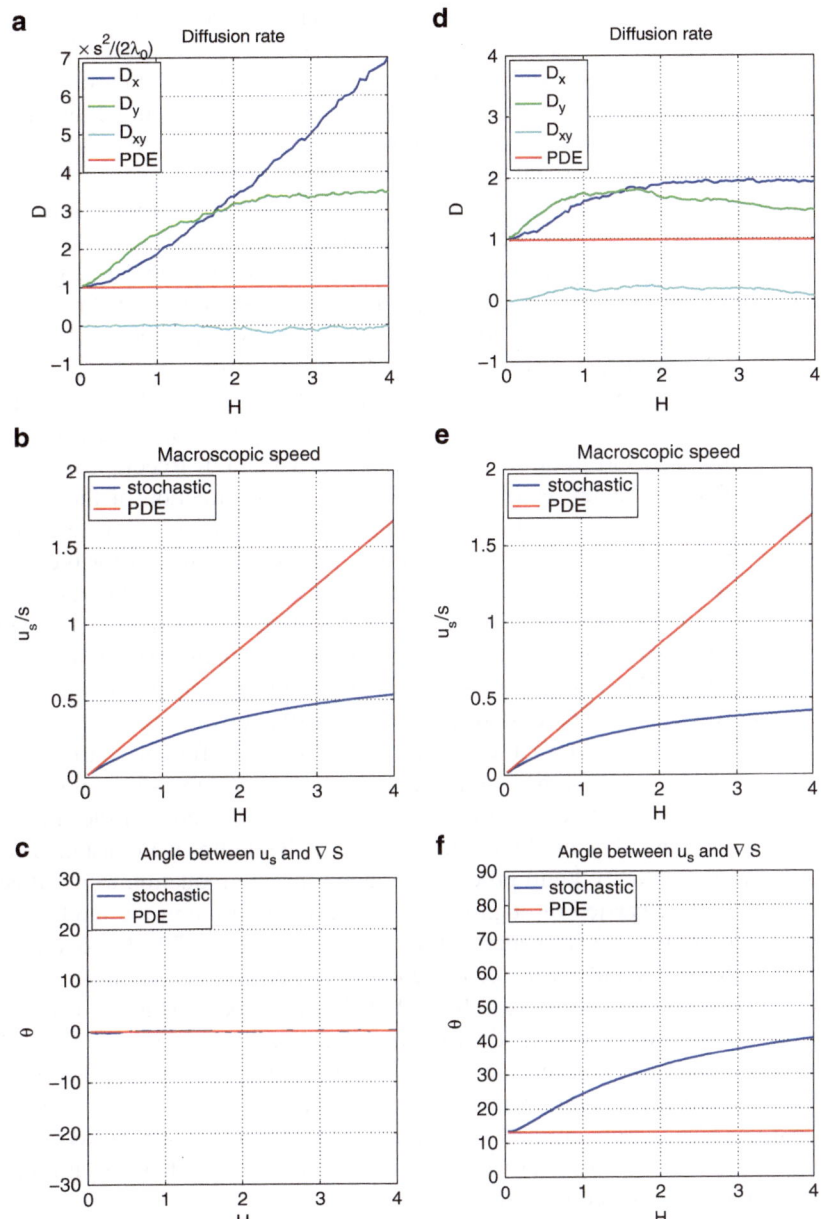

Fig. 9 A comparison of the cell-based and the macroscopic predictions of the chemotactic velocity and the diffusion coefficients in 2D. Here D_x and D_y are the diffusion coefficients perpendicular to and parallel to the signal gradient, resp., and D_{xy} is the cross diffusion coefficient. The *left column* is obtained with no swimming bias, and the *right column* is obtained with $\varepsilon_b = 0.04\pi$. Reproduced from [107] with permission

'swimming'. Leukocytes use this mode for movement through the extracellular matrix in the absence of adhesion sites [57]. Moreover, it has been shown that numerous cell types can sense the mechanical properties of their environment and adjust the balance between the modes appropriately [85]. Thus pure crawling and pure swimming are the extremes on a continuum of locomotion strategies, but many cells can choose the most effective strategy in a given context.

While 'run-and-tumble' organisms such as *E. coli* use temporal sensing to modulate their motile behavior, the motile program of eukaryotic cells such as Dd or leukocytes is more complicated. These cells are large enough to detect gradients in extracellular chemical and mechanical signals over the length of the cell, and can amplify small differences in the extracellular signal over the cell into large end-to-end intracellular differences that control the motile machinery [19, 78]. Given that these cells use spatial sensing, an individual-based model that incorporates direction sensing and movement cannot treat cells as points, but must allow for spatial variations in the finite cell volume (or area in 2D). Recent experiments show that cells in a steady gradient can polarize in the direction of the gradient without extending pseudopods [78], and thus must rely entirely on differences in the signal across the cell body for orientation. Analysis of a model for the cAMP relay pathway in Dd shows that a cell experiences a significant difference in the front-to-back ratio of cAMP when a neighboring cell begins to signal [23], which demonstrates that sufficient end-to-end differences for reliable orientation can be generated for typical extracellular signals; everything needed is that the direction-sensing pathways respond at least as fast as the cAMP pathway.

In addition to the fact that eukaryotes use spatial differences to measure signals, another major difference with 'run-and-tumble' swimmers lies in the force-generation machinery that drives the motion of eukaryotic cells. In the 'run-and-tumble' description of bacterial motion we assumed that jumps were instantaneous, which led to the velocity jump process. Furthermore, the reduction to a diffusion process can still be carried through if there is a finite lifetime in the tumble state, as long as the transitions are generated by a Poisson process [70]. In contrast, the directional changes in eukaryotic cells are much slower and depend directly on the signal location, and thus this has to be included in the model. This has been done at the single cell level, using a model for intracellular cAMP dynamics, and treating the cells as deformable viscoelastic ellipsoids that exert forces on the substrate and one another. This more complex model also produces realistic aggregation patterns [76], but there is a large gap between realistic, single-cell models and continuum descriptions. Thus far only relatively simple cell-based models have been used for the derivation of macroscopic descriptions.

One approach is to start with a Smoluchowski equation, and to postulate a relationship between the force and the chemotactic gradient. If one assumes that the motive force exerted by a cell is a function of the attractant concentration, one can compute the difference between the force at the leading and trailing edges, and then by a mean-value argument obtain a linear relation between this difference and the chemotactic gradient [79]. In this approach the chemotactic sensitivity is related to the rate of change of the force with attractant concentration. Support for this comes from experiments which show that as many pseudopods are produced down-gradient as up, but those up-gradient are more successful in generating cell movement [93]. However, Dd and perhaps other cells, adapt to the signal, and simplified models cannot capture this effectively [43]. Thus a different approach that incorporates signal transduction and internal dynamics is needed.

In [23, 30], a cell is described as a disk ($n = 2$) or a ball ($n = 3$) $B_\varrho = \{\xi \in \mathbf{R}^n |\; \|\; \xi\; \| \leq \varrho\}$, and the model is formulated in terms of the position of its center $\mathbf{x} \in \mathbf{R}^n$, its velocity $\mathbf{v} \in \mathbf{R}^n$, its internal state functions $\mathbf{y} : B_\varrho \to \mathbf{R}^{d_1}$ and its membrane state functions $\mathbf{z} : \partial B_\varrho \to \mathbf{R}^{d_2}$. Denote by $\overline{\mathbf{y}} = (\mathbf{y}, \mathbf{z}) \in \mathbb{Y}$ the combined internal and membrane state, where $\mathbb{Y}$ is, in general, an infinite-dimensional Banach space.

The internal state and the acceleration are assumed to evolve according to

$$\frac{d\overline{\mathbf{y}}}{dt} = \mathcal{G}(\overline{\mathbf{y}}, S), \tag{83}$$

$$\frac{d\mathbf{v}}{dt} = \mathcal{F}(\mathbf{x}, \mathbf{v}, \overline{\mathbf{y}}), \tag{84}$$

where $\mathcal{G} : \mathbb{Y} \times \mathbb{S} \to \mathbb{Y}$ is a mapping between Banach spaces and $\mathcal{F} : \mathbf{R}^n \times \mathbf{R}^n \times \mathbb{Y} \to \mathbf{R}^n$ is the force per unit mass on the centroid. Thus the acceleration depends on the internal state. In this formulation the combined internal state $\overline{\mathbf{y}}$ includes quantities that depend on the spatial location in the cell or on the membrane, and which may, for example, satisfy a reaction-diffusion equation such as

$$\frac{\partial \mathbf{y}}{\partial t} = D\Delta \mathbf{y} + \mathbf{f}(\mathbf{y}), \text{ in } B_\varrho, \tag{85}$$

$$B(\mathbf{y}, \mathbf{z}) = 0, \qquad \text{in } \partial B_\varrho. \tag{86}$$

Thus the boundary condition for $\mathbf{y}$ depends on the membrane state functions $\mathbf{z}$, perhaps to reflect binding or other processes such as scaffold formation. The boundary variables in turn evolve according to the equation

$$\frac{\partial \mathbf{z}}{\partial t} = \mathbf{g}(\mathbf{z}, S), \qquad \text{in } \partial B_\varrho, \tag{87}$$

where S is the external signal, and this could also incorporate diffusion on the boundary by suitably altering the equation.

Given the complexity of the single cell description, it is a formidable task to derive macroscopic equations for populations of eukaryotic cells. A simple model of the form (83–84) for a single cell was analyzed in [30]. This model captures the essential features of cell movement in response to traveling waves of chemoattractant. Moreover, in that context there is a mapping $\mathcal{P} : \mathbb{Y} \to \mathbf{R}^k$, $k < \infty$, satisfying $\mathcal{F}(\mathbf{x}, \mathbf{v}, \overline{\mathbf{y}}) = \mathbf{F}(\mathbf{x}, \mathbf{v}, \mathcal{P}(\overline{\mathbf{y}}))$ where $\mathbf{F} : \mathbf{R}^n \times \mathbf{R}^n \times \mathbf{R}^k \to \mathbf{R}^n$ such that a closed evolution equation for the variable $\overline{\mathbf{z}} = \mathcal{P}(\overline{\mathbf{y}})$ can be derived. Then the cellular random walk written in terms of $(\mathbf{x}, \mathbf{v}, \overline{\mathbf{y}})$ can be equivalently formulated in terms of the finite-dimensional state variables $(\mathbf{x}, \mathbf{v}, \bar{\mathbf{z}})$. In particular, one can formulate an equation for the probability distribution $p(\mathbf{x}, \mathbf{v}, \bar{\mathbf{z}})$ (cf. (76) written for $p(\mathbf{x}, \mathbf{v}, \mathbf{y})$ in the bacterial case). Asymptotic analysis of this transport equation leads to a system of macroscopic hyperbolic equations that accurately reflect the dynamics of the full system, but it is not known if that system can in turn be reduced to a PKS equation [30]. Other approaches have been used, e.g., generalized PKS equations have been derived beginning with a cellular Potts model [61], but the internal state plays no role in these formulations.

5 Discussion

How cells or organisms move about in space in response to signals, and how they coordinate their movement and form stationary or dynamic patterns is an important question in many biological processes, including embryonic development, cancer progression, wound healing and biofilm formation. These phenomena have been modeled in two ways in the literature. Firstly, there are continuum models based on phenomenological descriptions that lead to convection-diffusion equations such as the chemotaxis equation [42] for the evolution of the macroscopic cell density $n = n(\mathbf{x}, t)$. However, new experimental technology has advanced our knowledge on how cells detect, transduce, respond to, and propagate external stimuli, and this has led to the second approach, in which detailed cell-based models of collective cell movement towards chemical or mechanical signals [23,90,96,107] are incorporated. However, due to the complexity of intracellular dynamics and the large number of cells that are often involved, cell-based models are computationally expensive, and new techniques are needed to embed cell-level knowledge into macroscopic equations. This is a difficult problem, comparable to deriving the macroscopic rheological properties of a complex fluid such as the cytosol from knowledge of the molecular interactions, and thus not surprisingly, progress has been slow.

Here we have reviewed recent progress on deriving chemotaxis equations from space jump processes and velocity jump processes. When swimming bacteria such as *E. coli* move independently towards chemical signals, their movement can be described as independent velocity jump processes. When cells are well separated and the signal gradient is sufficiently small, chemotaxis equations are derived from the moment equations of the transport equation that describes the evolution of the cell in phase space [28–30, 41, 70, 104]. When the signal gradient is large, 1D stochastic simulations of a cell-based model show that the movement of cells is more persistent and cells run up the gradient with very little turning. Therefore statistically the diffusion rate decreases to zero, and the macroscopic velocity increases to the maximum cell speed, as the signal gradient increases. This shows that under extremely large signal gradients, the macroscopic equations for cells movement are more of a hyperbolic type, and also reflects the fact that the low-order moments in the internal dynamics cannot capture the strongly nonlinear dependence of the turning rate on the signal. This is similar to what is observed in eukaryotic cells, which suppress random movement in the presence of strong chemotactic signals. However in 2D stochastic simulations, bacteria moving roughly orthogonal to the signal gradient still run and tumble, and this leads to diffusion coefficients that do not approach zero, in contrast with the 1D case.

There are many open problems in this area, a few of which are listed below.

- A more complete analysis of the time scales and how they depend on the external signal and the internal dynamics is needed. For example, the second eigenvalue of the turning operator controls the rate at which the diffusion regime is approached, but little has been done to obtain better estimates of the second eigenvalue based on properties of the turning kernel.
- The formulation of VJ processes herein is based on the assumption that the velocity jumps are generated by a Poisson process, but there is some evidence mentioned earlier [64] that bacteria show abnormally long run lengths that are inconsistent with this assumption. In general the non-streaming component of the transport equation (54) is simply the time derivative of the stochastic process generating the jumps, which may change depending on the signal strength, and the use of other waiting time distributions in the VJ process should be explored.
- To date most derivations of macroscopic equations from a microscopic model have ignored density effects, but these are important in examples of bacterial movement and related problems. Most analyses of density effects begin with continuum descriptions and add forces due to active motile particles [44,58], but a more fundamental approach is needed.
- In many situations, cell-cell contact and contact-induced signaling is important for collective movement. To describe this one must include cell-cell mechanical interaction terms in $\mathbf{F}$, and cell-cell contact signaling terms in the internal dynamics. A suitable starting point for this may be to add the evolution of internal dynamics to the $(\mathbf{x}, \mathbf{v})$ evolution described by the Fokker-Planck-Kramers-Klein equation (15).
- As an adjunct to this, continuum-level descriptions of tissue movement based on microscopic models, should be formulated [35], but there are many difficult homogenization issues that arise here.
- The derivation of macroscopic equations for systems when the finite-dimensional reduction $\mathscr{P} : \mathbb{Y} \to \mathbf{R}^k$ is not possible is an open problem. In fact the entire formulation as a transport equation breaks down, and a new approach is needed.

References

1. W. Alt, Biased random walk models for chemotaxis and related diffusion approximations. J. Math. Biol. **9**, 147–177 (1980)
2. D. Applebaum, *Lévy Processes and Stochastic Calculus*, vol. 93 (Cambridge University Press, Cambridge, 2004)
3. R. Aris, *Vectors, Tensors and the Basic Equations of Fluid Mechanics* (Prentice-Hall, New York, 1962)

4. L. Arnold, *Stochastic Differential Equations, Theory and applications* (Wiley-Interscience, New York, 1974)
5. L. Bachelier, *Théorie de la spéculation* (Gauthier-Villars, Paris, 1900)
6. I.L. Bajec, F.H. Heppner, Organized flight in birds. Anim. Behav. **78**(4), 777–789 (2009)
7. M.N. Barber, B.W. Ninham, *Random and Restricted Walks: Theory and Applications*, vol. 10 (Gordon and Breach, New York, 1970)
8. H.C. Berg, D.A. Brown, Chemotaxis in esterichia coli analysed by three-dimensional tracking. Nature **239**, 500–504 (1972)
9. H.C. Berg, *Random Walks in Biology* (Princeton University Press, Princeton, 1983)
10. F. Binamé, G. Pawlak, P. Roux, U. Hibner, What makes cells move: requirements and obstacles for spontaneous cell motility. Mol. BioSystems **6**(4), 648–661 (2010)
11. L. Bocquet, J. Piasecki, Microscopic derivation of non-Markovian thermalization of a Brownian particle. J. Stat. Phys. **87**(5), 1005–1035 (1997)
12. M. Born, H.S. Green, A general kinetic theory of liquids. I. the molecular distribution functions. Proc. R. Soc. Lond. Ser. A. Math. Phys. Sci. **188**(1012), 10 (1946)
13. V. Capasso, D. Bakstein, *An Introduction to Continuous-Time Stochastic Processes: Theory, Models, and Applications to Finance, Biology, and Medicine* (Birkhauser, Basel, 2005)
14. V. Capasso, D. Morale, Asymptotic behavior of a system of stochastic particles subject to nonlocal interactions. Stoch. Anal. Appl. **27**(3), 574–603 (2009)
15. J.A. Carrillo, M. Fornasier, G. Toscani, F. Vecil, Particle, kinetic, and hydrodynamic models of swarming, in *Mathematical Modeling of Collective Behavior in Socio-Economic and Life Sciences*, ed. by G. Naldi, L. Pareschi, G. Toscani. Modelling and Simulation in Science and Technology, Birkhauser (2010), pp. 297–336
16. C. Cercignani, *Mathematical Methods in Kinetic Theory*, 2nd edn. (Plenum, New York, 1969)
17. C. Cercignani, R. Illner, M. Pulvirenti, *The Mathematical Theory of Dilute Gases* (Springer, New York, 1994)
18. S. Chandrasekhar, Stochastic problems in physics and astronomy. Rev. Mod. Phys. **15**, 2–89 (1943)
19. C.Y. Chung, S. Funamoto, R.A. Firtel, Signaling pathways controlling cell polarity and chemotaxis. Trends Biochem. Sci. **26**(9), 557–566 (2001). Review
20. R.V. Churchill, *Operational Mathematics* (McGraw-Hill, New York, 1958)
21. E.A. Codling, M.J. Plank, S. Benhamou, Random walk models in biology. J. R. Soc. Interface **5**(25), 813 (2008)
22. F. Cucker, S. Smale, On the mathematical foundations of learning. Bull. Am. Math. Soc. **39**(1), 1–49 (2001)
23. J.C. Dallon, H.G. Othmer, A continuum analysis of the chemotactic signal seen by *Dictyostelium discoideum*. J. Theor. Biol. **194**(4), 461–483 (1998)
24. B. Davis, Reinforced random walks. Probab. Theory Relat. Fields **84**(2), 203–229 (1990)
25. W.R. DiLuzio, L. Turner, M. Mayer, P. Garstecki, D.B. Weibel, H.C. Berg, G.M. Whitesides, Escherichia coli swim on the right-hand side. Nature **435**(7046), 1271–1274 (2005)
26. A.M. Edwards, R.A. Phillips, N.W. Watkins, M.P. Freeman, E.J. Murphy, V. Afanasyev, S.V. Buldyrev, M.G.E. da Luz, E.P. Raposo, H.E. Stanley et al., Revisiting Lévy flight search patterns of wandering albatrosses, bumblebees and deer. Nature **449**(7165), 1044–1048 (2007)
27. A. Einstein, Über die von der molekularkinetischen Theorie der Wärme geforderte Bewegung von in ruhenden Flussigkeiten suspendierten Teilchen. Ann. der Physik **17**, 549–560 (1905)
28. R. Erban, H. Othmer, From signal transduction to spatial pattern formation in *E. coli*: a paradigm for multi-scale modeling in biology. Multiscale Model. Simul. **3**(2), 362–394 (2005)
29. R. Erban, H.G. Othmer, From individual to collective behavior in bacterial chemotaxis. SIAM J. Appl. Math. **65**(2), 361–391 (2004)
30. R. Erban, H.G. Othmer, Taxis equations for amoeboid cells. J. Math. Biol. **54**, 847–885 (2007)
31. W. Feller, *An Introduction to Probability Theory* (Wiley, New York, 1968)

32. R. Ford, D.A. Lauffenburger, A simple expression for quantifying bacterial chemotaxis using capillary assay data: application to the analysis of enhanced chemotactic responses from growth-limited cultures. Math. Biosci. **109**(2), 127–150 (1992)
33. M. Franceschetti, When a random walk of fixed length can lead uniformly anywhere inside a hypersphere. J. Stat. Phys. **127**(4), 813–823 (2007)
34. R. Fürth, Die Brownische Bewegung bei Berücksichtigung einer Persistenz der Bewegungsrichtung. Zeitsch. f. Physik **2**, 244–256 (1920)
35. J. Galle, M. Hoffmann, G. Aust, From single cells to tissue architecture-a bottom-up approach to modelling the spatio-temporal organisation of complex multi-cellular systems. J. Math. Biol. **58**(1–2), 261–283 (2009)
36. G.W. Gardiner, *Handbook of Stochastic Processes for Physics, Chemistry and Natural Sciences*, 2nd edn. (Springer, Berlin, 1985)
37. S. Goldstein, On diffusion by discontinuous movements, and on the telegraph equation. Quart. J. Mech. Appl. Math. **VI**, 129–156 (1951)
38. S.-Y. Ha, E. Tadmor, From particle to kinetic and hydrodynamic descriptions of flocking. Kinetic Relat. Model **1**(3), 415–435 (2008)
39. R.L. Hall, Amoeboid movement as a correlated walk. J. Math. Biol. **4**, 327–335 (1977)
40. C.R. Heathcote, J.E. Moyal, The random walk [in continuous time] and its application to the theory of queues. Biometrika **46**(3–4), 400 (1959)
41. T. Hillen, H.G. Othmer, The diffusion limit of transport equations derived from velocity jump processes. SIAM J. Appl. Math. **61**, 751–775 (2000)
42. T. Hillen, K.J. Painter, A users guide to PDE models for chemotaxis. J. Math. Biol. **58**(1), 183–217 (2009)
43. T. Höfer, J.A. Sherratt, P.K. Maini, Cellular pattern formation during dictyostelium aggregation. Physica D **85**(3), 425–444 (1995)
44. C. Hohenegger, M.J. Shelley, Stability of active suspensions. Phys. Rev. E **81**(4), 046311 (2010)
45. D. Horstmann, From 1970 until present: the Keller-Segel model in chemotaxis and its consequences I. Jahresbericht der DMV **105**(3), 103–165 (2003)
46. J.Hu, H.G. Othmer, A theoretical analysis of filament length fluctuations in actin and other polymers. J. Math. Biol. (2011, to appear)
47. J.M. Hutchinson, P.M. Waser, Use, misuse and extensions of "ideal gas" models of animal encounter. Biol. Rev.-Camb. **82**(3), 335 (2007)
48. J.O. Irwin, The frequency distribution of the difference between two independent variates following the same Poisson distribution. J. R. Stat. Soc. **100**(3), 415–416 (1937)
49. M. Kac, *Some Stochastic Problems in Physics and Mathematics* (Field Research Laboratory, Magnolia Petroleum Company, Dallas, 1956)
50. N. van Kampen, *Stochastic Processes in Physics and Chemistry*, 3rd edn. (North-Holland, Amsterdam, 2007)
51. S. Karlin, H. Taylor, *A First Course in Stochastic Processes* (Academic, New York, 1975)
52. E. Keller, L. Segel, Initiation of slime mold aggregation viewed as an instability. J. Theor. Biol. **26**, 399–415 (1970)
53. V.M. Kenkre, The generalized master equation and its applications, in *Statistical Mechanics and Statistical Methods in Theory and Application* (Plenum, New York, 1977)
54. V.M. Kenkre, E.W. Montroll, M.F. Shlesinger, Generalized master equations for continuous-time random walks. J. Stat. Phys. **9**(1), 45–50 (1973)
55. J.C. Kluyver, A local probability theorem. Ned. Akad. Wet. Proc. A **8**, 341–350 (1906)
56. D.E. Koshland, *Bacterial Chemotaxis as a Model Behavioral System* (Raven Press, New York, 1980)
57. T. Lämmermann, B.L. Bader, S.J. Monkley, T. Worbs, R. Wedlich-Söldner, K. Hirsch, M. Keller, R. Förster, D.R. Critchley, R. Fässler et al., Rapid leukocyte migration by integrin-independent flowing and squeezing. Nature **453**, 51–55 (2008)
58. J. Lega, T. Passot, Hydrodynamics of bacterial colonies: a model. Phys. Rev. E Stat. Nonlinear Soft Matter Phys. **67**(3 Pt 1), 031906 (2003)

59. L. Li, S.F. Nørrelykke, E.C. Cox, Persistent cell motion in the absence of external signals: a search strategy for eukaryotic cells. PLoS One **3**(5), e2093 (2008)

60. R.L. Liboff, *Kinetic Theory: Classical, Quantum, and Relativistic Descriptions* (Springer, Berlin, 2003)

61. P.M. Lushnikov, N. Chen, M. Alber, Macroscopic dynamics of biological cells interacting via chemotaxis and direct contact. Phys. Rev. E **78**(6), 061904 (2008)

62. R.M. Macnab, Sensing the environment: bacterial chemotaxis, in *Biological Regulation and Development*, ed. by R. Goldberg (Plenum Press, New York, 1980), pp. 377–412

63. U.M.B. Marconi, P. Tarazona, Nonequilibrium inertial dynamics of colloidal systems. J. Chem. Phys. **124**, 164901 (2006)

64. F. Matthaus, M. Jagodic, J. Dobnikar, E. coli superdiffusion and chemotaxis–search strategy, precision, and motility. Biophys. J. **97**(4), 946–957 (2009)

65. R. Metzler, J. Klafter, The random walk's guide to anomalous diffusion: a fractional dynamics approach. Phys. Rep. **339**(1), 1–77 (2000)

66. E.W. Montroll, G.H. Weiss, Random walks on lattices. II. J. Math. Phys. **6**, 167 (1965)

67. G. Naldi, *Mathematical Modeling of Collective Behavior in Socio-Economic and Life Sciences* (Springer, Berlin, 2010)

68. K. Oelschläger, A fluctuation theorem for moderately interacting diffusion processes. Probab. Theor. Relat. Field **74**, 591–616 (1987)

69. A. Okubo, *Diffusion and Ecological Problems: Mathematical Models* (Springer, New York, 1980)

70. H. Othmer, T. Hillen, The diffusion limit of transport equations 2: chemotaxis equations. SIAM J. Appl. Math. **62**, 1222–1250 (2002)

71. H.G. Othmer, Interactions of Reaction and Diffusion in Open Systems, PhD thesis, University of Minnesota, 1969

72. H.G. Othmer, A. Stevens, Aggregation, blowup, and collapse: The ABC's of taxis in reinforced random walks. SIAM J. Appl. Math. **57**(4), 1044–1081 (1997)

73. H.G. Othmer, S.R. Dunbar, W. Alt, Models of dispersal in biological systems. J. Math. Biol. **26**, 263–298 (1988)

74. H.G. Othmer, K. Painter, D. Umulis, C. Xue, The intersection of theory and application in biological pattern formation. Math. Mod. Nat. Phenom. **4**, 3–79 (2009)

75. K.J. Painter, P.K. Maini, H.G. Othmer, Development and applications of a model for cellular response to multiple chemotactic cues. J. Math. Biol. **41**(4), 285–314 (2000)

76. E. Palsson, H.G. Othmer, A model for individual and collective cell movement in *Dictyostelium discoideum*. Proc. Natl. Acad. Sci. **97**, 11448–11453 (2000)

77. G.C. Papanicolaou, Asymptotic analysis of transport processes. Bull. AMS **81**, 330–392 (1975)

78. C.A. Parent, P.N. Devreotes, A cell's sense of direction. Science **284**(5415), 765–770 (1999). Review

79. E. Pate, H.G. Othmer, Differentiation, cell sorting and proportion regulation in the slug stage of *Dictyostelium discoideum*. J. Theor. Biol. **118**, 301–319 (1986)

80. C.S. Patlak, Random walk with persistence and external bias. Bull. Math. Biophys. **15**, 311–338 (1953)

81. K. Pearson, The problem of the random walk. Nature **72**(1865), 294–294 (1905)

82. R. Pemantle, A survey of random processes with reinforcement. Probab. Surv. **4**, 1–79 (2007)

83. B. Perthame, Mathematical tools for kinetic equations. Bull. Am. Math. Soc. **41**(2), 205–244 (2004)

84. L. Rayleigh, On the resultant of a large number of vibrations of the same pitch and of arbitrary phase. Phil. Mag. **10**(73), 491 (1880)

85. J. Renkawitz, K. Schumann, M. Weber, T. Lämmermann, H. Pflicke, M. Piel, J. Polleux, J.P. Spatz, M. Sixt, Adaptive force transmission in amoeboid cell migration. Nat. Cell Biol. **11**(12), 1438–1443 (2009)

86. K.I. Sato, *Lévy Processes and Infinitely Divisible Distributions* (Cambridge University Press, London, 1999)

87. P.A. Spiro, J.S. Parkinson, H.G. Othmer, A model of excitation and adaptation in bacterial chemotaxis. Proc. Natl. Acad. Sci. **94**(14), 7263–7268 (1997)
88. H. Spohn, *Large Scale Dynamics of Interacting Particles* (Springer, New York, 1991)
89. A. Stevens, The derivation of chemotaxis equations as limit dynamics of moderately interacting stochastic many-particle systems. SIAM J. Appl. Math. **61**, 183–212 (2000)
90. M.A. Stolarska, Y. Kim, H.G. Othmer, Multi-scale models of cell and tissue dynamics. Phil. Trans. R. Soc. A **367**(1902), 3525 (2009)
91. D.W. Stroock, Some stochastic processes which arise from a model of the motion of a bacterium. Probab. Theor. Relat. Field **28**, 305–315 (1974)
92. G.I. Taylor, Diffusion by continuous movements. Proc. Lond. Math. Soc. **20**, 196–212 (1920)
93. B.J. Varnum-Finney, E. Voss, D.R. Soll, Frequency and orientation of pseudopod formation of *Dictyostelium discoideum* amebae chemotaxing in a spatial gradient: further evidence for a temporal mechanism. Cell Motil. Cytoskeleton **8**(1), 18–26 (1987)
94. K. Kang, B. Perthame, A. Stevens, J.J.L. Velázquez, An integro-differential equation model for alignment and orientational aggregation. J. Differ. Equat. **246**(4), 1387–1421 (2009)
95. T. Vicsek, A. Zafiris, Collective motion (2010). arXiv preprint arXiv:1010.5017
96. D.C. Walker, G.Hill, S.M. Wood, R.H. Smallwood, J. Southgate, Agent-based computational modeling of wounded epithelial cell monolayers. IEEE Trans. Nanobiosci. **3**(3), 153–163 (2004)
97. Q.D. Wang, The global solution of the n-body problem. Celestial Mech. Dynam. Astron. **50**, 73–88 (1991)
98. G.H. Weiss, *Aspects and Applications of the Random Walk*, vol. 121 (North-Holland, Amsterdam, 1994)
99. D. Widder, *The Laplace Transform* (Princeton University Press, Princeton, 1946)
100. G. Wilemski, On the derivation of Smoluchowski equations with corrections in the classical theory of Brownian motion. J. Stat. Phys. **14**(2), 153–169 (1976)
101. Y. Wu, A.D. Kaiser, Y. Jiang, M.S. Alber, Periodic reversal of direction allows myxobacteria to swarm. Proc. Natl. Acad. Sci. **106**(4), 1222 (2009)
102. X. Xin, H.G. Othmer, A trimer of dimers - based model for the chemotactic signal transduction network in bacterial chemotaxis. Bull. Math. Biol., 1–44 (2012)
103. C. Xue, Mathematical Models of Taxis-Driven Bacterial Pattern Formation, PhD thesis, University of Minnesota, 2008
104. C. Xue, H.G. Othmer, Multiscale models of taxis-driven patterning in bacterial populations. SIAM J. Appl. Math. **70**(1), 133–167 (2009)
105. C. Xue, H.G. Othmer, R. Erban, *From Individual to Collective Behavior of Unicellular Organisms: Recent Results and Open Problems*, vol. 1167 (AIP, Melville, NY, 2009), pp. 3–14
106. C. Xue, H.J. Hwang, K.J. Painter, R. Erban, Travelling waves in hyperbolic chemotaxis equations. Bull. Math. Biol. **73**(8), 1695–1733 (2011)
107. C. Xue, E.O. Budrene, H.G. Othmer, Radial and spiral stream formation in proteus mirabilis colonies. PLoS Comput. Biol. **7**(12), e1002332 (2011)

Hybrid Modelling of Individual Movement and Collective Behaviour

Benjamin Franz and Radek Erban

Abstract Mathematical models of dispersal in biological systems are often written in terms of partial differential equations (PDEs) which describe the time evolution of population-level variables (concentrations, densities). A more detailed modelling approach is given by individual-based (agent-based) models which describe the behaviour of each organism. In recent years, an intermediate modelling methodology—hybrid modelling—has been applied to a number of biological systems. These hybrid models couple an individual-based description of cells/animals with a PDE-model of their environment. In this chapter, we overview hybrid models in the literature with the focus on the mathematical challenges of this modelling approach. The detailed analysis is presented using the example of chemotaxis, where cells move according to extracellular chemicals that can be altered by the cells themselves. In this case, individual-based models of cells are coupled with PDEs for extracellular chemical signals. Travelling waves in these hybrid models are investigated. In particular, we show that in contrary to the PDEs, hybrid chemotaxis models only develop a transient travelling wave.

1 Introduction

There are two fundamentally different approaches to the mathematical modelling of systems of interacting individuals (cells, animals) in biology. If the number of individuals is large, one often uses a continuum population-level approach, which yields partial differential equations (PDEs) for the spatially-distributed densities of individuals [39]. The advantage of PDE-based modelling is a well-developed mathematical theory and a number of existing numerical solvers which can be used

B. Franz · R. Erban (✉)
Mathematical Institute, University of Oxford, 24-29 St Giles', Oxford OX1 3LB, UK
e-mail: franz@maths.ox.ac.uk; erban@maths.ox.ac.uk

M.A. Lewis et al. (eds.), *Dispersal, Individual Movement and Spatial Ecology*,
Lecture Notes in Mathematics 2071, DOI 10.1007/978-3-642-35497-7_5,
© Springer-Verlag Berlin Heidelberg 2013

to efficiently simulate the system behaviour. However, continuum approximation becomes inaccurate if smaller groups of individuals are studied, and agent-based (individual-based) models become the method of choice [13, 50]. Examples can be found in zoological applications, like behaviour of fish schools, bird flocks and locust groups [10,51]. The individual behaviour of the agents is modelled as well as the interaction (e.g. attraction or repulsion) between them [12]. A number of these agents are then simulated on the computer and their collective behaviour is analysed. This approach allows for a more detailed description of the individual behaviour and does not discount various stochastic effects caused by a finite number of individuals. On the other hand mathematical analysis is often hard to achieve and simulations can be computationally intensive.

Another problem with purely agent-based models is that it is challenging to incorporate influences the agents might have on their environment. This is important whenever agents interact indirectly by modifying their (evolving) environment. A classical example is modelling chemotaxis where individual cells modify (secrete, consume) extracellular chemical signals which diffuse in the extracellular space [14, 19]. In this case, a *hybrid modelling* framework that seeks to combine the advantages of continuum and agent-based models is often used. The main idea of this modelling approach is to describe some species as a continuum and some species as a set of agents. For example, Schweitzer and Schimansky-Geier [46] studied a system of "active" walkers (individuals) that can secrete and interact through a (chemical) signal described by a reaction-diffusion PDE. One application of their abstract framework included ants which lay a pheromone into the ground to use it for their orientation. A more specific chemotactic example can be found in Dallon and Othmer [14] who developed a hybrid model for chemotaxis of slime mold *Dictyostelium discoideum* in which the cells are treated as individuals in a continuum field of the chemoattractant which again evolves according to a reaction-diffusion PDE. A similar hybrid modelling framework has also been applied to chemotaxis of bacteria [15,55] and leukocytes [26]. The use of the hybrid approach allows for faster simulations than the purely agent-based model which would treat extracellular chemicals as another set of agents. Extracellular signalling molecules are much smaller and more abundant than cells. This property is often used to justify that extracellular chemicals can be described as a continuum [14].

The use of hybrid models is becoming more widespread especially with the growing computational power that allows to consider more complex systems in this manner, including modelling tumour growth [44] and forest dynamics [37]. In cancer biology, several hybrid cellular automaton models have been proposed in the literature [45,47]. For example, Smallbone et al. [47] coupled a two-dimensional cellular automaton model (describing cells) with continuum (PDE-based models) of glucose, H^+ and oxygen concentrations, building on the previous work of Patel et al. [44] and Alarcón et al. [3]. A similar hybrid approach has been used in a number of other studies in cancer biology [5,24,42]. A hybrid forest model with trees modelled as agents and a continuum approach used for oxygen and other atmospheric gases is presented in [37]. In economical research hybrid models are used to estimate prices in the petrol market [29] and in general markets with a non uniform spatial demand of products [30, 31]. In these models the demand is described as a continuous function of space whereas the retailers are considered as agents.

The term hybrid modelling is sometimes applied for models which use both individual-based and continuum description for the same physical quantity. For example, a "hybrid" model for the spread of an epidemic disease is presented in [9]. It initially considers infected individuals as agents, but switches to a continuum model when the number of infected people in an area rises above a threshold. Coupling reaction-diffusion models with a different level of detail in different parts of the computational domain is presented in [17, 22]. "Hybrid" models of this type are useful because they can lead to computational savings. However, in this chapter, we will focus on hybrid models which describe some system components (e.g. cells or animals) as individual agents and some components (e.g. external chemicals) as continuum fields. The choice which description is used for each species is made at the beginning and will not change during the course of the simulation. We will summarise the progress in hybrid models which satisfy this definition, and clarify some of the problems and difficulties that arise from their use.

The outline of this chapter is as follows. Section 2 will give a short overview of the PDE-based and agent-based modelling approaches before the general mathematical framework for hybrid models is introduced in Sect. 3. Hybrid models can be considered as extensions of (purely) agent-based models. Therefore, their computer implementation often forms an integral part of the model. We will discuss it in detail in Sect. 4 where we describe the numerical simulation of hybrid models drawing special attention to the different treatment of the continuum and the agent-based subsystems as well as the problem of matching the two parts. In order to give a more practical insight into the topic we will perform a case study of a hybrid chemotactic model in Sect. 5. This case study will also be used to show some qualitative and quantitative differences that can occur when using a hybrid model instead of the corresponding continuum model.

2   Continuum vs. Agent-Based Models

Hybrid modelling is an intermediate approach between continuum (PDE-based) models and agent-based models of systems of interacting individuals. In this section we briefly review these common modelling approaches in mathematical terms. We will make use of our notation later in Sect. 3 when hybrid models are considered.

Continuum (mean-field) models give rules for the evolution of the spatially dependent concentration vector $\mathbf{c} \equiv \mathbf{c}(\mathbf{x}, t)$ where $\mathbf{x} \in \Omega \subset \mathbb{R}^m$, $m = 1, 2$ or 3, and t is the simulation time. The components of the vector $\mathbf{c}$ can be densities of individuals (cells, animals) and concentrations of extracellular signals. As the concentration vector $\mathbf{c}$ can change both with position $\mathbf{x}$ and time t, a general continuum model takes the form

$$\frac{\partial \mathbf{c}}{\partial t} = \mathscr{L}(\mathbf{c}, \mathbf{x}, t) \qquad \mathbf{x} \in \Omega \,, \tag{1}$$

where $\mathscr{L}$ is an operator on $\mathbf{c}$, which in most practical cases will be a differential or integral operator. To uniquely describe the time evolution of (1), one also has to specify suitable initial and boundary conditions.

Example 1 (Keller-Segel model). Continuum modelling is used in many areas of mathematical biology [39]. In chemotaxis modelling (which will be the subject of Sect. 5), a classical example of (1) is the Keller-Segel model of chemotaxis [35]. Here, $\Omega \subset \mathbb{R}$ and the vector $\mathbf{c}$ has two components, i.e. $\mathbf{c} = [c_1, c_2] = [n, S]$ where $n \equiv n(x, t)$ is the density of cells and $S \equiv S(x, t)$ is the concentration of the chemoattractant. The evolution equation (1) is a coupled system of two PDEs for n and S:

$$\frac{\partial n}{\partial t} = D_n \frac{\partial^2 n}{\partial x^2} - \frac{\partial}{\partial x}\left(n\chi(S)\frac{\partial S}{\partial x}\right), \tag{2}$$

$$\frac{\partial S}{\partial t} = D_S \frac{\partial^2 S}{\partial x^2} - k(S)n, \tag{3}$$

where D_n and D_S are diffusion constants of cells and chemoattractant, respectively. The strength of chemotaxis is controlled by chemotactic sensitivity $\chi(S)$ and therefore by the concentration of substrate S which is consumed by cells with the rate $k(S)$.

The applicability of continuum modelling depends on the number of particles in the studied system. In Example 1, the interacting "particles" are unicellular microscopic organisms (n) and molecules of chemical signal (S). As there are often more signalling molecules than cells, the validity of mean-field assumptions is dictated by the number of cells in the system and the interaction between them [25]. If the system only consists of a few cells, it is more accurate to use an individual-based approach which is introduced in the next section.

2.1 Agent-Based Modelling

In contrary to the continuum models the so-called *agent-based models* treat every particle as an individual that follows an inherent set of rules. This means in particular that individual behaviour and interactions between different agents account for the possibly complex behaviour of the system. Agent-based models are commonly used for systems with a small number of individuals that follow non-trivial behavioural rules, for example in modelling of collective animal behaviour [12] or human crowds in panic situations [27]. While continuum models have a well-developed mathematical theory, agent-based models are sometimes written as computer routines which are difficult to theoretically analyse. The literature also fails to agree on a general definition of an agent. In this chapter, we use a definition which is slightly adapted from [54] and used in [23].

Definition 1. An agent is a system that uses a fixed set of rules based on communication with other agents and information about the environment in order to change its internal state and fulfil its design objective.

This definition, however, is only a formal description, which now has to be put into a more rigorous context. Following from Definition 1, the mathematical description of an agent has to incorporate the behavioural rules of an agent as well as the possibility of communication between them. Therefore, we assume a finite number N of agents numbered from 1 to N. In general N can depend on time, taking into account birth or death of agents. We define the current state of an agent by its internal state variable $y_i(t)$, $i = 1, \ldots, N$, which can describe its position, velocity and internal memory. It is this internal state and its time evolution that describes the rules of an agent. Since these agents represent different individuals, we assume that other agents generally have no means to access all internal state variables. In order to allow for communication between the agents, we define a set of external states $\mathbf{w}_i(t)$, which are observable by other agents. The observable states $\mathbf{w}_i(t)$ of every agent are in principle available to every other agent, which is ensured by creating the set of external states $\mathscr{X}$. The general agent-based model following these definitions then takes the form

$$\mathbf{y}_i(t + \Delta t) = \mathbf{f}_i(\mathbf{y}_i(t), t, \Delta t, \mathscr{X}), \qquad i = 1, \ldots, N, \qquad (4)$$

$$\mathbf{w}_i(t) = \mathbf{g}_i(\mathbf{y}_i(t)), \qquad i = 1, \ldots, N, \qquad (5)$$

$$\mathscr{X} = \{\mathbf{w}_1, \ldots, \mathbf{w}_N\}. \qquad (6)$$

We can see that the evolution of $\mathbf{y}_i$ is given by the function $\mathbf{f}_i$, which notably depends on the time step Δt. This general description can entail discretised versions of ordinary differential equations (ODEs) as well as stochastic differential equations (SDEs). Additionally, agent-based systems that only change discretely can be written in the form (4)–(6).

We understand the external states of an agent merely as an observable representation of the internal states, which is why $\mathbf{w}_i(t)$ directly depends on $\mathbf{y}_i(t)$ through the function $\mathbf{g}_i$. The distinction between observable and non-observable states is often used to represent internal memories that cannot be perceived by other agents [23].

Example 2 (Animal behaviour). Agent-based models have been successfully used for the modelling of collective animal behaviour [51]. Couzin et al. [12] showed that a relatively simplistic model can yield complex collective behaviour and can be used to model fish schools and bird flocks.

In this model, the internal states of an agent $\mathbf{y}_i$ are defined to be its position $\mathbf{x}_i \in \mathbb{R}^m$ ($m = 2, 3$) and its velocity $\mathbf{v}_i \in \mathbb{R}^m$. Since both the position and velocity of an agent potentially influence the motion of other agents, both are observable and hence $\mathbf{w}_i = \mathbf{y}_i = [\mathbf{x}_i, \mathbf{v}_i] \in \mathbb{R}^{2m}$, which means that $\mathbf{g}_i = \text{Id}$. The update rules $\mathbf{f}_i$, $i = 1, \ldots, N$, in this example are equivalent for each agent and incorporate the different rules for the different zones in the model (zone of attraction, orientation and repulsion).

Example 3 (Chemotactic movement under a stationary signal). A simple agent-based model for chemotaxis in one dimension can be written as follows [28]: the internal state $y_i(t)$ of an agent is defined as its current position in $\mathbb{R}$. Additionally, we assume that the signal $S(x)$ is fixed and there is no interaction between agents, hence no observable states are required. All agents start at some initial position $y_{0,i} \in \mathbb{R}$ and move according to the stochastic differential equation

$$\mathrm{d}y_i(t) = \chi(S)\frac{\partial S}{\partial x}\,\mathrm{d}t + \sqrt{2D_n}\,\mathrm{d}W\,, \qquad\qquad i = 1,\ldots,N\,, \qquad (7)$$

where $\chi(S)$ is the chemotactic sensitivity function introduced in Example 1, D_n is the diffusion constant of the bacteria and $\mathrm{d}W$ is the Wiener-process, also known as Brownian motion [33]. We can discretise (7) to obtain an update rule equivalent to (4) as follows

$$y_i(t + \Delta t) = y_i(t) + \chi(S)\frac{\partial S}{\partial x}\Delta t + \sqrt{2D_n\Delta t}\,\xi\,,$$

where ξ is a normally distributed random variable with zero mean and unit variance. In the limit of infinitely many particles, this agent-based description is equivalent to the PDE (2), which is written for the density of cells [49]. However, if we considered a time-evolving signal which is consumed by cells as in Example 1, a purely agent-based model would have to simulate the trajectories of all signal molecules. This would be computational intensive and a hybrid model which combines agent-based simulations with PDEs can then be used to optimize computational efficiency and accuracy.

3 Hybrid Modelling: Theoretical Framework

Because of their hybrid nature the general framework for these models necessarily combines the two frameworks presented in Sect. 2. We define a vector of continuous variables $\mathbf{c}(\mathbf{x}, t)$ on a domain $\Omega \subset \mathbb{R}^m$, $m = 1, 2$ or 3. The update rule for $\mathbf{c}$ is again governed by an operator $\mathscr{L}$, which now also depends on the current states of the agents. The N agents are represented by their internal state variables $\mathbf{y}_i(t)$ and their set of observable states $\mathbf{w}_i(t)$ defined in (5). To allow interactions between the agents and the continuous variables $\mathbf{c}$, the set of observable states $\mathscr{X}$ as defined in (6) is used. The update rules for the system are

$$\frac{\partial \mathbf{c}}{\partial t} = \mathscr{L}(\mathbf{c}, \mathbf{x}, t, \mathscr{X})\,, \qquad\qquad \mathbf{x} \in \Omega\,, \qquad (8)$$

$$\mathbf{y}_i(t + \Delta t) = \mathbf{f}_i(\mathbf{y}_i(t), t, \Delta t, \mathscr{X}, \mathbf{c})\,, \qquad\qquad i = 1,\ldots,N\,, \qquad (9)$$

Fig. 1 Concept of a hybrid model. *Arrows* symbolise direction of influence

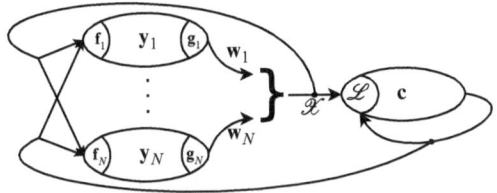

where $\mathscr{X}$ is given in (6). In (8) we see that the agents can influence the continuous variables **c** through the set of observable states $\mathscr{X}$. Similarly, the behaviour of the agents can be altered by the continuous variables, as the operator $\mathbf{f}_i$ now also depends on **c**. Figure 1 shows a graphical representation of the hybrid model. It contains the N agents represented by the internal states $\mathbf{y}_i$ on the left. Through the function $\mathbf{g}_i$ the observable states $\mathbf{w}_i$ are generated which then influence the update of the continuous variables **c** as well as the agents' behaviour themselves. We, however, encounter a problem in this definition, as the continuous variables are defined for every time t, while the internal agent states are only defined for discrete times. To overcome this problem we can consider (9) in the limit $\Delta t \to 0$, where it takes the general form of an SDE

$$\mathrm{d}\mathbf{y}_i = \mathbf{f}_i^{(1)}(\mathbf{y}_i(t), t, \mathscr{X}, \mathbf{c})\, \mathrm{d}t + \mathbf{f}_i^{(2)}(\mathbf{y}_i(t), t, \mathscr{X}, \mathbf{c})\, \mathrm{d}W \ ,$$

where $\mathbf{f}_i^{(1)}$ and $\mathbf{f}_i^{(2)}$ respectively represent the stochastic and the deterministic part of the SDE.

Example 4 (Hybrid cellular automaton model for carcinogenesis). In [47] Smallbone et al. present a hybrid cellular automaton model for the formation of cancer. This model uses reaction-diffusion equations to calculate the concentration of oxygen, glucose and hydrogen ions in the environment of the cells. The concentrations of these chemicals therefore constitute the continuous variables **c**. Each cell of the cellular automaton is represented by an agent with the internal state $y_i \in \mathbb{N}$ defining which of the finite number of possible phenotypes the cell at this position has (including the "phenotype" empty). As these phenotypes are observable by neighbouring cells, we have $w_i = y_i$. This cellular automaton model has a generation-based update rule, which means that the states y_i are only updated once every time step. The rules of the model then represent the probabilistic functions f_i in (4), where the change depends on the current phenotype, the neighbouring cells and the concentrations of the considered chemicals at the cell position.

Example 5 (Hybrid model for chemotaxis of Dictyostelium discoideum*).* Dallon and Othmer developed a hybrid model for the chemotaxis of *Dictyostelium discoideum* [14] that combines individual cell movement with a continuous extracellular concentration of cAMP modelled by a PDE. The internal states of the agents are the position of the individual $\mathbf{x}_i$, as well as the variables representing the intracellular processes. Only the position and one of the intracellular variables influence the external field and therefore form the observable states $\mathbf{w}_i$. The update rules $\mathbf{f}_i$ are given through ODEs for the internal dynamics and rules of motion for the position.

3.1   A Position-Based Hybrid Model

So far, we have defined a general framework for a hybrid model that allows for a great freedom in the choice of internal and external states of the agents. In the next step we want to refine this framework for the more specific models used in chemotaxis modelling [14, 18, 49]. In order to be able to interpret the agents as part of a species situated inside the domain Ω, we need to introduce the notion of an agent's position in Ω. Moreover, we assume that all agents are equal for an external observer except for their position, or in other words the set of observable states of the agents $\mathbf{w}_i(t)$ is the position $\mathbf{x}_i(t)$ of the agents inside Ω, i.e.

$$\mathbf{w}_i(t) \equiv \mathbf{x}_i(t) .$$

This definition excludes Couzin et al. models for animal behaviour [12] as well as cellular automaton models [47], but it is sufficient for the chemotaxis example studied in Sect. 5.

Because of the agents' similarity, we no longer need to define an abstract set $\mathscr{X}$, but can instead define a density function ϱ_δ on Ω through

$$\varrho_\delta(\mathbf{x}, t) = \sum_{i=1}^{N} \delta\left(\mathbf{x} - \mathbf{x}_i(t)\right) , \qquad \mathbf{x} \in \Omega . \tag{10}$$

When discussing numerical simulations of hybrid models, we will see that this definition of ϱ_δ is already a first step towards obtaining a continuous density function for the agents. With this definition we can redefine the operator $\mathscr{L}$, which governs the behaviour of the continuous variables $\mathbf{c}$ and (1) reads as follows

$$\frac{\partial \mathbf{c}}{\partial t} = \mathscr{L}(\mathbf{c}, \mathbf{x}, t, \varrho_\delta) .$$

For the evolution of the internal agent states $\mathbf{y}_i$ we assume now that every agent can only perceive information about the continuous variables $\mathbf{c}$ at its current position. Hence, the operator $\mathbf{f}_i$ no longer depends on $\mathbf{c}$ on the whole domain, but only on $\mathbf{c}(\mathbf{x}_i)$ and the first spatial derivative in this point, i.e. $\mathbf{f}_i$, $i = 1, \ldots, N$, are functions for all further considerations. Equation (9) therefore becomes

$$\mathbf{y}_i(t + \Delta t) = \mathbf{f}_i(\mathbf{y}_i, t, \Delta t, \varrho_\delta, \mathbf{c}(\mathbf{x}_i, t), \nabla \mathbf{c}(\mathbf{x}_i, t)) . \tag{11}$$

This special type of hybrid systems still allows for a wide range of flexibility and can therefore be used to model a variety of different processes. In Sect. 5 we study position-based models for chemotaxis in more depth.

3.2 Initial and Boundary Conditions

An important aspect of modelling is the incorporation of initial and boundary conditions. Hybrid models necessarily combine the conditions from the two different approaches. For the continuous variables one usually has an initial value $c_0(\mathbf{x})$, while for the agents an initial distribution of their position and internal states is given, which is then used to generate each agents' position at the beginning of the simulation. In some applications the agents can be born during the course of the simulation. In this case, we have to ensure the appropriate initialisation of its internal variables.

A similar idea of independent conditions for the continuum and the agent-based parts of the hybrid model is used for the boundary conditions. The values of the continuous variables on the boundary usually have to satisfy an equation of the type

$$\mathscr{G}(\mathbf{c}, \mathbf{x}, t) = 0, \qquad \mathbf{x} \in \partial\Omega, \tag{12}$$

where $\mathscr{G}$ is a general operator. In the most commonly used cases (12) enforces certain values on $\mathbf{c}$ or its gradient on the boundary. For the agents the boundary conditions are often given in a more descriptive manner. For example, agents can leave the domain through one end and automatically reappear on the other end. This *periodic boundary condition* implies that the number of agents in the system is conserved. Periodic boundaries are widely used because of their simplicity and because they effectively shape an infinite domain. *Reactive boundaries* absorb agents with a probability p, while reflecting them with probability $1 - p$ [16]. If $p = 0$, one often speaks of a *reflecting boundary*, while for $p = 1$ the condition is called an *absorbing boundary*.

4 Hybrid Modelling: Numerical Implementation

For similar reasons as in purely agent-based models it is often very hard to obtain analytic results for hybrid models. This increases the importance of numerical simulations for gaining insight into the behaviour of the system. The mixture of different modelling frameworks, however, renders the process of setting up a numerical simulation non-trivial. Each part of the model has to be considered differently and a way of matching the two parts has to be developed. In this section we discuss a numerical framework and evaluate difficulties one has to overcome when implementing a hybrid model.

The general task for the numerical simulation of a hybrid model is to calculate approximations for both $\mathbf{c}$ and $\mathbf{y}_i$ at times $t_j = j\Delta t$, $j = 1, 2, \ldots$ given initial data for each of these variables according to Sect. 3.2. We additionally assume that the domain Ω can (for the continuous part of the hybrid model) be adequately represented by the points $\mathbf{r}_1, \ldots, \mathbf{r}_L \in \Omega$, which means that we seek to compute approximate values for $\mathbf{c}(t_j, \mathbf{r}_l)$, $j = 1, 2, \ldots, l = 1, \ldots, L$ and $\mathbf{y}_i(t_j)$, $i = 1, \ldots, N$. In order to simplify the notation, we introduce

$$\underline{C}_j = [\mathbf{c}(t_j, \mathbf{r}_1), \ldots, \mathbf{c}(t_j, \mathbf{r}_L)] \quad j = 0, 1, \ldots .$$

Due to the different characters of the continuous and the agent-based subsystems, different approaches have to be used for their numerical solutions. For each of the subsystems one tries to answer the question of how to get from t_j to t_{j+1} still guaranteeing an accurate approximation of the system. For the continuous variables this means, we seek a solver that generates the values of $\underline{C}_{j+1}$ using the values $\underline{C}_0, \ldots, \underline{C}_j$ and the current distribution of the agents $\varrho_\delta(\cdot, t_j)$ given by (10), which can be symbolised as

$$\left\{ \underline{C}_0, \ldots, \underline{C}_j, \varrho_\delta(\cdot, t_j) \right\} \overset{\mathscr{L}_d}{\longmapsto} \underline{C}_{j+1}. \tag{13}$$

In (13) we introduced the operator $\mathscr{L}_d$, which is a discretised version of the continuous operator $\mathscr{L}$ used in (8). In the most common case, where $\mathscr{L}$ is a differential operator, $\mathscr{L}_d$ could be a finite element or finite difference approximation of $\mathscr{L}$. Note that in (13) we have made the implicit assumption that the solver used for (8) only takes the positions of the agents at time t_j into account. For the agents equation (11) is already given in a time-discrete way and can therefore be used directly to update the internal states.

The introduction of this general scheme raises some immediate problems, which we will discuss in the remainder of this section. The first difficulty are the differing spatial resolutions for the two subsystems, which we address in Sect. 4.1. Other problems like time stepping, choices of solvers and the influence of stochastic effects are presented in Sect. 4.2.

4.1 Spatial Matching in Numerical Simulations

A spatial matching between the continuous variables and the agents is required during a numerical simulation of a hybrid system, because different spatial resolutions are applied. The agents can be positioned at an arbitrary point inside the domain Ω, while the data for $\mathbf{c}$ is only calculated at the points $\mathbf{r}_l$. This triggers a two-way matching problem, as one has to generate estimates for the agent distribution at the points $\mathbf{r}_l$ as well as for the continuous variables $\mathbf{c}$ everywhere inside Ω.

First, let us consider estimating the agent density distribution throughout Ω and especially at the points $\mathbf{r}_l$, which is necessary for the update relation (13). So the general mapping we are trying to achieve is

$$\varrho_\delta(\mathbf{x}) = \sum_{i=1}^{N} \delta(\mathbf{x} - \mathbf{x_i}) \overset{\psi}{\longmapsto} \varrho(\mathbf{x}) \in C^0(\Omega).$$

The requirements for the estimated density function $\varrho(\mathbf{x})$ can alter for different applications, but here we require it to be at least a continuous function in Ω. One way to achieve such a mapping is the so-called *kernel density estimation* [52]. In general the kernel density estimation can be used to estimate the probability density function of a random process, if one has been given a number of realisations of this process. The name stems from the use of a kernel $K(\mathbf{x})$, which is typically a continuous, symmetric and normalised function. Let us for simplicity assume a one dimensional random process, in which case these conditions take the form

$$K(x) \in C^0(\mathbb{R}), \quad K(-x) = K(x), \quad \int_\mathbb{R} K(x)\mathrm{d}x = 1. \tag{14}$$

Additionally, $K(x)$ is often required to be non-negative in order to generate a non-negative estimate. Most commonly used kernels include a Gaussian kernel and a piecewise linear kernel with compact support. In practice a scaled version of K is used, which leads to the introduction of a bandwidth parameter h. We define

$$K_h(x) = \frac{1}{h} K\left(\frac{x}{h}\right),$$

which still satisfies the conditions (14). With given positions $x_1, \dots, x_N$, an estimate of the probability density function is then given by

$$\varrho(x) = \sum_{i=1}^{N} K_h(x - x_i) = K_h(x) * \varrho_\delta(x). \tag{15}$$

Figure 2 shows an example of a kernel density estimation for 100 normally distributed random variables using a Gaussian kernel with different bandwidths h. In Fig. 2a we can see that the choice of a very small h leads to a highly oscillating estimate, while a very big h can lead to the estimate being too wide as shown in Fig. 2c. An optimal choice for the parameter h and the kernel itself always depends on the nature of the problem and the number of samples N.

The second spatial matching problem that occurs when simulating a combined continuous and agent-based system is the need to estimate the values of the continuous variables (and possibly their derivatives) at an arbitrary position inside Ω. The operator we are looking for can be symbolised through

$$(\mathbf{r}_1, \mathbf{c}(\mathbf{r}_1)), \dots, (\mathbf{r}_L, \mathbf{c}(\mathbf{r}_L)) \overset{\Theta}{\longmapsto} \hat{\mathbf{c}}(\mathbf{x}) \in C^0(\Omega).$$

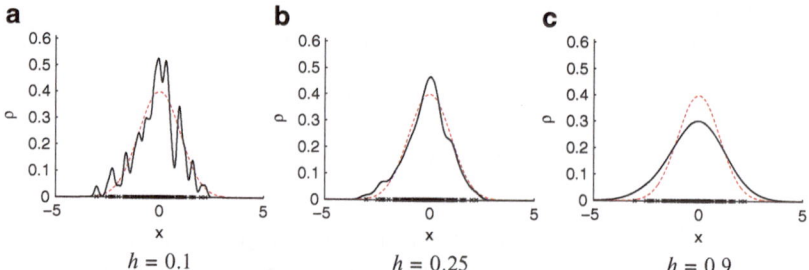

Fig. 2 Kernel density estimate for $N = 100$ agents, which are placed according to a normal distribution with different bandwidths h. *Crosses* along the x-axis represent the agents, the *dashed line* is the underlying Gaussian probability density function and the *solid line* is the generated estimate according to (15)

Though similar to the operator Ψ, we here have the advantage that we know the positions of the points $\mathbf{r}_1, \ldots, \mathbf{r}_L$ beforehand and that we know they give an adequate representation of the domain Ω. With this additional information, one can argue that the problem at hand represents an interpolation problem from the grid points $\mathbf{r}_l$ onto the whole domain Ω. This result allows for the use of approaches from the well-studied fields of interpolation and approximation theory [53]. In some cases the interpolation regime is already implicitly incorporated in the numerical solution of the update equation for the continuous variables, for example if one chooses to use a finite element approach.

Example 6 (Numerical realisation of Example 5). In Example 5 we presented a hybrid model for chemotaxis of slime mold *Dictyostelium discoideum* developed by Dallon and Othmer [14]. To generate a discretised operator $\mathscr{L}_d$ they used the particle-in-cell method [41]. For the kernel density estimation Ψ they use a piecewise linear kernel and for the interpolation operator Θ a fifth order spline interpolation was employed.

4.2 Other Aspects of Numerical Simulations

The spatial matching between the two parts is the biggest additional challenge posed by the use of a hybrid model. Here, we discuss some other problems that occur during this process. The first problem is the choice of a solver both for the continuous variables and for the internal states of the agents. One can choose from a wide range of standard approaches for both problems. The way the two parts are interwoven, however, sets some restrictions. It is, for example, almost always impossible to use a fully implicit solver for both parts, especially if the functions $\mathbf{f}_i$ for the internal agent states contain random variables. Additionally, one has to consider the accuracy of the different solvers and should ideally try to match these to prevent unnecessary computational effort that does not lead to more accurate results.

The discrete nature of the agent-based parts automatically introduces stochastic effects into the system. Various examples of these effects will be discussed in Sect. 5. It is important to consider these effects when choosing the time stepping and the spatial resolution for the simulation. In particular, these choices will depend on the number of agents in the system. It is generally possible to allow different time steps for different parts of the system, for example the agents could be simulated with a finer time stepping than the continuous variables or vice-versa. For each part of the system the time steps have to be chosen in a way that ensures an accurate solution depending on the spatial resolution and the solver that is used. In Sect. 5 we study one application area of hybrid systems in more detail and analyse the effect of some of these choices on the system.

5 Case Study: Hybrid Modelling of Chemotaxis

In Sect. 3 we introduced a general framework for hybrid models that combine agent-based models with mean-field equations and we now concentrate on one application area for hybrid modelling: cell migration. In particular we focus on the movement of cells induced by gradients in the concentration of extracellular chemicals, a process that is known as *chemotaxis*. Chemotaxis is one of the main forms of cell migration and is used in a variety of cells, including bacteria cells [8]. Hybrid models of chemotaxis have been successfully used in the literature [14, 15, 26, 55].

The first notion of chemotaxis goes back to the late nineteenth century, when Engelmann and Pfeffer detected the process. In the late 1960s it was Adler [1, 2] who performed experiments with the bacteria *E. coli* that helped understanding and quantifying the process and that were later used as comparison for the early mathematical models. Adler placed a colony of *E. coli* at one end of a long thin pipe that was filled with oxygen and an additional energy source. Through the process of chemotaxis the colony started to move with a constant speed away from the closed end forming a narrow band of bacteria. The band was visible to the naked eye and Adler was able to measure the speed with which it moved forward.

In the 1970s the first mathematical descriptions of chemotaxis were formulated, with the Keller-Segel model, which we will discuss in Sect. 5.1, as one of the early breakthroughs. A review of the impact this first model had on the modelling of chemotaxis is given in [32]. Section 5.2 will introduce a hybrid version of this model, which we will further investigate and analyse in Sect. 5.3.

5.1 The Keller-Segel Model

As mentioned above, Keller and Segel developed the first mathematical model to describe the process of chemotaxis in 1971 [35]. The original model considers both the bacteria and the chemotactic substrate in a continuum limit, which therefore results in a coupled system of two PDEs. The original form of the system only

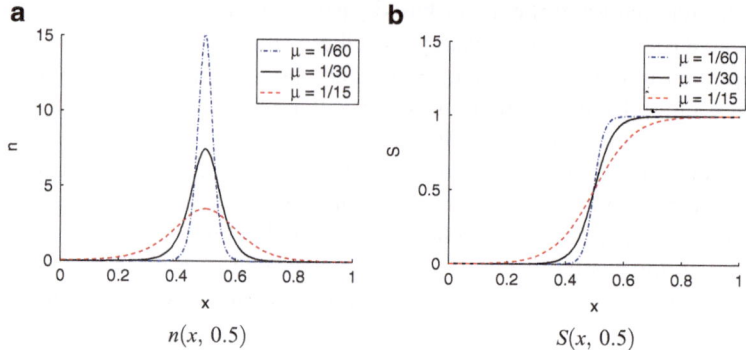

Fig. 3 Travelling wave solution of the Keller-Segel model (16) and (17) for different values of the parameter μ, where $\kappa = 2$

considers one spatial dimension and gives a way to compute the concentration of bacteria denoted by $n(x, t)$ and the concentration of substrate $S(x, t)$ through the PDEs (2) and (3), introduced in Example 1. In (2) we can see that the behaviour of the bacteria is governed by two independent effects and therefore takes the form of a general advection-diffusion equation. The diffusion of the bacteria occurs with the diffusion constant D_n, while the advection is governed by the chemotactic sensitivity $\chi(S)$. The substrate, as seen in (3), diffuses with the diffusion constant D_S and is consumed by the bacteria with a consumption rate $k(S)$ that depends on the concentration of substrate itself.

In a follow-up to the paper [35], Keller and Segel showed that under certain conditions the developed system of partial differential equations yields travelling wave solutions [36]. In particular they were able to proof that travelling wave solutions can only exist if $\chi(S)$ has a singularity at some critical value S_{crit}. For reasons of simplicity they concentrated on the simplest such functions $\chi(S) = \frac{\kappa}{S}$ with the critical concentration at $S_{crit} = 0$. In their analysis Keller and Segel made some additional assumptions for the various parameter values and simplified (2) and (3) to the nondimensionalised PDEs

$$\frac{\partial n}{\partial t} = \mu \frac{\partial}{\partial x} \left(\frac{\partial n}{\partial x} - n \frac{\kappa}{S} \frac{\partial S}{\partial x} \right) , \tag{16}$$

$$\frac{\partial S}{\partial t} = -n . \tag{17}$$

The nondimensionalised system is set up for $x \in [0, 1]$ with an initial value of $S(x, 0) = 1$ and no-flow boundary conditions. As initial distribution of the agents we choose $n(x, 0) = \delta(x)$, which corresponds to the initial state of Adler's experiments where all bacteria were inserted at one end of the tube. We consider reflective boundary conditions for the bacteria at $x = 0$ and $x = 1$.

In order to investigate the influence of the two dimensionless parameters μ and κ on the travelling wave, Figs. 3 and 4 show the concentration of n and S at $t = 0.5$

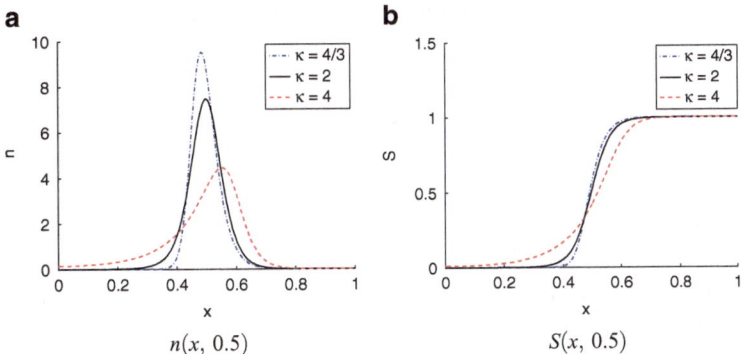

a   **b**

$n(x, 0.5)$ $S(x, 0.5)$

Fig. 4 Travelling wave solution of the Keller-Segel model for different values of the parameter κ, where $\mu = 1/30$

for various values of μ and κ. In Fig. 3 we can see that the parameter μ influences the width of the wave while leaving its general shape untouched. Increasing μ leads to a wider wave and a decrease in the maximum of n. Accordingly, the gradient in S is higher for the narrower bands caused by smaller values of μ. As can be seen in Fig. 4, the parameter κ influences the general shape of the wave. In the case $\kappa = 2$ the travelling band of bacteria is symmetric, while a κ bigger than two leads to a wave that is steeper in the front (right) and falls slowly in the back (left) of the wave. Choosing κ smaller than two causes an opposite effect with the wave being bent backwards.

5.2 Hybrid Models of Chemotaxis

One of the assumptions made by Keller and Segel in their original model is to consider the bacteria as a continuum rather than explicitly describe their individual behaviour. For systems that do not satisfy this assumption hybrid chemotaxis models have been developed in the literature [14, 15, 26, 55]. In this section we present three of them. The bacteria are modelled as agents with varying numbers of internal states and their position $x_i \in \Omega$, as the only observable state. All three models consider the substrate in a continuum limit and the PDE (17) takes the role of (8) in our description of the hybrid modelling framework.

Model I

The first approach to design a hybrid version of the Keller-Segel model, is to interpret the evolution equation for n as a Fokker-Planck equation for a number of randomly moving particles similarly to the idea presented in Example 3. The movement of each of the agents is described by the stochastic differential equation

$$\mathrm{d}x_i = \mu \frac{\kappa}{S(x_i)} \frac{\partial S(x_i)}{\partial x} \mathrm{d}t + \sqrt{2\mu}\mathrm{d}W .$$ (18)

The parameters used in (18) correspond to the ones in the dimensionless Keller-Segel equations (16) and (17). This particle-based description of (16) shows one of the weaknesses in the original Keller-Segel model. According to (18) an agent can theoretically jump any given distance in one time step, implying that some of them can move with a speed that is not achievable for bacteria.

A hybrid model which uses (18) for computation of cell trajectories is analysed in [40]. They use perturbation theory and methods from statistical physics to investigate the non-mean-field-behaviour of a hybrid model where cells produce a chemoattractant that diffuses and degrades on its own. Another approach to analyse a chemotaxis model similar to (18) is presented in [49]. Here, an individual-based description is used for both cells and chemical signal. Then macroscopic PDEs (similar to (2) and (3)) are derived in the limit of infinitely many individuals for appropriately rescaled interactions between individuals.

Model II

Driven by weaknesses of the first model, a different type of random walk, known as velocity-jump process, seems a more realistic choice for the bacterial behaviour. The motion of bacteria *E. coli* consists of two phases [8]. During a run-phase the bacterium moves with a constant speed straight into a chosen direction. This run lasts for a randomly distributed time before the bacterium enters the tumble-phase in which it chooses a new direction randomly [7]. As we are considering a one-dimensional model, there are only two possible directions of motion: to the left and to the right. A right-moving agent continues to the right for a time that is given by an exponentially distributed random variable before it switches its direction. In order to incorporate the bias of bacteria towards higher concentrations of chemoattractants, Othmer et al. [43] introduced a biased velocity-jump process. In this biased random walk the duration for the run phase depends on information gathered at the current position of the individual. In particular, the model in [43] allows the agents to directly measure the gradient of the substrate concentration at their current position. The run-phase then tends to be longer, if the concentration increases in the current direction of motion, while for a decreasing signal, the turning probability is increased.

The turning frequency λ is therefore adjusted according to the current movement direction, the value and the gradient of S. To represent the direction of motion, the velocity $v_i(t) = \pm s$ is introduced, where s denotes the constant speed. In terms of the hybrid modelling framework introduced in Sect. 3, the internal variable is $\mathbf{y}_i = [x_i, v_i]$. The agent-based description of the bacteria can be written in the form

$$x_i(t + \Delta t) = x_i(t) + v_i(t)\Delta t \,,$$

$$v_i(t + \Delta t) = \begin{cases} -v_i(t) & \text{with probability } \lambda^{\pm}\Delta t \\ v_i(t) & \text{otherwise} \end{cases} \,,$$

where

$$\lambda^{\pm} = \lambda_0 \mp \frac{\kappa s}{2S}\frac{\partial S}{\partial x} \,.$$

In a continuum limit this velocity-jump process is equivalent to the hyperbolic chemotaxis equation [21]

$$\frac{1}{2\lambda_0}\frac{\partial^2 n}{\partial t} + \frac{\partial n}{\partial t} = \frac{s^2}{2\lambda_0}\frac{\partial}{\partial x}\left(\frac{\partial n}{\partial x} - n\frac{\kappa}{S}\frac{\partial S}{\partial x}\right) \,, \tag{19}$$

where n is the concentration of bacteria. This shows that changing the type of random-walk used for the agents can influence the corresponding continuum equation. Nevertheless (19) can be used to adjust the parameters of the agent-based model to match the parameters of the Keller-Segel model, as the large time behaviour of (19) is given by the classical chemotaxis equation (16), where we have $\mu = s^2/(2\lambda_0)$ [34]. Lui et al. [38] showed that coupling the hyperbolic chemotaxis equation (19) with (3) for the substrate also yields travelling wave solutions similar to the original Keller-Segel system. An investigation of this case for a more general dependence of the turning frequency is given in [56].

Model III

More accurate descriptions of the individual behaviour of bacteria incorporate the sensing and processing of extracellular signals [6, 48]. Hybrid models with descriptions of these intracellular processes have been used by Dallon and Othmer [14] as well as Xue et al. [57]. Erban and Othmer [18, 19] used an agent with a toy version of the internal dynamics that includes two main features of the sensing process: a fast excitation and a slower adaptation. We will use a simple model with one additional internal variable z_i that acts as a memory and allows the agent to identify increasing or decreasing signal concentrations [18]. The model is based on a velocity-jump process with a turning frequency λ, which depends on z_i. This internal variable is chosen to follow the value of a sensing function $g(S)$ with the adaptation time t_a. Thus, the model can be written in the hybrid form presented in Sect. 3, using $\mathbf{y}_i = [x_i, v_i, z_i]$ as follows:

$$x_i(t + \Delta t) = x_i(t) + v_i(t)\Delta t \,,$$

$$v_i(t + \Delta t) = \begin{cases} -v_i(t) & \text{with probability } \lambda\Delta t \,, \\ v_i(t) & \text{otherwise} \,, \end{cases}$$

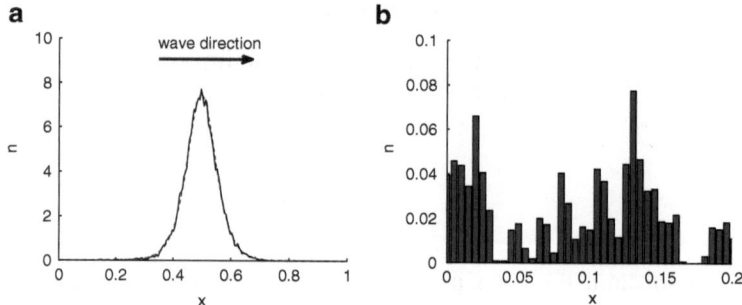

Fig. 5 Numerical simulation of the hybrid Keller-Segel model with internal dynamics (Model III). Parameters are $N = 10^4$, $t_a = 1/\lambda_0 = 1.5 \times 10^{-3}$, $s = 133.33$, $g(S) = 4.5 \times 10^{-3} \log(S)$, $\Delta t = 10^{-4}$. (**a**) Distribution of agents at time 0.5 (*solid line*) and the results given by the Keller-Segel model (16) and (17) (*dashed line*, which is almost indistinguishable from the *solid line*). (**b**) Histogram of agent positions in subinterval $[0, 0.2]$

$$z_i(t + \Delta t) = z_i(t) + \frac{g(S(x_i(t))) - z_i(t)}{t_a} \Delta t \, ,$$

where

$$\lambda = \lambda_0 + z_i - S(x_i) \, .$$

In the limit $\Delta t \to 0$ and $N \to \infty$ this process can be described by the chemotaxis equation

$$\frac{\partial n}{\partial t} = \frac{s^2}{2\lambda_0} \frac{\partial}{\partial x} \left(\frac{\partial n}{\partial x} - \frac{2t_a}{1 + 2\lambda_0 t_a} \frac{dg}{dS} \frac{\partial S}{\partial x} \right) \, , \qquad (20)$$

provided that t is large ($t \gg 1/\lambda_0$) and the gradient of S is shallow [18]. Choices for the parameters of this model can be made by matching (20) with the classical chemotaxis equation (2), which especially indicates that g is given through $dg/dS \sim \chi(S)$.

In Fig. 5a a simulation of the hybrid model of type III is shown. Simulations of the other two models were also performed, with results almost identical to the one seen in Fig. 5a. We simulate $N = 10^4$ agents with the dimensionless model parameters $t_a = 1/\lambda_0 = 1.5 \times 10^{-3}$, $s = 133.33$, $g(S) = 4.5 \times 10^{-3} \log(S)$ and $\Delta t = 10^{-4}$. These parameters were chosen in such a way that they match the global parameters $\mu = 1/30$ and $\kappa = 2$ used for the classical Keller-Segel model. On a first impression, it looks as though the resulting agent distribution at $t = 0.5$ matches the predicted concentration of the Keller-Segel system well except for some stochastic effects. In Fig. 5b, however, we show the agent distribution in the region behind the travelling band. Further analysis of this region showed that here the extracellular signal is completely exploited. Some agents are left in this zone and undergo an unbiased random walk without a chemotactic signal to guide them. This means that these agents do not necessarily manage to catch up with the travelling wave again but instead stay in the exploited region. In the remainder of this section, we study this effect, which we refer to as *dropout* in more detail. We will show that it significantly influences the system dynamics for large times.

5.3 Analysis of the Dropout

In Fig. 5b we saw that the hybrid model, in contrast to the original Keller-Segel model, creates a region behind the wave where the substrate is completely exploited. The main assumptions for a mean-field approach are violated in this region, namely the number of bacteria and the concentration of extracellular material are very small, which renders a continuum approach here not applicable [25]. Stochastic effects due to the small number of bacteria then lead to the complete exploitation of S, which causes the dropout of some of the agents. These agents can no longer sense any gradient in extracellular substrate and are therefore moving completely randomly, which makes it very unlikely for them to become part of the travelling band again. Due to the constant loss of agents, the velocity and the height of the wave will decrease as the wave moves along. Note that a complete exploitation in these models is only possible under the assumption that S does not diffuse, which was made by Keller and Segel and is incorporated in the PDE (17). The dropout effect is interesting for us, because it shows a qualitative difference between the hybrid model and the original Keller-Segel model, as the hybrid model only yields transient travelling wave solutions. In this section we create measures for this dropout in order to get an estimate of the number of lost agents from the simulations. We will then move on to analyse the effect of some system parameters on the dropout. Finally, some theoretical results about the loss of agents are presented and compared to numerical results.

Dropout Measures

In order to be able to quantify the dropout of agents from the travelling wave, we need to investigate certain conditions that render an agent as dropped out. A condition of this form allows us to define an index set $\Gamma(t)$ that contains the agents who are currently part of the wave.

However, before defining and comparing different conditions for the dropout, we investigate some global statistical values of the agent set. The first measure to indicate the fact that agents have dropped out is the position of the centre of the wave $c(t)$. From [36] we know that the theoretical wave speed of the nondimensionalised Keller-Segel system is 1 and therefore the predicted position of the centre of the wave is $c_{mf}(t) = t$. In comparison to that the actual position of the wave can be measured from the agents' positions via

$$c_1(t) = \frac{1}{N} \sum_{i=1}^{N} x_i(t).$$ (21)

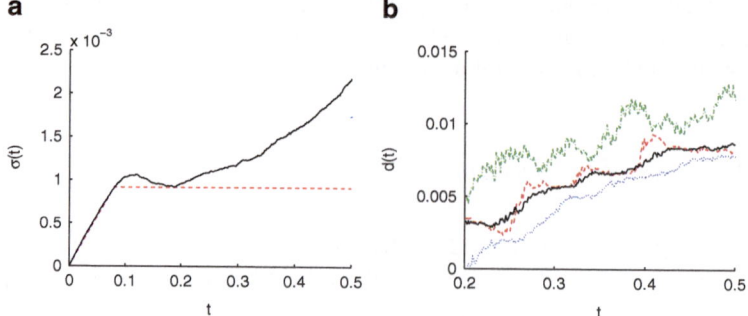

Fig. 6 Simulation results of the variance and dropout for short times, where the same parameter values as in Fig. 5 are used. (**a**) Variance $\sigma(t)$ estimated from the simulation (*solid line*) and variance of the stationary wave given by the mean-field model σ_{mf} (*dashed line*). (**b**) Dropout given by (25): $d_1(t; 0.1)$ (*dash-dotted line*), $d_1(t; 0.15)$ (*solid line*), $d_1(t; 0.2)$ (*dotted line*) and $d_2(t)$ (*dashed line*)

The problem with this option is that it includes dropped out agents for the calculation of the wave centre, which can bias the calculation. To overcome this problem, a second option for finding the centre of the wave is given through

$$c_2(t) = \frac{1}{|\Gamma|} \sum_{i \in \Gamma} x_i(t), \tag{22}$$

which implies that the found centre position depends on the choice for the index set Γ. For short times $c_1(t)$ and $c_2(t)$ give similar results, but will differ for large times. Using this wave centre $c_1(t)$, we can calculate the variance of the agent positions as an indicator for the width of the wave and therefore for the dropout. In Fig. 6a this variance is compared to the variance of the travelling wave solution found by Keller and Segel, which is $\sigma_{mf} = (\pi \mu)^2/3$. Initially the measured variance increases linearly towards the theoretical value, which is caused by the start of the agents on the boundary $x = 0$. After the wave is fully developed, the variance starts to rise over the theoretical value, which indicates a significantly wider wave and therefore dropout of agents.

With these statistical values for the agent set we have now different options to define an agent as dropped out from the wave and therefore to define the index set Γ. The first option is to allow an agent to have a certain distance r from the centre of the wave. Agents with a distance bigger than r are therefore considered to be dropped out. Hence,

$$\Gamma_1 \equiv \Gamma_1(t; r) = \{i \in \{1, \dots, N\} \mid x_i(t) \geq c_1(t) - r\} . \tag{23}$$

Because of the non-finite support of the travelling wave solution for the original Keller-Segel system, the measure defined in (23) is strongly dependent on r, which makes the choice of r important. One should choose r in a way that the solution of the original Keller-Segel model only predicts a very small number of dropout agents. One way to pick r is to use a multiple of the theoretical standard deviation of the wave.

A second option of defining an agent as dropped out is to use the observation that S is exploited behind the wave. An agent is then considered to be dropped out of the wave if the value of S at its current position is 0. Thus,

$$\Gamma_2 \equiv \Gamma_2(t) = \{i \in \{1, \ldots, N\} \mid S(x_i(t)) = 0\} . \tag{24}$$

Using the sets Γ_1 and Γ_2 we can now define 2 dropout measures $d_1(t; r)$ and $d_2(t)$ by

$$d_1(t; r) = 1 - \frac{1}{N} |\Gamma_1(t; r)|, \quad \text{and} \quad d_2(t) = 1 - \frac{1}{N} |\Gamma_2(t)| . \tag{25}$$

Figure 6b shows plots of the behaviour of $d_1(t; r)$ and $d_2(t)$. We can see that after the initial period of adjustment due to the start on the boundary $x = 0$, all measures have an increasing trend with some fluctuations around it. The measure $d_1(t; 0.15)$ matches well with $d_2(t)$, but has less fluctuations.

Large Time Behaviour

In this section we investigate the large time behaviour of the travelling wave in the hybrid chemotaxis Model III. We study the behaviour of the bacteria and the signal in the half-line $[0, \infty]$. For large times the definitions $c_1(t)$ and $c_2(t)$ given by (21) and (22) differ significantly because many agents drop behind the wave. Therefore, $c_2(t)$ is more meaningful to describe the centre of the wave in this case. However, as $c_2(t)$ depends on Γ, we can no longer use $\Gamma \equiv \Gamma_1$ to find the agents that have dropped out, because Γ_1 depends on the definition of the centre of the wave. We therefore use $d_2(t)$ given by (25) as measure for the dropout in the analysis of large time behaviour, where we are particularly interested in the slowing down of the wave. Hence, we define the velocity of the wave $v(t)$ through

$$v(t) = \frac{c_2(t + \Delta T) - c_2(t)}{\Delta T} , \tag{26}$$

where ΔT is chosen to be much larger than Δt in order to minimise the fluctuations in $v(t)$. We simulate $N = 10^4$ agents with the same parameters as before. The results of one simulation are shown in Fig. 7. We see that after $t = 50$ about 40% of the agents have dropped out from the wave. The predicted slowing down of the wave is demonstrated in Fig. 7b, where we plot $v(t)$ as a function of time. We use $\Delta T = 0.1$ in the definition (26). As the velocity shrinks with the number of agents in the wave, we have $v(t) \approx 1 - d_2(t)$, which is also demonstrated in Fig. 7b.

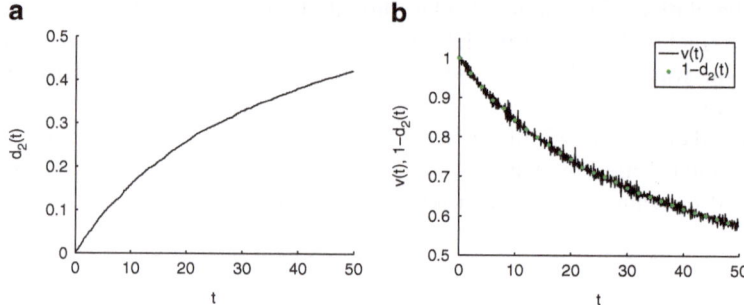

Fig. 7 Dropout and velocity of the travelling wave for large time, where the same parameter values as in Fig. 5 are used. (**a**) Dropout $d_2(t)$ given by (25). (**b**) Velocity of the wave $v(t)$ given by (26) (*solid line*) compared with $1 - d_2(t)$ (*circles*)

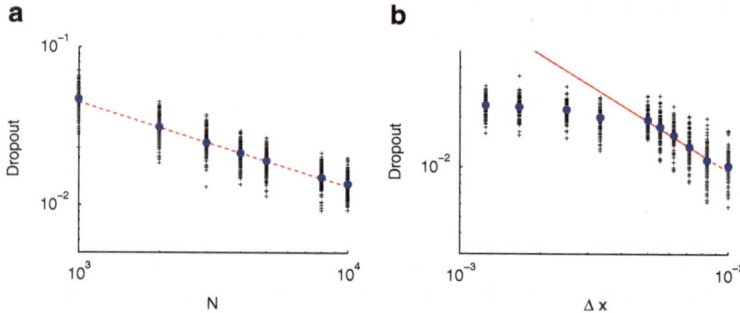

Fig. 8 Dropout $d_1(0.5; 0.15)$ given by (25) as a function of (**a**) N and (**b**) Δx. In each figure we show results given by individual simulations (*crosses*), average values of $d_1(0.5; 0.15)$ estimated from simulations (*circles*) and linear fits explained in the text (*dashed-line*)

Dropout in Dependence on N and Δx

In the next step we use the derived measure (25) in order to analyse the influence of certain system parameters on the dropout. In particular, we are interested in the dependence of the dropout on the number of agents N and the gridsize Δx. The variation of the number of agents N in the system is a way of comparing the hybrid with the continuum model. One would expect that the dropout goes to 0 as N goes to infinity. On the other hand the Δx dependence is a problem of the hybrid model, as one would ideally want the dropout to be independent of the chosen grid. We performed a number of simulations for various values of N (200 simulations for each value) and Δx (100 simulations for each value) and in each case measured the value of $d_1(0.5; 0.15)$. The results are plotted in Fig. 8. In Fig. 8a we plot the average values of $d_1(0.5; 0.15)$ estimated from simulations as circles. The best linear fit in the double logarithmic plot, shown as the dashed line, has a gradient of -0.53, which indicates that $d_1 \sim 1/\sqrt{N}$. This relationship can be explained through the central limit theorem, which predicts that the noise in the system should shrink with $\sqrt{N}$.

The plot in Fig. 8b shows a more complicated dependence. For larger values of Δx the dashed line with gradient -1 can be fitted indicating that a finer grid leads to an increase in dropout, which seems slightly surprising at first glance, as one expects a finer grid to allow for a more accurate representation of the original PDE. This effect can, however, be explained by the lower number of agents per gridpoint and therefore the higher noise expected at each gridpoint. As Δx decreases the dropout seems to level off, meaning that the choice of a finer grid at this point does not influence the dropout drastically. Bearing in mind that we ideally wanted the dropout to be independent of Δx, this levelling off effect seems to indicate the region of choice for Δx in order to get an accurate solution.

Theoretical Analysis

More theoretical insight into the dropout effect can be obtained by considering a simplified system, where the concentration of extracellular material S is a given function that does not change over time. A natural choice for the function $S(x)$ is the travelling wave solution found by Keller and Segel [36]. Using the knowledge of the exploited region behind the wave, we can adjust this function slightly to allow for the analysis of the dropout effect. We therefore define S to be equal to 0 for x smaller than some critical position x_c and to take the form of the travelling wave solution everywhere else. In this section we will use $\kappa = 2$, thus we put

$$S(x) = \begin{cases} (1 + \exp(-x/\mu))^{-1}, & x \geq x_c, \\ 0, & x < x_c. \end{cases} \tag{27}$$

To be able to use a time-independent function for S we have to make adjustments to the movement of the agents, as they would otherwise follow the increasing gradient towards the right of the real axis. Therefore, we subtract the expected wave speed of 1 from the movement velocity of the agents in order to keep them in a position that is realistic for the travelling wave. In other words, we use a coordinate system that moves with the travelling wave solution. For example, for an agent of Model I the evolution equation becomes

$$dx_i = \left(\mu \frac{2}{S(x_i)} \frac{dS(x_i)}{dx} - 1 \right) dt + \sqrt{2\mu} dW.$$

With the help of this simplified system we can now make further analytic and simulative investigations into the effect of different x_c on the quantity of the dropout. If an agent enters the exploited region $x < x_c$, two behaviours are considered. In the first case, the agent would be considered dropped out and is absorbed by

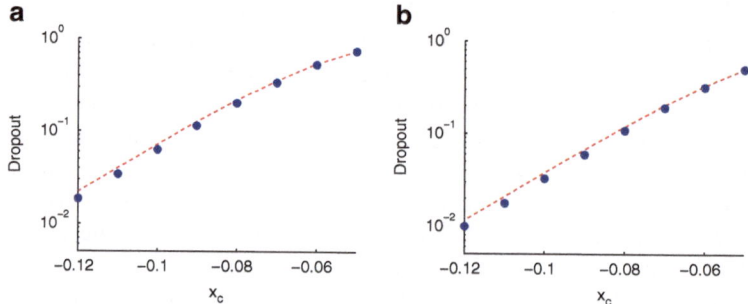

Fig. 9 Dropout $d_1(0.5; 0.15)$ defined by (25) as a function of x_c for static signal given by (27) where the same parameter values as in Fig. 5 are used. In each figure we show average values of $d_1(0.5; 0.15)$ estimated from 100 simulations (*circles*). (**a**) Simulations where no comeback from $x = x_c$ is allowed. The *dashed line* is a result of the theoretic analysis given by (30). (**b**) Simulations where dropout agents can return. The *dashed line* is 50 % of the dropout predicted by (30)

the boundary, so that it has no chance of becoming part of the wave again. The second case allows the agent to randomly move around in the exploited area and therefore allows the agent to enter the non-exploited region again. For both cases we performed 100 simulations for each of the considered values of x_c and measured the value of $d_1(0.5; 0.15)$ as defined before, this time using 0 as the mean position. The average values of $d_1(0.5; 0.15)$ estimated from the simulations are shown in Fig. 9 as circles. To analyse the case of an absorbing boundary at $x = x_c$ we consider the system in the limit $N \to \infty$, which is described by the following equation (compare to (16))

$$\frac{\partial n}{\partial t} - \frac{\partial n}{\partial x} = \mu \frac{\partial}{\partial x} \left(\frac{\partial n}{\partial x} - n \frac{2}{S} \frac{dS}{dx} \right). \tag{28}$$

The boundary condition on the left-hand boundary can be written in the form $n(x_c) = 0$. Further conditions for $x \to \infty$ can be introduced. We look for a separable solution of the form

$$n(x, t) = \exp(-\lambda t) M(x),$$

where λ is a positive constant. Plugging this ansatz into (28) leads to

$$\mu M'' + M' - 2\mu \left(M \frac{S'}{S} \right)' + \lambda M = 0, \tag{29}$$

where primes denote derivatives with respect to x. For the ODE (29) a non-negative solution is sought that satisfies $M(x_c) = 0$ and $M(x) \to 0$ as $x \to \infty$. The general solution for (29) is

$$M(x) = C_1 \frac{2\lambda\mu \exp\left(x\frac{3+\gamma}{2\mu}\right) + (2\lambda\mu - 1 - \gamma)\exp\left(x\frac{1+\gamma}{2\mu}\right)}{\left(\exp\left(\frac{x}{\mu}\right) + 1\right)^2}$$

$$-C_2 \frac{2\lambda\mu \exp\left(x\frac{3-\gamma}{2\mu}\right) + (2\lambda\mu - 1 + \gamma)\exp\left(x\frac{1-\gamma}{2\mu}\right)}{\left(\exp\left(\frac{x}{\mu}\right) + 1\right)^2},$$

where $\gamma = \sqrt{4\lambda\mu + 1}$. The integration constants C_1 and C_2 have to be chosen to satisfy the boundary conditions. Because of the nature of (29) as an eigenvalue problem, only the quotient C_1/C_2 can be determined uniquely, which also means that the condition $M(x) \to 0$ as $x \to \infty$ is satisfied for all values $C_1, C_2 \in \mathbb{R}$. Taking a closer look at the limit $x \to \infty$, we can see that the direction of the approach changes in dependence of λ, in particular, a non-negative solution can only be obtained for λ smaller than a critical value $\lambda_c(x_c)$. This critical value $\lambda_c(x_c)$ is achieved for the case where C_1/C_2 turns out to be 0. Applying the left-hand boundary condition $M(x_c) = 0$ for this case yields to the unique value $\lambda_c(x_c)$ given through

$$\lambda_c(x_c) = -\frac{1}{\mu}\exp\left(-\frac{x_c}{\mu}\right)\left(1 + \exp\left(-\frac{x_c}{\mu}\right)\right)^{-2} = -S'(x_c^+).$$

This value $\lambda_c(x_c)$ can now be used to get a predicted value of the dropout $d_{pred}(x_c, t)$ via

$$d_{pred}(x_c, t) = 1 - \exp(\lambda(x_c)t). \tag{30}$$

The function $d_{pred}(x_c, 0.5)$ is plotted as the dashed line in Fig. 9a. We can see that it matches well with the simulation results. The slight overestimation given by $d_{pred}(x_c, 0.5)$ can be explained through the time it takes before the first agents start reaching the critical position x_c from the starting position at $x = 0$. For the situation with comeback, we choose a value $\lambda = \alpha\lambda_c(x_c)$ to predict the dropout, where α is a constant. Matching this with the simulation results as shown in Fig. 9b we found that $\alpha \approx 0.5$, which indicates that about 50 % of agents come back into the wave after they have dropped out. This effect could be modelled by using a reactive boundary [16] instead of the free diffusion zone behind the wave.

6 Discussion

In this chapter, we reviewed the advances that have been made in the field of hybrid modelling of collective behaviour. Hybrid models combine agent-based models with mean-field concentration models and allow a more accurate description of certain systems than the general mean-field approach. Compared to purely agent-based models hybrid models have the advantage of a reduced computational complexity and a wider range of applicability. As hybrid models explicitly consider individual behaviour as well as interactions between individuals, stochastic effects are incorporated which can alter the behaviour from that of the corresponding continuum model. This became especially clear during the studies of hybrid chemotaxis models in Sect. 5. We showed that the hybrid models do not produce a travelling wave in the classical sense, as agents are dropping out behind the wave. This effect leads to a decrease in the number of agents in the wave, which also slows down the wave, as demonstrated in Fig. 7. We also discussed some of the problems and difficulties related to the use of hybrid models. In particular the spatial matching between the discrete agents and the continuous variables has to be considered. We showed in Fig. 8 that the choice of the gridsize can have a significant effect on the behaviour of hybrid models and has to be handled with care.

Throughout this chapter, we mainly focused on hybrid models which include agents with internal variables. These models are particularly useful whenever individuals describe living cells (e.g. Model III introduced in Sect. 5.2) where internal variables model important intracellular processes. In particular, the mathematical framework (8) and (9) covers a rich class of complex biological systems. One disadvantage of models with internal dynamics is that they are more complicated to analyse [20]. In the case of models without internal variables (e.g. Model I introduced in Sect. 5.2), one can apply approaches which were developed for analysis of many-particle systems in statistical physics. Perturbation methods and closure approximations have been used for analysis of variants of Model I in the literature [25, 40]. Kinetic and hydrodynamic descriptions of hybrid models based on velocity jump processes without internal dynamics are derived in [11]. Mathematical analysis becomes more challenging whenever cells are not described as pointwise objects. For example, the cellular Potts model is a lattice-based approach which takes into account the finite size of biological cells. Using Kramers-Moyal expansion, Alber et al. [4] derived a continuous macroscopic description of a two-dimensional cellular Potts model. Models of cells which take into account both the finite size and internal variables are even more difficult to analyse. A continuum model for chemotaxis of disk-like cells with internal variables was derived for stationary signals in [20]. Analysis of a hybrid model of the finite-sized cells with internal dynamics remains an open problem.

Acknowledgements The research leading to these results has received funding from the European Research Council under the *European Community's* Seventh Framework Programme *(FP7/2007–2013)* / ERC *grant agreement* No. 239870. This publication was based on work supported in part by Award No KUK-C1-013-04, made by King Abdullah University of Science and Technology (KAUST). RE would also like to thank Somerville College, University of Oxford, for a Fulford Junior Research Fellowship; Brasenose College, University of Oxford, for a Nicholas Kurti Junior Fellowship; the Royal Society for a University Research Fellowship; and the Leverhulme Trust for a Philip Leverhulme Prize.

References

1. J. Adler, Chemotaxis in bacteria. Science **153**, 708–716 (1966)
2. J. Adler, Chemoreceptors in bacteria. Science **166**, 1588–1597 (1969)
3. T. Alarcón, H. Byrne, P. Maini, A cellular automaton model for tumour growth in inhomogeneous environment. J. Theor. Biol. **225**, 257–274 (2003)
4. M. Alber, N. Chen, P. Lushnikov, S. Newman, Continuous macroscopic limit of a discrete stochastic model for interaction of living cells. Phys. Rev. Lett. **99**, 168102 (2007)
5. A. Anderson, A hybrid mathematical model of solid tumour invasion: the importance of cell adhesion. Math. Med. Biol.: J. IMA **22**, 163–186 (2005)
6. N. Barkai, S. Leibler, Bacterial chemotaxis - united we sense . . . Nature **393**, 18–21 (1998)
7. H. Berg, How bacteria swim. Sci. Am. **233**, 36–44 (1975)
8. H. Berg, D. Brown, Chemotaxis in Esterichia coli analysed by three-dimensional tracking. Nature **239**, 500–504 (1972)
9. G. Bobashev, M. Goedecke, F. Yu, J. Epstein, A hybrid epidemic model: combining the advantages of agent-based and equation-based approaches. In *Proceedings of the 2007 Winter Simulation Conference*, ed. by S. Henderson (IEEE, New York, 2007), pp. 1532–1537
10. J. Buhl, D. Sumpter, I. Couzin, J. Hale, E. Despland, E. Miller, S. Simpson, From disorder to order in marching locusts. Science **312**, 1402–1406 (2006)
11. P. Chavanis, Nonlinear mean field Fokker-Planck equations. Application to the chemotaxis of biological populations. Eur. Phys. J. B **62**, 179–208 (2008)
12. I. Couzin, J. Krause, R. James, G. Ruxton, N. Franks, Collective memory and spatial sorting in animal groups. J. Theor. Biol. **218**(1), 1–11 (2002)
13. I. Couzin, J. Krause, N. Franks, S. Levin, Effective leadership and decision-making in animal groups on the move. Nature **433**, 513–516 (2005)
14. J. Dallon, H. Othmer, A discrete cell model with adaptive signalling for aggregation of dictyostelium discoideum. Phil. Trans. R. Soc. B: Biol. Sci. **352**(1351), 391–417 (1997)
15. R. Erban, From individual to collective behaviour in biological systems, Ph.D. thesis, University of Minnesota, 2005
16. R. Erban, S.J. Chapman, Reactive boundary conditions for stochastic simulations of reaction-diffusion processes. Phys. Biol. **4**(1), 16–28 (2007)
17. R. Erban, S.J. Chapman, Time scale of random sequential adsorption. Phys. Rev. E **75**(4), 041116 (2007)
18. R. Erban, H. Othmer, From individual to collective behaviour in bacterial chemotaxis. SIAM J. Appl. Math. **65**(2), 361–391 (2004)
19. R. Erban, H. Othmer, From signal transduction to spatial pattern formation in *E. coli*: a paradigm for multi-scale modeling in biology. Multiscale Model. Simul. **3**(2), 362–394 (2005)
20. R. Erban, H. Othmer, Taxis equations for amoeboid cells. J. Math. Biol. **54**(6), 847–885 (2007)
21. R. Erban, I. Kevrekidis, H. Othmer, An equation-free computational approach for extracting population-level behavior from individual-based models of biological dispersal. Physica D **215**(1), 1–24 (2006)
22. M.Flegg, J. Chapman, R. Erban, Two Regime Method for optimizing stochastic reaction-diffusion simulations. J. R. Soc. Interface. **9**, 859–868 (2012)
23. B. Franz, Synchronisation Properties of an Agent-Based Animal Behaviour Model, Master's thesis, University of Oxford, 2009
24. P. Gerlee, A. Anderson, An evolutionary hybrid cellular automaton model of solid tumour growth. J. Theor. Biol. **246**, 583–603 (2007)
25. R. Grima, Multiscale modelling of biological pattern formation. Curr. Top. Dev. Biol. **81**, 435–460 (2008)
26. Z. Guo, P. Sloot, J.C. Tay, A hybrid agent-based approach for modeling microbiological systems. J. Theor. Biol. **255**(2), 163–175 (2008)
27. D. Helbing, I. Farkas, T. Vicsek, Simulating dynamical features of escape panic. Nature **407**, 487–490 (2000)

28. F. Hellweger, V. Bucci, A bunch of tiny individuals: individual-based modeling for microbes. Ecol. Model. **220**, 8–22 (2009)
29. A. Heppenstall, A. Evans, M. Birkin, A hybrid multi-agent/spatial interaction model system for petrol price setting. Trans. GIS **9**, 35–51 (2005)
30. A. Heppenstall, A. Evans, M. Birkin, Using hybrid agent-based systems to model spatially-influenced retail markets. J. Artif. Soc. Soc. Simul. **9**, 2 (2006)
31. A. Heppenstall, A. Evans, M. Birkin, Genetic algorithm optimisation of an agent-based model for simulating a retail market. Environ. Plann. B: Plann. Des. **34**, 1051–1070 (2007)
32. D. Horstmann, From 1970 until present: the Keller-Segel model in chemotaxis and its consequences i. Jahresbericht der DMV **105**(3), 103–165 (2003)
33. N. van Kampen, *Stochastic Processes in Physics and Chemistry*, 3rd edn. (North-Holland, Amsterdam, 2007)
34. G. Karch, Selfsimilar profiles in large time asymptotics of solutions to damped wave equations. Stud. Math. **143**, 175–197 (2000)
35. E. Keller, L. Segel, Model for chemotaxis. J. Theor. Biol. **30**, 225–234 (1971)
36. E. Keller, L. Segel, Traveling bands of chemotactic bacteria: a theoretical analysis. J. Theor. Biol. **30**, 235–248 (1971)
37. J. Landsberg, R. Waring, A generalised model of forest productivity using simplified concepts of radiation-use efficiency, carbon balance and partitioning. Forest Ecol. Manag. **95**, 209–228 (1997)
38. R. Lui, Z.A. Wang, Traveling wave solutions from microscopic to macroscopic chemotaxis models. J. Math. Biol. **61**(5), 739–761 (2010)
39. J. Murray, *Mathematical Biology* (Springer, Berlin, 2002)
40. T. Newman, R. Grima, Many-body theory of chemotactic cell-cell interactions. Phys. Rev. E **70**, 051916 (2004)
41. P. O'Rourke, J. Brackbill, On particle-grid interpolation and calculating chemistry in particle-in-cell methods. J. Comput. Phys. **103**, 37–52 (1993)
42. J. Osborne, A. Walter, S. Kershaw, G. Mirams, A. Fletcher, P. Pathmanathan, D. Gavaghan, O. Jensen, P. Maini, H. Byrne, A hybrid approach to multi-scale modelling of cancer. Phil. Trans. R. Soc. A: Math. Phys. Eng. Sci. **368**, 5013–5028 (2010)
43. H. Othmer, S. Dunbar, W. Alt, Models of dispersal in biological systems. J. Math. Biol. **26**, 263–298 (1988)
44. A. Patel, E. Gawlinski, S. Lemieux, R. Gatenby, A cellular automaton model of early tumor growth and invasion. J. Theor. Biol. **213**, 315–331 (2001)
45. B. Ribba, T. Alarcón, K. Marron, P. Maini, Z. Agur, The use of hybrid cellular automaton models for improving cancer therapy, in *ACRI 2004*, ed. by P. Sloot, B. Chopard, A. Hoekstra. Lecture Notes in Computer Science, vol. 3305 (Springer, Berlin, 2004), pp. 444–453
46. F. Schweitzer, L. Schimansky-Geier, Clustering of "active" walkers in a two-component system. Physica A **206**, 359–379 (1994)
47. K. Smallbone, R. Gatenby, R. Gillies, P. Maini, D. Gavaghan, Metabolic changes during carcinogenesis: potential impact on invasiveness. J. Theor. Biol. **244**, 703–713 (2007)
48. P. Spiro, J. Parkinson, H. Othmer, A model of excitation and adaptation in bacterial chemotaxis. Proc. Natl. Acad. Sci. USA **94**, 7263–7268 (1997)
49. A. Stevens, The derivation of chemotaxis equations as limit dynamics of moderately interacting stochastic many-particle systems. SIAM J. Appl. Math. **61**, 183–212 (2000)
50. D. Sumpter, The principles of collective animal behaviour. Phil. Trans. R. Soc. B: Biol. Sci. **361**, 5–22 (2006)
51. D. Sumpter, *Collective Animal Behavior* (Princeton University Press, Princeton, 2010)
52. M. Wand, M. Jones, *Kernel Smoothing* (Chapman & Hall/CRC, London, 1994)
53. N. Wiener, *Extrapolation, Interpolation, and Smoothing of Stationary Time Series* (MIT, Cambridge, 1964)

54. M. Wooldridge, *An Introduction to Multi-Agent Systems* (Wiley, New York, 2002)
55. C. Xue, H. Othmer, Multiscale models of taxis-driven patterning in bacterial populations. SIAM J. Appl. Math. **70**(1), 133–167 (2009)
56. C. Xue, H.J. Hwang, K. Painter, R. Erban, Travelling waves in hyperbolic chemotaxis equations. Bull. Math. Biol. **1**, 1–29 (2010)
57. C. Xue, E. Budrene, H. Othmer, Radial and spiral stream formation in *Proteus mirabilis* colonies. PLoS Comput. Biol. e1002332 (2011)

From Individual Movement Rules to Population Level Patterns: The Case of Central-Place Foragers

Hsin-Hua Wei and Frithjof Lutscher

Abstract We consider a model for the dynamics of a consumer-resource population where the foraging behavior of the central-place forager is explicitly modeled as a random walk. The model consists of a discrete map between generations and a partial differential equation within a season. We determine analytically the conditions under which the consumer can stay in the system. We then explore numerically how different assumptions on the foraging strategy affect the stability of the coexistence equilibrium. We find a number of ways in which foraging behavior destabilizes the coexistence equilibrium and leads to population cycles. We also find an instability resembling a flip bifurcation even though the model has compensatory dynamics. This modeling framework can serve in the future to explore the evolution of foraging strategies, thereby complementing previous ecological theory of central-place foraging.

1 Introduction

The question of how individuals move in space to forage, traditionally, falls into the field of behavioral ecology. Foraging strategies are optimized to maximize resource intake, but the dynamics on the population level are usually not considered. Models for population dynamics, on the other hand, focus on birth, and survival while keeping behavioral aspects at a minimum [9]. In some sense, this dichotomy results from the different time scales involved, with foraging happening on a faster time scale than population growth and survival. However, when evolutionary aspects of foraging behavior are considered, the dynamics of populations need to be included [4]. The goal of this study is to develop and analyze population dynamic models

H.-H. Wei · F. Lutscher (✉)
Department of Mathematics and Statistics, University of Ottawa, 585 King Edward Ave.,
Ottawa, ON, Canada K1N6N5
e-mail: whh9@hotmail.com; flutsche@uottawa.ca

M.A. Lewis et al. (eds.), *Dispersal, Individual Movement and Spatial Ecology*,
Lecture Notes in Mathematics 2071, DOI 10.1007/978-3-642-35497-7_6,
© Springer-Verlag Berlin Heidelberg 2013

that explicitly include the spatial foraging behavior of individuals, specifically for central-place foragers.

Central-place foragers are individuals who have a place from which they depart and to which they must return after a foraging bout [14, 16]. This central place can be a nest where offspring need to be fed or a refuge where predation pressure is relieved. Many species are central-place foragers, for example, ants with their colonies; many birds, in particular sea birds; many cave-dwelling species such as crickets and bats; mammals with their den site. Central-place foraging theory considers optimal foraging strategies based on maximizing fitness, i.e., food intake or rate of delivery to the central place. It is often assumed that individuals have full, or at least partial, knowledge of the size or quality and distance of resource items, and there is no re-stocking of resources. The optimal strategy then is to order resource items according to their energy content per time necessary to acquire the resource. Some predictions of this theory include that individuals should concentrate their foraging efforts near the central place and should consider far-away items only if they are large [5]. This optimization process does not consider the dynamics of a population over multiple generations.

Models for population dynamics between generations focus on birth and death of individuals. These processes may depend on age structure, interaction with conspecifics, with competitors or with predators, but behavioral aspects of foraging are kept at a minimum. For example, some of the simplest models for the dynamics of a single species collapse foraging behavior into a single parameter that describes the amount of "clumping" of the species on the locations of their resource. A robust prediction of these models is that increased clumping enhances the stability of the positive population equilibrium; population cycles are less likely. The mechanism of this stabilizing effect is that as more individuals concentrate on few resource patches, most mortality occurs there, whereas the few individuals in low-density patches buffer the population against cycles [9].

The number of studies linking behavioral aspects of central-place foraging with population dynamics is fairly limited. Some consider the impact of central-place foragers on the community surrounding the central place (e.g. [2]). Some considered effects on the population itself, for example on resource division between foragers [8], or on nestling survival [20]. Only recently was the effect of foraging strategies on the stability of the corresponding consumer-resource system considered [6]. The authors there used spatial distributions to describe foraging locations, and then considered a population-level optimization scheme to adjust the distribution to the resources in a given year. Some of these spatial distributions were based on random-walk models for individual movement. It was found that stability of the consumer-resource equilibrium sensitively depends on the spatial distribution of the consumer and, hence, on foraging behavior.

The present study takes the connection between individual movement behavior and population dynamics one step further by including more complex behavior into the random walks that generate the foraging distribution. It is, thus, possible to examine the effect of each aspect of foraging behavior on the population dynamic patterns such as stability versus population cycles.

In the next section, we formulate our model in two steps: first the population dynamic processes of growth, reproduction, and survival are formulated in a discrete-time system. Then the foraging process is modeled as a random walk in continuous time and space as a partial differential equation. This framework is quite general; we give three particular functions that can describe the choice of foraging locations, and we consider their impact on the system. More specifically, we consider the cases that individuals make their choice based on resource availability alone; on resource availability and travel time; or on resource availability and density of other foragers. In Sect. 3 we briefly consider persistence conditions of the consumer population. We find simple analytical rules. In Sect. 4 we explore how the different foraging rules affect the stability of the population equilibrium. We conclude with a discussion of the patterns that emerge and suggestions as to where this modeling framework can be applied.

2 Model Derivation

The model consists of two parts: the discrete map for the consumer and resource density from one generation to the next, and the continuous description of individual movement to the foraging location within a season. In a non-spatial setting, a similar approach of between and within season dynamics was used by Geritz and Kisdi [7] to give mechanistic interpretations of depensatory and overcompensatory population maps.

2.1 Population Dynamics

We begin with a consumer-resource model in discrete time, where we denote the resource and consumer densities in generation n by F_n, C_n, respectively. We assume for simplicity that the resource grows before the consumer emerges to forage, see [3] for a more general model. We denote by G the growth of the resource and by P the probability that a given resource is consumed. Then the equations between generations are

$$
\begin{aligned}
F_{n+1} &= G(F_n)(1 - P(C_n)), \\
C_{n+1} &= sC_n + bG(F_n)P(C_n).
\end{aligned}
\tag{1}
$$

The parameter b is the conversion coefficient from resource to consumer biomass. The survival probability of consumers from one generation to the next is denoted by s. For $s = 0, b = 1$ we obtain a standard host parasitoid model [12]. The dynamics of this non-spatial model have been analyzed for a variety of growth functions and capture probabilities [10], though typically with $s = 0$. In general, the survival probability could depend on the amount of acquired resources, i.e., $s = s(G(F)P(C))$. Here, we include the case of constant $s > 0$ since adult survival is often much slower to decrease than reproductive success when food is scarce [1]. We limit ourselves to the Beverton-Holt growth function and the negative exponential capture probability, i.e.,

$$G(F) = \frac{rF}{1 + \frac{r-1}{E}F}, \qquad P(C) = 1 - \exp(-aC). \qquad (2)$$

All parameters are assumed positive, and $r > 1$. The dynamics of (1), (2) allow for three scenarios: the consumer may become extinct while the resource approaches its carrying capacity E; consumer and resource may stably coexist; or stable consumer-resource cycles can be observed [19].

Next, we introduce the spatial variable X, and write $F_n(X)$ for the spatial distribution of the resource in generation n. We assume that the central place of the consumer is located at $X = 0$. We write the spatial distribution of foragers in generation n as the product $C_n K_n(X)$ of the total number of consumers C_n with a probability density of their locations $K_n(X)$. In particular, K_n is a non-negative function whose integral over the entire space is one. In reality, consumers will only forage in some suitable region, Ω, for example an island or otherwise limited habitat patch. If consumers leave this patch, then the integral of K_n over Ω may be less than unity. For simplicity, we will consider the one-dimensional interval $\Omega = [-L/2, L/2]$, corresponding to a patch of length L with the central place in the middle of the patch. We assume that individuals who forage inside the patch return to the central place whereas the ones who leave the region do not return. Then the spatial model reads

$$F_{n+1}(X) = G(F_n)(1 - P(C_n K_n(X))),$$

$$C_{n+1} = sI_n C_n + b \int_\Omega G(F_n(X)) P(C_n K_n(X)) dx, \qquad (3)$$

where $I_n = \int_\Omega K_n(X) dx \leq 1$ denotes the percentage of foragers returning to the central place. We call the functions K_n the *foraging kernels* in analogy with dispersal kernels that describe the movement of individuals between two generations [13]. In (3) we have assumed that the dispersal of the resource between generations is negligibly small compared to the foraging area of the consumer. A variety of alternative assumptions about consumer mortality during foraging and outside of Ω were discussed by Fagan et al., as was the case that resource redistribution occurs on the same scale as consumer foraging [6]. Only the model details change, the overall approach is the same.

When $K_n = K = 1/|\Omega|$ is constant in space and between generations, and if F_0 is constant in space, then model (3) is equivalent to the non-spatial model (1). The assumption of spatially constant K implies that individuals search equally throughout the domain. In contrast, optimal foraging theory for central-place foragers states that individuals should concentrate their search efforts near the central place [2]. Such a concentrated effort near the central place could be described by a spatial distribution K that is peaked near zero, for example the Gaussian or Laplace distribution, as was done in [6]. Then, however, the resource near the central place becomes more and more depleted over time, so that individuals should concentrate their efforts further away. Fagan et al. used a population-level optimization scheme to model changes in foraging locations in response to resource availability [6]. Here, we use individual-based movement rules to let consumers adjust to resource distribution in every generation. We introduce a random walk model with appropriate foraging rules in the next section.

2.2 Individual Movement

In this section, we model individual movement to the foraging location by a partial differential equation, similar to the approach in [13, 17]. We assume that individuals move randomly across the landscape after they emerge from the central location, and they settle at a location with a given rate A. The distribution of settling locations will then be the foraging kernel K. The equations for mobile (u) and settled (v) individuals are

$$\frac{\partial u}{\partial t} = D \frac{\partial^2 u}{\partial X^2} - Au, \qquad u(0, X) = \delta(X),$$

$$\frac{\partial v}{\partial t} = Au, \qquad v(0, X) = 0, \tag{4}$$

where the Dirac delta for the initial condition indicates that each individual begins the foraging process at the central location. Formally, the foraging kernel is given by

$$K(X) = v(T, X) = \int_0^T Au(t, X)dt. \tag{5}$$

The time that the movement process takes is denoted by T. If movement happens on a fast time scale, we can assume $T = \infty$, which we will do from here on. If individuals change their movement behavior at the boundary of the patch, then the equation for u has to be supplied with appropriate boundary conditions. For example, if we assume that the boundary is hostile so that individuals at the boundary are removed and never return, then the condition is

$$u(t, x) = 0 \qquad \text{for} \qquad x \in \partial \Omega. \tag{6}$$

When the settling rate is a constant, independent of all external influences, and the domain Ω is the entire real line, then the resulting foraging kernel is the double-exponential or Laplace kernel [13]

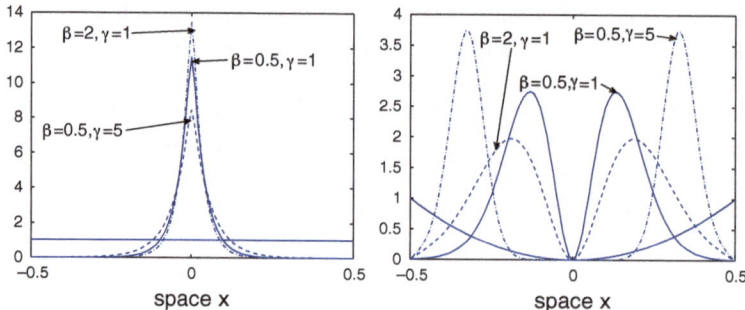

Fig. 1 Foraging kernels with resource-dependent settling rate (8). *Left:* The resource profile is the constant one and the resulting kernel is the Laplace kernel. *Right:* The resource profile is a parabola and the resulting kernels have peaks away from the central place. The (scaled) diffusion coefficient is $D = 0.01$ and the maximal settling rate is $\alpha = 10$

$$K(X) = \frac{1}{2}\sqrt{A/D}\exp\left(-\sqrt{A/D}|X|\right). \qquad (7)$$

If the boundary of a finite patch is permeable for individuals to some degree, then the resulting kernels can be obtained as some infinite series [18]. The effect of the Laplace and other kernels, fixed in time, on the dynamics of system (3) has been studied in [6]. The interesting part here is, of course, to let the settling rate A depend on various factors, such as resource density, forager density, or time. We consider several possibilities here. We illustrate the effect of each set of assumptions with two scenarios: (i) spatially constant resource distribution and (ii) resources depleted near the central place. In Sect. 4 we study how the different assumptions affect the dynamics of the entire system.

2.2.1 Resource-Dependent Settling

We begin with the assumption that the settling rate is an increasing bounded function of resource density. We choose Hill's function

$$A(F) = \frac{\alpha F^{\gamma}}{\beta^{\gamma} + F^{\gamma}}. \qquad (8)$$

The parameter α denotes the maximum settling rate, β is the half-saturation constant, and γ controls the steepness of the curve. For $\gamma = 1$, the curve is concave down, for $\gamma > 1$ the curve is sigmoidal. For spatially constant resource distribution, this settling rate is constant, and hence the resulting kernel is a Laplace kernel. When the resource at the central place is depleted, however, then no individuals settle there, so that the resulting foraging kernel has two maxima away from the central place, see Fig. 1. For all simulations in this section, we chose the simple quadratic function $F(x) = 4x^2$ on $[-1/2, 1/2]$ as a hypothetical resource distribution, depleted near the central place. We found that with this function, we can capture all phenomena

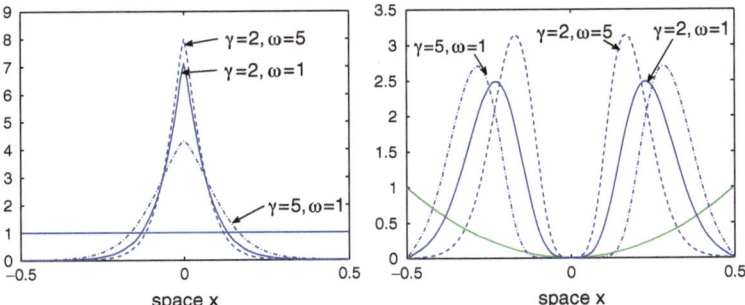

Fig. 2 Foraging kernels with time-dependent settling rate (8). *Left:* The resource profile is the constant one and the resulting kernel is similar to the Laplace kernel. *Right:* The resource profile is a parabola and the resulting kernels have peaks away from the central place. The (scaled) diffusion coefficient is $D = 0.01$ and the maximal settling rate is $\alpha = 10$, and $\beta = 2$

that we found in the simulations of the full consumer-resource system, see also Fig. 4 in [6] for typical steady state distributions of the resource.

The effect of the various parameters on the resulting distributions are indicated in the figure as well. We did calculate the mean absolute distance moved and the variance in move length, and we found that both are increasing functions of the half saturation constant, β, and the shape parameter, γ, and decreasing functions of the maximum settling rate α, (plots not shown).

2.2.2 Time-Dependent Settling

The longer an individual moves before consuming resources, the more energy it spends. Hence, as time goes on, an individual should be more likely to settle and forage even if resources are not optimal. We modify the previous case by replacing F with $(1 + \omega t)F$, where $\omega > 0$ indicates how fast a forager accepts low-density resource locations. Hence, the settling rate is

$$A_{\text{time}}(F, t) = A((1 + \omega t)F) = \frac{\alpha[(1 + \omega t)F]^{\gamma}}{\beta^{\gamma} + [(1 + \omega t)F]^{\gamma}}. \tag{9}$$

Increasing the parameter ω gives shorter travel distances and a smaller variance of the foraging kernel for both, a spatially constant and spatially varying resource distribution F. The profiles are given in Fig. 2. The mean absolute distance moved and the variance in move length are decreasing functions of ω, (plot not shown).

2.2.3 Prospect-Dependent Settling

In the final example, we assume that an individual settles at a location depending on its prospective return. We use the ratio of resource availability to previously

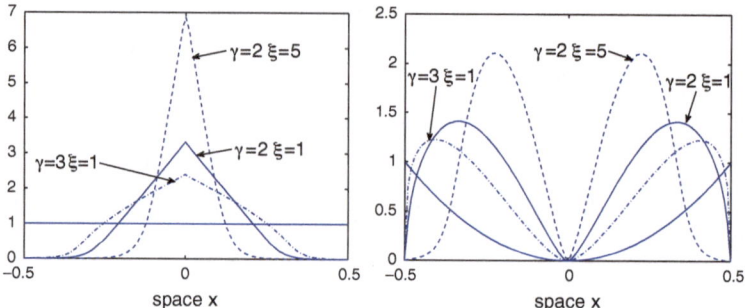

Fig. 3 Foraging kernels with prospect-dependent settling rate (8). *Left:* The resource profile is the constant one and the resulting kernel clearly differs from the Laplace kernel, at least for small ξ. *Right:* The resource profile is a parabola and the resulting kernels have peaks away from the central place. The (scaled) diffusion coefficient is $D = 0.01$ and the maximal settling rate is $\alpha = 10$, and $\beta = 2$

settled individuals (F/v) as a measure for prospective return. In other words, even if resource density is high the individual might choose to not settle there if there are already many settled individuals foraging. We set

$$A_{\text{prospect}} = A(\xi F/v) = \frac{\alpha(\xi F/v)^{\gamma}}{\beta^{\gamma} + (\xi F/v)^{\gamma}}. \tag{10}$$

Here, $\xi > 0$ indicates how strongly the moving individuals react to their expected return. We observe that the settling rate defined here has the unrealistic property that $A_{\text{prospect}} = \alpha$ wherever $v = 0$, independent of the resource density F. Therefore, we slightly modify the rate and replace v with $v + \epsilon$ for some small value of ϵ. (In all simulations, we used $\epsilon = 0.01$.)

The effects of ξ on the resulting forager distribution is depicted in Fig. 3. Increasing ξ clearly decreases the distance moved and the variance of the foraging kernel, (plots not shown).

2.2.4 Nondimensionalization

Before we close this section, we non-dimensionalize the model by setting $F_n = E f_n$, $C_n = a/L c_n$, $X = Lx$. then the equations read

$$f_{n+1}(x) = \frac{r f_n(x)}{1 + (r - 1) f_n(x)} \exp(-c_n k_n(x))),$$

$$c_{n+1} = s I_n c_n + \tilde{b} \int_{-1/2}^{1/2} \frac{r f_n(x)}{1 + (r - 1) f_n(x)} [1 - \exp(-c_n k_n(x))] dx, \tag{11}$$

where now $\tilde{b} = bE/a$ and $k(x) = LK(Lx)$. The diffusion coefficient in (4) scales by L^2, the parameters β in (8) and ξ in (10) scale with E. In all the simulations, we choose hostile boundary conditions at $x = \pm 1/2$. We write b instead of $\tilde{b}$ when no confusion can arise.

3 Effects of Movement Rules on Persistence

We begin our analysis with the question of whether or not the consumer can persist. Hence, we linearize at the resource-only steady state $(F^*(X), 0)$ and check the stability condition. After linearizing (3) the equation for the consumers decouples and in its dimensional form reads

$$C_{n+1} = \left(sI_n + b \int G(F^*(X))K_n(X)dX \right) C_n. \tag{12}$$

If there is no consumer, then the resource is homogeneously distributed throughout the domain, i.e., F^* is independent of X. In addition, since it is a steady state for the resource, we have $G(F^*) = F^*$. Hence, the condition for $(F^*(X), 0)$ to be unstable so that the consumer can persist in the system is

$$\int_\Omega K(X)dX > \frac{1}{bF^* + s}, \quad \text{or} \quad \int_{-1/2}^{1/2} k(x)dx > \frac{1}{b+s} \tag{13}$$

in the dimensional and non-dimensional form, respectively. This persistence condition differs slightly from the one found in [6] since we included the loss of consumers into the survival term here.

The effects of movement rules on persistence is fairly easy to see, since we only need to consider the case of spatially constant resource distribution. The resource-dependent settling rate (8) becomes a constant that increases with α and decreases with β^γ. The integral in (13) therefore also increases with α and decreases with β^γ. For the time-dependent settling rate, persistence is more likely with increasing ω and in the prospect-dependent rate, persistence is more likely with increasing ξ.

The persistence condition for the consumer is fairly easy to determine. Since the resource distribution at the resource-only steady state is constant in space, the invasion condition of the consumer reduces to knowing the value of the integral (13). In the next section, we assume that parameters are such that the consumer can persist, and we ask how movement rules affect the shape and stability of the coexistence steady state.

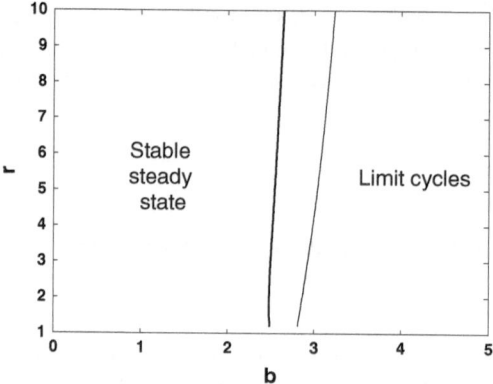

Fig. 4 Bifurcation diagram for the non-spatial consumer resource model in the r-b plane. The *line* indicates a Neimark Sacker bifurcation that separates stable coexistence (to the *left*) from invariant loops (to the *right*). The *thick line* corresponds to $s = 0.5$, the *thinner line* to $s = 0.2$. The region labeled "stable steady state" includes the region where the consumer is extinct and the resource is at its stable equilibrium

4   Effects of Movement Rules on Stability

In this section we assume that the consumer can persist, and we ask whether the resulting consumer-resource system shows stable coexistence or cyclic behavior. For the purpose of comparison, we begin with the non-spatial model, or, equivalently, with a spatial model where dispersal is uniform in the domain. The three remaining parameters in the model are r, b, and s. The bifurcation diagram in the r-b plane for different values of s is given in Fig. 4. Roughly, these results can be summarized by saying that increasing the values of b and s may destabilize the system and lead to cycles whereas increasing r tends to stabilize the coexistence state. For spatially varying but temporally fixed kernels, some of these relationships may change. For example, Fagan et al. showed that with a fixed kernel invariant loops may appear for some intermediate range of s-values (using a top-hat kernel) or only for small values of s (using a Laplace kernel) [6] .

In the case that movement behavior depends on resource density, we now investigate how the various parameters affect stability.

4.1   Resource-Dependent Settling

We begin with the model where individuals choose their foraging site according to resource density as in the settling rate (8). In our extensive numerical simulations of the full spatial system, we found essentially three different behaviors with resource and consumer coexisting. There could be a stable steady state, invariant loops, or

Fig. 5 Snapshots of the forager distribution every six generations on a stable invariant loop in the consumer-resource system with resource dependent settling rate (8). The parameters are $s = 0.5$, $b = 3.5, r = 2, \alpha = 10$, $\beta = 0.2, \gamma = 1, D = 0.01$

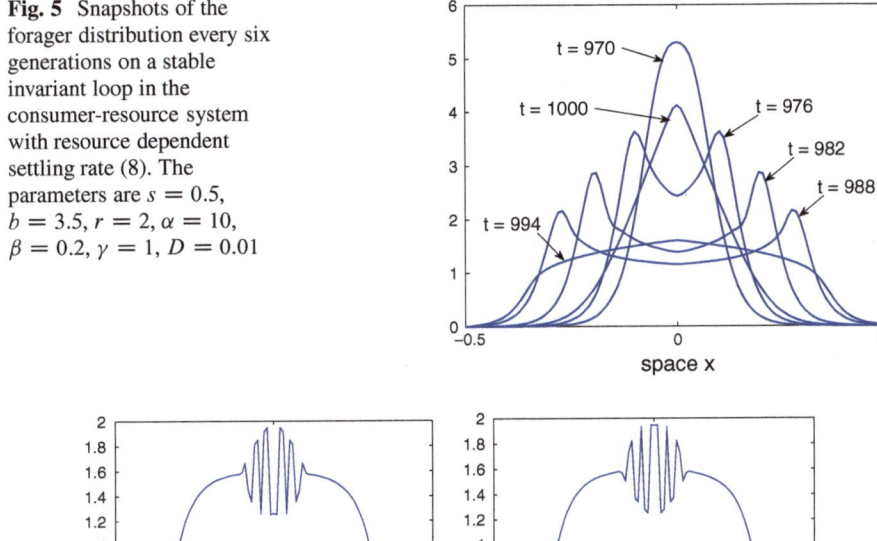

Fig. 6 Snapshots of the forager distribution at the two stages of a 'switch' in the consumer-resource system with resource dependent settling rate (8). The dynamics alter between these two states. The parameters are $s = 0.5, b = 3.5, r = 8, \alpha = 10, \beta = 0.2, \gamma = 1, D = 0.01$

'switches'. The changes in the consumer foraging locations over the course of a typical solution are depicted in Fig. 5. From any nonzero initial condition, solutions have stabilized at the invariant loop after around 800 generations. We begin our observations at $t = 970$, when foragers are concentrated near the central place. Six generations later, the resources near the central place are depleted so that consumers concentrate away from the central place. As consumers successively deplete the resources near the central place, the peaks of forager density move away from the central place for several generations. At $t = 994$, however, resources near the central place have recovered enough to attract consumers there. After several generations of foragers concentrated near the central place and the cycle starts over (through $t = 1,000$ until $t = 1,006$).

In the case of a 'switch', the forager distribution switches between two states as does the resource distribution, but the number of consumers remains relatively constant. An example is given in Fig. 6. We found that such patterns only occurred for high resource growth rate and high settling rate. Consumers locally deplete the resource in one year to such a low level that the probability of settling there the next year is significantly decreased. Then the resource grows back to very high levels

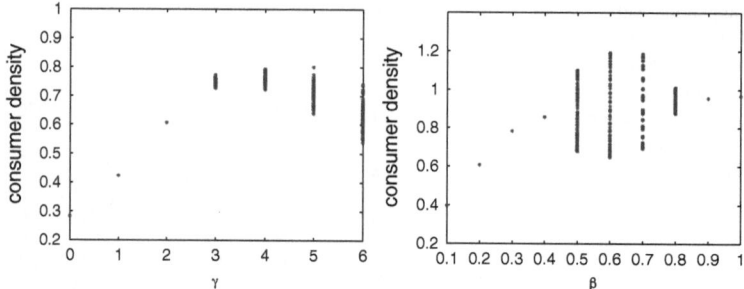

Fig. 7 Orbit diagrams and bifurcations for the resource-dependent settling rate. *Left:* Cycles appear as γ increases, for $\beta = 0.2$. *Right:* Intermediate values of β give rise to cycles for $\gamma = 2$. In both plots the other parameters are $s = 0.5, b = 3, r = 3, \alpha = 10, D = 0.01$

during the season when the consumer density is low. Because of the high maximum settling rate, consumers settle quickly and closely together as soon as there is a sufficiently high resource density. This mechanism leeds to the alternating, in space and time, high-low density pattern near the central place in Fig. 6.

The different behaviors that we observed are very similar to a Neimark-Sacker bifurcation (in the case of cycles) and a flip bifurcation (in the case of 'switches') for finite dimensional maps. However, the spatial system that we study is not finite dimensional. Its analytical study is even more complicated by the fact that the equation for the resource has no 'smoothing' effect, so that the spectrum of the linearized operator is not necessarily a point spectrum. Introducing dispersal of the resource will make the eigenvalue problem well posed [6]. Alternatively, we use numerical simulations here to study the stability of the coexistence state. At each time step, we used a central difference scheme to solve the partial differential equation of consumer movement in MATLAB. We present orbit diagrams, in which we plot the consumer density, c_n, for 50 generations after the initial transients have died out; typically after 2,000 generations. Because we need to solve a partial differential equation in each time step, numerical solutions require a good amount of computing power. For that reason, we have a fairly coarse discretization of parameter space. A detailed study of phenomena such as phase-locking, that can be found in discrete maps, is beyond the scope of this work.

We find the following behaviors. With respect to the population dynamics parameters, the behavior is qualitatively quite similar to the non-spatial case and the orbit diagrams are not shown. Increasing the conversion coefficient, b, or the survival rate, s, can destabilize the coexistence state and lead to cycles. Increasing the resource growth rate, r, can stabilize cycles. In contrast to the non-spatial situation, increasing r even further can destabilize the steady state again and lead to a 'switch' as described above.

For the movement parameters, a number of different bifurcations can occur. Increasing the shape parameter γ in the settling rate (8) can destabilize the steady state and lead to population cycles. Cycles can also occur for intermediate ranges of the half-saturation rate, β, see Fig. 7.

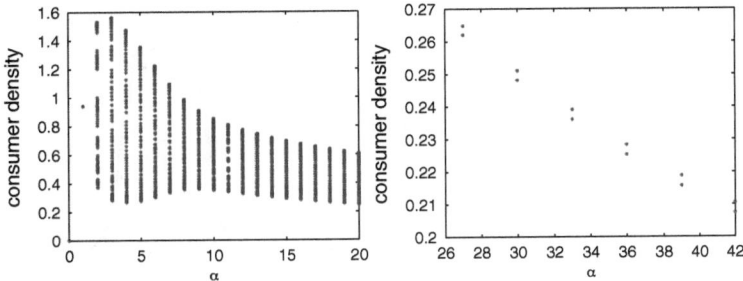

Fig. 8 Orbit diagrams and bifurcations with respect to α. *Left:* Cycles appear as α increases, cycle amplitude and mean decreases for large enough α. Parameters are $s = 0.5$, $b = 4$, $r = 3$, $\beta = 0.1$, $\gamma = 1$, $D = 0.01$. *Right:* Another bifurcation can destabilize the steady state and lead to 'switches' for large values of α. Here: $b = 2$, $\gamma = 2$, and the other parameters are as in the other plot

The behavior with respect to the maximum settling rate, α, depends on the other parameters. Typically, the coexistence state is stable for small enough α and loses stability to population cycles as α increases. This behavior is similar and related to the bifurcations with respect to the conversion coefficient, b. In both cases, higher values mean higher consumer growth rates. However, while the value of b does not affect the resource density in a given year, the value of α does since more resource is consumed. Consequently, the amplitude and the mean of these population cycles tend to decrease for large α, see Fig. 8. In some cases, the orbits may disappear altogether and the coexistence state is stable for high α (plots not shown). In some cases, there can be another bifurcation, similar to a flip bifurcation, and a 'switch' pattern can result, see Fig. 8.

4.2 Time-Dependent Settling

We now consider the behavior of the system with the time-dependent settling rate (9). The orbit diagram in Fig. 9 shows that increasing the parameter ω can destabilize the coexistence state and trigger the occurrence of invariant loops. For any given value of ω, the dynamic behavior with respect to the other parameters is qualitatively similar to the previous case of resource-dependent settling. We do not present other orbit diagrams here. We do, however, point out that the foraging distribution over the course of one cycle is much more concentrated near the central place than was the case for the resource-dependent settling (see the right plot in Fig. 9). As time increases, individuals tend to settle even at lower resource density so that they can obtain at least some resources. This increasing (in time) settling rate leads to a narrowly peaked distribution.

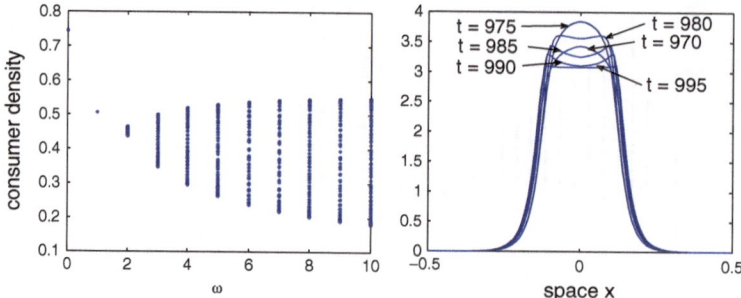

Fig. 9 Orbit diagram and forager distribution for the time-dependent settling rate. *Left:* As the parameter ω increases, a stable coexistence state can be destabilized and invariant loops can occur. *Right:* The distribution of foragers over the course of a cycle is much more narrowly peaked around the central place than was the case with resource-dependent settling. Parameters are: $s = 0.5$, $r = 3$, $b = 3.5$, $\alpha = 10$, $\beta = 0.2$, $\gamma = 2$ for the orbit diagram. When $\gamma = 1$, then the onset of cycles occurs for larger values of ω. The foraging distributions are plotted with the same parameters, except $\gamma = 1$ and $\omega = 15$

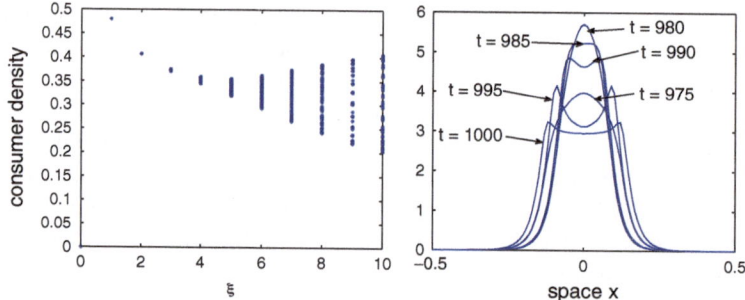

Fig. 10 Orbit diagram and forager distribution for the prospect-dependent settling rate. *Left:* As the parameter ξ increases, a stable coexistence state can be destabilized and invariant loops can occur. *Right:* The distribution of foragers over the course of a cycle is much more narrowly peaked around the central place than was the case with resource-dependent settling. Parameters are: $s = 0.5$, $r = 3$, $b = 3.3$, $\alpha = 10$, $\beta = 0.2$, $\gamma = 1$ for the orbit diagram. The foraging distributions are plotted with the same parameters and $\xi = 16$

4.3 Prospect-Dependent Settling

The results for prospect-dependent settling are quite similar to the case of time-dependent settling. For fixed ξ, the qualitative behavior of the system is the same as in the case of resource-dependent settling. As a function of ξ, the coexistence state can be destabilized for increasing ξ. Similarly to the time-dependent settling rate, the forager distribution tends to be more narrowly peaked around the central place. The results are illustrated in Fig. 10.

5 Discussion

Foraging theory typically considers a single individual and maximizes fitness. It assumes that the individual has knowledge of available resource items, and it largely ignores population dynamic aspects. Conversely, population dynamic models do not usually include foraging behavior. The main goal of this study was to develop and analyze a mathematical model for population dynamics, based on individual movement rules for central-place foragers. This framework allowed us to explore and elucidate the effect of different foraging strategies on the stability of the population, and, vice versa, the effect of population dynamics on spatial patterns in the resource via individual-based movement rules. All these movement rules were local in the sense that an individual decides at each step whether to settle and forage or to move and in which direction. We did not assume that the individual has prior knowledge over available resources.

We found that population dynamics parameters in the spatial model have similar effects on the stability of a population as in the non-spatial model. We also found that movement rules can have a significant effect on population dynamics. The most interesting observations are probably that sometimes intermediate parameter values can destabilize a steady state and lead to oscillations, and secondly that the system might exhibit two-cycles (reminiscent of flip bifurcations in overcompensatory dynamics, see [10]) even though the dynamics here are compensatory.

One of the classical results about stability of populations was that increased clumping of the consumer on resource patches increases stability of the steady state [9]. Instability there, under the assumption of constant resource, results in a flip-bifurcation and the appearance of stable two-cycles in the population. We do not have a parameter of clumping in our model; aggregation at certain spatial locations rather emerges as a result of a situation-specific selection of foraging sites. We can interpret a decreased variance in distance moved as increased clumping, see Sect. 2.2.1. Aggregation on the resource, which is dynamic in our model, causes a decrease in available resource at that location in the next year. We observe that increasing γ decreases the variance of distance moved, but also increases the likelihood of population cycles (Fig. 7). A similar pattern emerges for ω (Fig. 9). Hence, we find quite different behavior than the simple models that abstract much of the foraging process.

Our work also advances the theory and ideas for mechanistic models of dispersal processes and their resulting kernels as in [13, 15, 18]. In particular, we found mechanistic underpinnings for distributions that are quite different from the commonly used Gaussian or Laplace distributions. We claim that resource and prospect dependent settling should be quite likely and need to be incorporated into future models in spatial ecology. The situation of central-place foragers, where all individuals emerge from the same location, facilitates this inclusion of mechanisms. The challenge for the future is to incorporate such density-dependent movement and settling behavior into other models where foragers or dispersers are spatially distributed to begin with. The modeling framework will not be either discrete or continuous, but rather a hybrid of the two as we used here or, in a non-spatial setting, Geritz and Kisdi earlier [7].

We want to highlight two directions for future research that naturally extend our work. First, there are a number of analytical questions to be asked. For example, the question of a rigorous stability analysis of the coexistence state, potentially under the assumption of (small) resource dispersal; an exploration of the dynamics with overcompensatory resource growth; investigation of phenomena such as phase-locking; or the inclusion of resource dependent movement into the partial-differential equation, similar to prey-taxis [11]. The second topic for future exploration is to use our modeling framework to return to the question of optimal central-place foraging and consider the evolutionary dynamics of certain movement-related parameters. One would adopt the viewpoint of adaptive dynamics [4] and produce pairwise invasibility plots to find evolutionary stable states. This process will lead to insight about evolutionary mechanisms of population cycles. In particular, some of the predictions of classical foraging theory can now be explored in connection with their population dynamic consequences.

References

1. T.R. Birkhead, R. Furness, Regulation of seabird population, in *Behavioural Ecology*, ed. by R. Sibly, R. Smith (Blackwell Science, Oxford, 1985), pp. 145–168
2. J. Chase, Central-place forager effects on food web dynamics and spatial pattern in northern California meadows. Ecology **79**(4), 1236–1254 (1998)
3. C. Cobbold, M. Lewis, F. Lutscher, J. Roland, How parasitism affects critical patch size in a host-parasitoid system: application to Forest Tent Caterpillar. Theor. Popul. Biol. **67**(2), 109–125 (2005)
4. U. Dieckmann, Can adaptive dynamics invade? Trends Ecol. Evol. **12**, 128–131 (1997)
5. P. Elliot, Foraging behavior of a central-place forager: field tests of theoretical predictions. Am. Nat. **131**(2), 159–174 (1988)
6. W. Fagan, F. Lutscher, K. Schneider, Population and community consequences of spatial subsidies derived from central-place foraging. Am. Nat. **170**(6), 901–915 (2007)
7. S. Geritz, E. Kisdi, On the mechanistic underpinning of discrete-time population models with complex dynamics. J. Theor. Biol. **228**, 261–269 (2004)
8. T. Getty, Analysis of central-place space-use patterns: the elastic disc revisited. Ecology **62**(4), 907–914 (1981)
9. M. Hassell, R. May, From individual behavior to population dynamics, in *Behavioural Ecology*, ed. by R. Sibly, R. Smith (Blackwell Science, Oxford, 1985), pp. 3–32
10. M. Kot, *Elements of Mathematical Ecology* (Cambridge University Press, Cambridge, 2001)
11. J. Lee, T. Hillen, M. Lewis, Pattern formation in prey-taxis systems. J. Biol. Dynam. **3**(6), 551–573 (2009)
12. M. May, *Stability and Complexity in Model Ecosystems* (Princeton University Press, Princeton, 1973)
13. M. Neubert, M. Kot, M.A. Lewis, Dispersal and pattern formation in a discrete-time predator-prey model. Theor. Popul. Biol. **48**(1), 7–43 (1995)
14. G. Orians, N. Pearson, On the theory of central-place foraging, in *Analysis of Ecological Systems*, ed. by G. Horn, R. Mitchell, G. Stairs (Ohio State University Press, Columbus, 1979), pp. 154–177

15. T. Robbins, Seed Dispersal and Biological Invasion: A Mathematical Analysis, Ph.D. thesis, University of Utah, 2004
16. T. Schoener, Generality of the size-distance relation in models of optimal feeding. Am. Nat. **114**, 902–914 (1979)
17. P. Turchin, *Quantitative Analysis of Movement* (Sinauer Associates, Sunderland, 1998)
18. R.W. Van Kirk, M.A. Lewis, Edge permeability and population persistence in isolated habitat patches. Nat. Res. Model. **12**, 37–64 (1999)
19. H. Wei, Population Dynamics of Central-Place Foragers, Master's thesis, University of Ottawa, 2010
20. W. Wolff, An individual-oriented model of a wading bird nesting cology. Ecol. Model. **72**, 75–114 (1994)

Transport and Anisotropic Diffusion Models for Movement in Oriented Habitats

Thomas Hillen and Kevin J. Painter

Abstract A common feature of many living organisms is the ability to move and navigate in heterogeneous environments. While models for spatial spread of populations are often based on the diffusion equation, here we aim to advertise the use of transport models; in particular in cases where data from individual tracking are available. Rather than developing a full general theory of transport models, we focus on the specific case of animal movement in oriented habitats. The orientations can be given by magnetic cues, elevation profiles, food sources, or disturbances such as seismic lines or roads. In this case we are able to present and contrast the three most common scaling limits, (i) the parabolic scaling, (ii) the hyperbolic scaling, and (iii) the moment closure method. We clearly state the underlying assumptions and guide the reader to an understanding of which scaling method is used in what kind of situations. One interesting result is that the macroscopic drift velocity is given by the mean direction of the underlying linear features, and the diffusion is given by the variance-covariance matrix of the underlying oriented habitat. We illustrate our findings with specific applications to wolf movement in habitats with seismic lines.

T. Hillen (✉)
Centre for Mathematical Biology, University of Alberta, Edmonton, AB, Canada T6G2G1
e-mail: thillen@ualberta.ca

K.J. Painter
Heriot-Watt University, Edinburgh EH14 4AS, UK
e-mail: K.Painter@hw.ac.uk

M.A. Lewis et al. (eds.), *Dispersal, Individual Movement and Spatial Ecology,*
Lecture Notes in Mathematics 2071, DOI 10.1007/978-3-642-35497-7_7,
© Springer-Verlag Berlin Heidelberg 2013

1   Introduction

1.1   Biological Motivation

Successful navigation through a complicated and evolving environment is a fundamental task carried out by an enormous range of organisms. Migration paths can be
staggering in their length and intricacy: at the microscopic scale, nematode worms
can determine the shortest path through the intricate maze-like structure of the soil to
locate plant roots [38] while at the macroscopic scale salmon return from the ocean
upstream through bifurcating rivers and streams to spawn at their original birth site
[25]. Selecting a path requires the detection, processing and integration of a myriad
of cues drawn from the surrounding environment. In many instances the intrinsic
orientation of the environment provides a valuable navigational aid. The earth's
magnetic field provides one such example: species such as turtles and whales use
an inbuilt compass to navigate to breeding or feeding grounds [25], while butterflies
and other insects fly up slopes to local peaks in a mate locating strategy known as
"hilltopping" [36]. Pigeons [24] and cane toads [6] have been shown to fly or hop
in the direction of roads, while caribou and wolves move along the seismic lines cut
into forests by oil exploration companies [29]. An aligned environment also plays
a fundamental role in the migration of individual cells: many cell types, including
immune cells, fibroblasts and certain types of cancer cells migrate in alignment with
the fibre network constituting the surrounding extracellular matrix (ECM).

The above examples provide the motivation for the present paper where we focus
on mathematical models for movement in oriented habitats and their scaling limits.
The aim is to clarify some of the tools of the trade, allowing the reader to adapt the
methods to any given specific situation, such as those outlined above. In the case of
the present paper we shall use cell movement in collagen tissues to derive the model
equations, before demonstrating their adaption to wolf movement on seismic lines
and the motion of organisms in a stream. We note that these should be considered
illustrative examples rather than indepth studies, although we note that a detailed
application to glioma growth will be covered in a forthcoming paper [35].

1.2   Mathematical Modelling

Transport models (often referred to as kinetic models) form a powerful tool in
the analysis and modelling of animal and cell movement. Modern experimental
methods allow us to track an individual's movement in intricate detail, whether
by GPS tracking of mammals [29, 39] or through confocal microscopy of cells in
tissues [13, 14]. The wealth of data generated can be employed to extract precise
information on mean travel speeds, velocities, the distribution of turning angles, the
"choice" of new velocities amongst others. Within this context the transport model
fits naturally, relying on particle speed and turning distributions as key inputs.

Transport models have a long history in continuum mechanics. For example, the theory of dilute gases is entirely based on the kinetic Boltzmann equation of interacting gas particles [8]. Over the last few decades this theory has been transferred to the modelling of living entities, with the obvious advantage of shipping previously developed methodologies with it [4, 9, 18, 32, 33, 37]. However, wholesale removal from the shelf of continuum mechanics is inadvisable: methods must be carefully adjusted to reflect the biological situation.

A highly utilized tool in the study of transport models is a consideration of scaling limits, thus allowing approximation to a reduced (and typically simpler) model such as a diffusion- or drift-dominated partial differential equation. A variety of scaling limits have been considered, found under the general headings of *parabolic limit, diffusion limit, hyperbolic limit, Chapman-Enskog expansions, Hilbert expansions*, and *moment closures* [9, 10, 16, 18] (with, most likely, many further terminologies dispersed throughout the literature).

In the hope that we can make transport equations more broadly accessible for ecological and cellular processes, in this chapter we explore such systems as a means of modelling migration. We will open the following section with a presentation of the transport equation approach, as well as a specific formulation that incorporates guided movement due to a fixed and oriented environment. This relatively simple model will be used to motivate and illustrate the various scalings. Here, with our attention fixed on ecological applications, we restrict attention to the three most commonly used methods: (i) the parabolic scaling, (ii) the hyperbolic scaling, and (iii) the moment closure. We will not attempt to present the most abstract and general theory, rather we focus on a nontrivial and interesting case which retains enough simplicity to directly apply each of the scaling limits above. In particular, we will attempt to answer the following questions:

- Is there a better method among those three methods?
- How and when do we employ hyperbolic scaling, parabolic scaling or moment closure?
- What are the specific assumptions behind these three methods and how do they differ?
- In which cases do these scalings lead to the same results?

While all methods have been discussed individually, as far as we are aware there has not been a study which directly compares these methods in the ecological context. We find that each of the methods (i), (ii) and (iii) have their own range of applicability and there are situations when one is favourable over the other. As it turns out, the parabolic limit (i) plays a central role, as special cases of (ii) and (iii) both lead back to (i). To illustrate the findings and methodologies in a transparent manner we will explore some simple case studies and consider specific applications, including the movement of wolves and caribou along seismic lines in Western Canada. Finally, we will provide a brief discussion of the findings.

2 Transport Equations

The application of transport equations to biological processes grew from seminal work of the 1980s (see [1, 33]) as an approach for modelling biological movement, whether by cells or organisms. Transport equations typically refer to mathematical models in which the particles of interest are structured by their position in space, time and velocity. Here we will use $p(t, x, v)$ to describe the population density of cells/organisms at time $t \geq 0$, location $x \in \Omega \subset \mathbb{R}^n$ and velocity $v \in V \subset \mathbb{R}^n$. We will generally consider an unbounded spatial domain $\Omega = \mathbb{R}^n$ to avoid specifying boundary conditions and, given that we consider biological movement, the set of possible velocities V is taken to be compact. It is worth noting that this is a key distinction from the kinetic theory of gas molecules, where $V = \mathbb{R}^n$ permits (at least theoretically) individual molecules to acquire infinite momentum. Here we shall typically consider $V = [s_1, s_2] \times \mathbb{S}^{n-1}$, with $0 \leq s_1 \leq s_2 < \infty$.

The time evolution of $p(t, x, v)$ is described by the *transport equation*

$$p_t(t, x, v) + v \cdot \nabla p(t, x, v) = \mathscr{L} p(t, x, v), \tag{1}$$

where the index t denotes the partial time derivative and $\mathscr{L}$ is the *turning operator*: a mathematical representation for modelling the velocity changes of the particles. In many instances $\mathscr{L}$ could be described by a nonlinear interaction operator incorporating changes in velocity due to interactions between individuals. For example, the coherence of a fish school is maintained through an individual altering velocity in response to that of an immediate neighbour, while certain populations of cells migrate as a cohort by forming strong adhesive bonds with their neighbours. Here we will ignore such scenarios, thus allowing us to focus our attention on the simpler case of linear operators $\mathscr{L}$.

Typically, $\mathscr{L}$ is defined via an integral operator representation

$$\mathscr{L}\varphi(v) = -\mu\varphi(v) + \mu \int_V T(x, v, v')\varphi(v')dv', \tag{2}$$

where the first term on the right hand side gives the rate at which particles switch away from velocity v and the second term denotes the switching into velocity v from all other velocities. The parameter μ is the *turning rate*, with $1/\mu$ the *mean run time* between individual turns. The kernel $T(x, v, v')$ denotes the probability density of switching velocity from v' to v, given that a turn occurs at location x. The mathematical properties of T set the stage for much of the theory that follows and it is certainly possible to set down a general theory for transport equations (see for example [7, 18, 23, 37]). However, the resulting burden of advanced functional analysis would overwhelm the aims of the present paper. Rather, we focus on a simple yet non-trivial case which allows us to present the scaling methods in a transparent manner. Specifically, we restrict to the case in which the turning operator does not depend on the incoming velocity v':

$$T(x, v, v') = q(x, v)$$

where q satisfies $q \geq 0$. This assumption limits the applicability, since animals as well as cells have a tendency to maintain a particular direction (persistence) such that the incoming and outgoing velocities show a strong correlation. Here we ignore this form of persistence, and we assume that the dominating directional cue is given by the oriented environment. As mentioned already, a general treatment is possible, but it would deter from our purpose to present the theory in a relatively transparent way.

2.1 Movement in an Oriented Environment

Here we present a dedicated and simple model based on the transport equation (1) with turning operator (2) to describe movement in an oriented environment. We follow the modelling approach developed by Hillen [17] and extended in [34] to describe contact-guided movement of cells within a network, for example an extracellular matrix (ECM) predominantly composed of collagen fibres. We motivate the model by briefly describing its derivation in relation to cell movement, as in the above articles, while noting that the model itself is quite general and can easily be adapted to model the movement of organisms in an oriented landscape, as shown in later sections.

The ECM imparts orienteering cues to cells through their tendency to follow fibres, a process known as contact guidance [12, 15]. More generally, contact guidance describes the oriented motility response of cells to anisotropy in the environment, whether it arises from collagen fibres, muscle fibres, neuronal axons, arteries and so forth. Contact guidance is believed to play important roles in tissue development, homeostasis and repair, from patterning of the pectoral fin bud of the teleost embryo [43] to immune cell guidance [41,42] and fibroblast-mediated tissue repair following injury [15]. Particular interest in contact-guided migration of cells further stems from its influence in directing the pathways of invasive cancer cells [13, 14].

Following the approach in [17] and [34], we represent the oriented structure of the environment by defining a directional distribution $\tilde{q}(x, \theta)$ for $\theta \in \mathbb{S}^{n-1}$, with $\tilde{q} \geq 0$ and $\int_{\mathbb{S}^{n-1}} \tilde{q}(x, \theta) d\theta = 1$. In the case of cell migration, the fibres along which cells migrate do not provide a particular direction to movement (i.e. there is no "up" or "down" a collagen fibre) and in such instances we would assume symmetry $\tilde{q}(x, -\theta) = \tilde{q}(x, \theta)$ for all $\theta \in \mathbb{S}^{n-1}$. For more on distinct forms for the directional distribution, see below.

To model contact-guided migration, we assume that cells choose their new direction according to the given fibre network, hence $q(x, v) \sim \tilde{q}(x, \hat{v})$, where $\hat{v} = v/||v||$ denotes the corresponding unit vector. Note that this assumes that cells only take guidance information from the directional distribution: there is no explicit component for random migration or orientation to chemical signalling cues built directly into the turning operator, although these can be built into the directional distribution as we demonstrate later. Since q is a probability distribution on V, and $\tilde{q}$ a probability distribution on $\mathbb{S}^{n-1}$, we need to scale appropriately:

$$q(x,v) := \frac{\tilde{q}(x,\hat{v})}{\omega}, \quad \text{with} \quad \omega = \int_V \tilde{q}(x,\hat{v})dv = \begin{cases} \frac{1}{n}(s_2^n - s_1^n) & \text{for } s_1 < s_2 \\ s^{n-1} & \text{for } s_1 = s_2 = s. \end{cases}$$

For this choice of turning kernel, (2) simplifies to

$$\mathcal{L}\varphi(v) = \mu(q(x,v)\bar{\varphi} - \varphi(v)), \quad \text{with} \quad \bar{\varphi} := \int_V \varphi(v)dv.$$

We make one final simplification, which is to assume individuals have a fixed speed s, i.e. $V = s\mathbb{S}^{n-1}$. While the extension to $V = [s_1, s_2] \times \mathbb{S}^{n-1}$ is trivial, it introduces some cumbersome integration constants that blur the analytical details.

To summarise, the transport model we study in this paper is given by

$$p_t(t,x,v) + v \cdot \nabla p(t,x,v) = \mu(q(x,v)\bar{p}(t,x) - p(t,x,v)) \tag{3}$$

on $\mathbb{R}^n \times s\mathbb{S}^{n-1}$, where $q(x,v)$ is the direction distribution that represents the external network structure.

It is worth noting that different cell types adopt distinct migration strategies, with correspondingly variable degrees of interaction with the surrounding network. For individually migrating cells the two principle migration strategies are amoeboid and mesenchymal. While the former is characterised by fleeting contact between cells and the ECM, and correspondingly minimal distortion of the network [42], the latter involves extensive structural modification of the ECM via a processes of cell-mediated proteolytic degradation. Consequently the stand-alone equation (3) is more appropriately a model for amoeboid rather than mesenchymal migration. The latter would require augmentation of (3) with an evolution equation for varying $q(t,x,v)$ due to cell-matrix interactions: while such extensions have been extensively considered in detail in [17] and [34], we do not consider this further here.

As mentioned earlier, while originally developed in the context of cell migration the above transport equation can easily be adapted to ecological applications. For example, to model the population movements of hilltopping butterflies we would reinterpret p as the density of butterflies, q as a spatially varying directional distribution with a maximum corresponding to the local direction of increasing slope, with parameters s and μ for butterfly speed and frequency of turns to be estimated from tracking of individual flights.

2.2 Environmental Distributions

Representing anisotropy of the environment through the directional distribution q provides the means to describe a wide range of oriented landscapes. Here we briefly consider some potential forms for q.

Strictly-Aligned Environments

A strictly aligned environment with local direction $\gamma \in \mathbb{S}^{n-1}$ can be modelled by choosing the singular q-distribution:

$$q(x,v) = \frac{1}{\omega}\delta_0(\hat{v} - \gamma).$$

The above effectively forces an individual to choose γ as a movement direction following a turn. A full mathematical solution theory of (3) for such q requires a notion of measure valued solutions, which was developed in [20]. We discuss this case in connection to applications in Sect. 7.3.

Regularly-Aligned Environments

For many landscapes, while oriented structures provide a directional cue, the individuals will typically move over a wide range of directions. For example, while wolves preferentially follow the seismic lines cut into forested areas they also move off the lines and into surrounding forest. Similarly, butterflies do not take the steepest route during hilltopping, rather their flight pattern fluctuates [36]. Such behaviours can be accounted for by allowing q to take the form of a regular probability distribution over V.

In summary, we assume that q has the general form

$$q(x,.) \in L^2(V), \qquad q(x,v) \geq 0, \qquad \int_V q(x,v)dv = 1. \tag{4}$$

With the above assumptions in place it is noteworthy to mention two statistical quantities later revealed to be of importance, the expectation and the variance-covariance matrix:

$$\mathbb{E}_q(x) = \int_V vq(x,v)dv, \qquad \mathbb{V}_q(x) = \int_V (v - \mathbb{E}_q(x))(v - \mathbb{E}_q(x))^T q(x,v)dv.$$

The product vv^T denotes the dyade product of two vectors and it defines a matrix. Other authors prefer to use tensorial notation such as $vv^T = v \otimes v$ [9, 10].

Furthermore, we consider potential restrictions on q that could result from distinct forms of environmental anisotropy. While information provided by magnetic cues, the sun and ocean currents could provide a unidirectional movement cue, topographical information in the form of roads, seismic lines and collagen fibres may only provide bidirectional anisotropy, i.e. animals or cells choose both directions with equal probability. In this latter case, we would assume symmetry of q,

$$q(x, -v) = q(x, v).$$

A direct consequence of this symmetry is

$$\mathbb{E}_q = 0 \quad \text{and} \quad \mathbb{V}_q(x) = \int_V v v^T q(x, v) dv.$$

3 The Parabolic Scaling

In this and the following two sections we discuss the three principal scalings: (i) parabolic scaling, (ii) hyperbolic scaling and (iii) moment closure. We will show that each of the methods have their own range of applicability, and that there are situations when one is favourable over the other. With the aim of making this methodology broadly accessible, we aim for transparent presentation by revealing all steps in the analysis, noting that such details are often omitted in the literature. To illustrate the structuring of what follows, the graphic (Fig. 1) outlines the relationships between the scalings as they will be discussed in this manuscript. The parabolic limit (i) is found to play a pivotal role, as special cases of (ii) and of (iii) both lead back to (i). For readers less motivated by the technical aspects of what follows, we would like to note that each section is concluded with a summarising box and a comparison between the scalings is presented in Sect. 6.

3.1 Motivation of the Parabolic Limit

As illustrated in Fig. 1, the parabolic limit marks a full-stop for scaling in our analyses, with all paths eventually leading to it. Given its obvious and considerable importance to the modelling community, we therefore discuss this case first. Two ways to motivate the parabolic limit are (a) an appropriate scaling of space and time, and (b) large turning rates and large speeds of the particles. These two approaches are, in fact, equivalent, as we next illustrate.

E. coli bacteria on a petri-dish display an average turning rate of $\mu \approx 1/s$ and an average speed of $s \approx 10^{-2}$ mm/s (see [18]), whereas durations for experiments that investigate population level dynamics are typically of the order of hours or days.

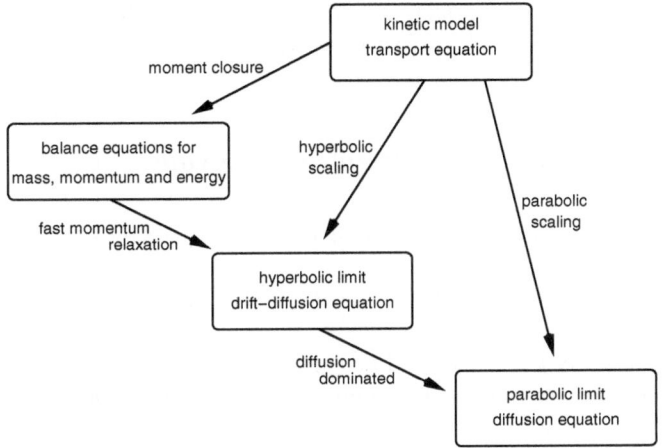

Fig. 1 Relations between the scalings and limit equations as discussed in the text

Taking a unit $U = 10,000\,\text{s}$ ($\approx 3\,\text{h}$), the turning rate and speed on this timescale become

$$\mu = 10^4 \frac{1}{U} \quad \text{and} \quad s = 10^2 \frac{mm}{U}.$$

Hence, introducing a small parameter

$$\varepsilon = 10^{-2},$$

we have

$$\mu = O(\varepsilon^{-2}), \quad \text{and} \quad s = O(\varepsilon^{-1}).$$

By writing $\mu = \varepsilon^{-2}\tilde{\mu}$ and $s = \varepsilon^{-1}\tilde{s}$ we obtain the equation

$$p_t + \varepsilon^{-1}\tilde{s}\theta \cdot \nabla p = \varepsilon^{-2}\tilde{\mu}(q\bar{p} - p),$$

where, for $v \in V$, we write $s\theta = v$ with $\theta \in \mathbb{S}^{n-1}$. Removing the ~'s on the scaled parameters and rearranging we obtain

$$\varepsilon^2 p_t + \varepsilon v \cdot \nabla p = \mu(q\bar{p} - p). \tag{5}$$

Alternatively, we can simply introduce macroscopic time and space scales

$$\tau = \varepsilon^2 t, \qquad \xi = \varepsilon x$$

and rescale model (3) accordingly to obtain

$$\varepsilon^2 p_\tau + \varepsilon v \cdot \nabla_\xi p = \mu(q\bar{p} - p). \tag{6}$$

Formally, (5) and (6) are identical, though we note that we shall employ the second formulation with the new time and space coordinates (τ, ξ).

3.2   Parabolic Limit in an Oriented Landscape

We first study the properties of the turning operator defined on $L^2(V)$:

$$\mathscr{L}\varphi(v) := \mu(q(x,v)\bar{\varphi} - \varphi).$$

The kernel of $\mathscr{L}$ is given by the linear space $\langle q(x,.)\rangle$. Hence we work in the weighted Lebesgue space $L^2_{q^{-1}}(V)$ where the inner product of a function f with q is given by

$$\int_V f(v)q(x,v)\frac{dv}{q(x,v)} = \int_V f(v)dv = \bar{f}.$$

On the complement set, $\langle q \rangle^\perp$, we can define a pseudo-inverse by solving the resolvent equation. Given a function $\psi \in \langle q \rangle^\perp$ we solve for $\phi \in \langle q \rangle^\perp$ such that

$$\mathscr{L}\phi = \psi. \tag{7}$$

Since $\phi, \psi \in \langle q \rangle^\perp$, we have $\bar{\phi} = \bar{\psi} = 0$ and the resolvent equation (7) reduces to

$$\phi = -\frac{1}{\mu}\psi. \tag{8}$$

where the pseudo-inverse appears as multiplication with $-\mu^{-1}$.

To analyse the scaled equation (6) we take the scaled coordinates (τ, ξ) and make a regular expansion in ε, called a *Hilbert expansion*:

$$p(\tau, \xi, v) = p_0(\tau, \xi, v) + \varepsilon p_1(\tau, \xi, v) + \varepsilon^2 p_2(\tau, \xi, v) + h.o.t.$$

Substituting into (6) and comparing orders of magnitude of ε:

- ε^0: The terms of leading order are $\mathscr{L}p_0(\tau, \xi, v) = 0$ which implies

$$p_0(\tau, \xi, v) = q(\xi, v)\bar{p}_0(\tau, \xi).$$

- ε^1: The terms of order one are

$$(\nabla \cdot v)p_0 = \mathscr{L}p_1.$$

This equation can be solved on $\langle q \rangle^\perp$, if the right hand side satisfies the solvability condition $(\nabla \cdot v)p_0 \in \langle q \rangle^\perp$. This condition reads

$$\int_V (\nabla \cdot v) q(\xi, v) p_0(\tau, \xi, v) \frac{dv}{q(\xi, v)} = \nabla \cdot \int_V v q(\xi, v) \, dv \, \bar{p}_0(\tau, \xi).$$

Crucially, this term is only equal to zero for arbitrary p_0 when we impose the following extra condition on q:

$$\mathbb{E}_q = \int_V v q(\xi, v) dv = 0. \tag{9}$$

We can then solve for the first order term and find

$$p_1(\tau, \xi, v) = -\frac{1}{\mu} \nabla \cdot v \, p_0(\tau, \xi, v).$$

- ε^2: The second order terms are

$$p_{0,\tau} + v \cdot \nabla p_1 = \mathscr{L} p_2.$$

From assumption (4) it follows that $\int_V \mathscr{L}\phi(v) dv = 0$ for all $\phi \in L^2_{q^{-1}}$. Hence we integrate the above equation and use index notation for summation over repeated indices:

$$0 = \int_V (p_{0,\tau} + v \cdot \nabla p_1) dv,$$

$$= \bar{p}_{0,\tau} - \frac{1}{\mu} \int (v \cdot \nabla)(\nabla \cdot v) \, (\bar{p}_0(\tau, \xi) q(\xi, v)) \, dv,$$

$$= \bar{p}_{0,\tau} - \frac{1}{\mu} \partial_i \partial_j \left(\int_V v^i v^j q(\xi, v) \, dv \, \bar{p}_0(\tau, \xi) \right).$$

This last equation can be written as a diffusion equation for the macroscopic density $\bar{p}_0(\tau, \xi)$:

$$\bar{p}_{0,\tau}(\tau, \xi) = \nabla\nabla(D(\xi) \bar{p}_0(\tau, \xi)). \tag{10}$$

with a macroscopic *diffusion tensor*

$$D(\xi) = \frac{1}{\mu} \int_V v v^T q(\xi, v) dv. \tag{11}$$

Since we assumed $\mathbb{E}_q = 0$ in (9), we find that the diffusion tensor for the particles is given by the variance-covariance matrix of the underlying fibre network:

$$D(\xi) = \frac{1}{\mu} \mathbb{V}_q(\xi).$$

With q assumed to be non-singular, the variance-covariance matrix $\mathbb{V}_q$ (and hence the diffusion tensor D) is positive definite and symmetric and (11) is uniformly parabolic.

We summarise this limit in the following result:

The Parabolic Scaling. In addition to (4) we make the following assumptions:

(A1)
$$\mathbb{E}_q = \int_V vq(x,v)dv = 0. \tag{12}$$

(A2) There exists a small parameter $\varepsilon > 0$ such that either

$$\mu = \varepsilon^{-2}\tilde{\mu}, \qquad s = \varepsilon^{-1}\tilde{s},$$

or

$$\tau = \varepsilon^2 t, \qquad \xi = \varepsilon x,$$

where $\tilde{\mu}, \tilde{s}, \tau, \xi$ are of order one.

Let $p(\tau, \xi, v)$ be a solution of the scaled kinetic equation

$$\varepsilon^2 p_\tau + \varepsilon v \cdot \nabla_\xi p = \mu(q\bar{p} - p). \tag{13}$$

Then the leading order term p_0 of a regular expansion $p = p_0 + \varepsilon p_1 + \varepsilon^2 p_2 + \dots$ satisfies

$$p_0(\tau, \xi, v) = \bar{p}_0(\tau, \xi)q(\xi, v),$$

where $\bar{p}_0(\tau, \xi)$ is solution of the parabolic limit equation

$$\bar{p}_{0,\tau}(\tau, \xi) = \nabla\nabla(D(\xi)\bar{p}_0(\tau, \xi)) \tag{14}$$

with *diffusion tensor*

$$D(\xi) = \frac{1}{\mu}\int_V vv^T q(\xi, v)dv. \tag{15}$$

4   The Hyperbolic Scaling

The parabolic limit of the previous section considered macroscopic time and space scales, where time is scaled quadratically in ε and space linearly. For the parabolic limit to work it was necessary to specify $\mathbb{E}_q = 0$, with a diffusion equation arising.

In the hyperbolic scaling we will observe that $\mathbb{E}_q$ corresponds to a drift term, which dominates when nonzero, and that in the hyperbolic limit we derive both a drift term and a diffusion correction.[1] For that, we assume that macroscopic time and space scales are both linear in ε, i.e.

$$\sigma = \varepsilon t, \quad \xi = \varepsilon x.$$

Under this rescaling, the transport equation (1) becomes

$$\varepsilon p_\sigma + \varepsilon(v \cdot \nabla)p = \mathscr{L} p. \tag{16}$$

Again, we use the operator properties of $\mathscr{L}$ on the space $L^2_{q^{-1}}(V)$ and split the solution into two parts (the *Chapman-Enskog expansion*):

$$p(\sigma, \xi, v) := \bar{p}(\sigma, \xi)q(\xi, v) + \varepsilon p^\perp(\sigma, \xi, v) \tag{17}$$

$$\text{with} \quad \int_V p^\perp(\sigma, \xi, v)q(\xi, v)\frac{dv}{q(\xi, v)} = \int_V p^\perp(\sigma, \xi, v)dv = 0.$$

Substituting the expansion (17) into (16) gives:

$$\varepsilon \bar{p}_\sigma q + \varepsilon^2 p^\perp_\sigma + \varepsilon(v \cdot \nabla)(\bar{p}q) + \varepsilon^2(v \cdot \nabla)p^\perp = \mathscr{L}\left(\bar{p}q + \varepsilon p^\perp\right)$$
$$= \varepsilon \mathscr{L} p^\perp. \tag{18}$$

Integrating (18) over V and dividing by ε yields

$$\bar{p}_\sigma + \nabla \cdot \left(\int_V vq \, dv \, \bar{p} + \varepsilon \int vp^\perp dv\right) = 0, \tag{19}$$

where we used

$$\int_V p^\perp_\sigma dv = \frac{\partial}{\partial \sigma}\int_V p^\perp dv = 0.$$

Once again, the expectation of q appears

$$\bar{p}_\sigma + \nabla \cdot \left(\mathbb{E}_q \bar{p} + \varepsilon \int_V vp^\perp dv\right) = 0, \tag{20}$$

and to leading order this is the drift-dominated model

$$\bar{p}_\sigma + \nabla \cdot (\mathbb{E}_q \bar{p}) = 0, \tag{21}$$

where the drift velocity is given by the expectation of q.

---

[1]This section is an adaptation of Sect. 4.1.3 from [17]. It was inspired by Dolak and Schmeiser [11] who apply this scaling to chemotactic movement and, while their results do not directly apply here, the methods are the same.

We determine the next order correction term by constructing an approximation to $p^\perp$. From (20) we obtain $\bar{p}_\sigma$, substitute into (18) and divide by ε:

$$\mathscr{L}p^\perp = -q\nabla \cdot \left(\mathbb{E}_q \bar{p} + \varepsilon \int_V v p^\perp dv\right) + \varepsilon p_\sigma^\perp + (v \cdot \nabla)(\bar{p}q) + \varepsilon(v \cdot \nabla)p^\perp$$

$$= (v \cdot \nabla)(\bar{p}q) - q\nabla \cdot (\mathbb{E}_q \bar{p}) + O(\varepsilon). \tag{22}$$

Hence to leading order we have:

$$\mathscr{L}p^\perp \approx q(v - \mathbb{E}_q) \cdot \nabla\bar{p} + (v \cdot \nabla q - q\nabla \cdot \mathbb{E}_q)\bar{p}. \tag{23}$$

To apply the pseudo-inverse of $\mathscr{L}$ on $\langle q \rangle^\perp$, we must check the solvability condition

$$\int_V \mathscr{L}p^\perp q \frac{dv}{q} = \int_V \mathscr{L}p^\perp dv,$$

$$\approx \nabla\bar{p} \cdot \int_V (q(v - \mathbb{E}_q)dv + \bar{p}\int_V (v \cdot \nabla q - q\nabla \cdot \mathbb{E}_q)dv,$$

$$= \nabla\bar{p} \cdot (\mathbb{E}_q - \mathbb{E}_q) + \bar{p}\nabla \cdot (\mathbb{E}_q - \mathbb{E}_q),$$

$$= 0.$$

Hence we can apply the pseudo-inverse of $\mathscr{L}$ and find

$$p^\perp \approx -\frac{1}{\mu}\left(q(v - \mathbb{E}_q) \cdot \nabla\bar{p} + (v \cdot \nabla q - q\nabla \cdot \mathbb{E}_q)\bar{p}\right). \tag{24}$$

Substituting (24) into (20) we obtain

$$\bar{p}_\sigma + \partial_j(\mathbb{E}_q^j \bar{p})$$

$$= \frac{\varepsilon}{\mu}\partial_j\left(\int_V v^j\left[q(v^i - \mathbb{E}_q^i)\partial_i\bar{p} + (v^i\partial_i q - q\partial_i \mathbb{E}_q^i)\bar{p}\right]dv\right)$$

$$= \frac{\varepsilon}{\mu}\partial_j\int_V v^j(v^i - \mathbb{E}_q^i)q\, dv\, \partial_i\bar{p}$$

$$+ \frac{\varepsilon}{\mu}\partial_j\left(\left[\int_V v^j(v^i\partial_i q - q\int_V v'^i\partial_i q\, dv')dv\right]\bar{p}\right).$$

The two integrals inside the square brackets can be written as

$$\int_V v^j(v^i\partial_i q - q\int_V v'^i\partial_i q\, dv')dv$$

$$= \int_V v^j v^i\partial_i q\, dv - \int_V v^j q\, dv\int_V v'^i\partial_i q\, dv'$$

$$= \int_V (v^j - \mathbb{E}_q^j) v^i \partial_i q \, dv$$

$$= \int_V (v - \mathbb{E}_q) v \cdot \nabla q \, dv,$$

Hence we obtain

$$\bar{p}_\sigma + \nabla \cdot (\mathbb{E}_q \bar{p}) = \frac{\varepsilon}{\mu} \nabla \cdot \int_V v(v - \mathbb{E}_q)^T q \, dv \cdot \nabla \bar{p}$$

$$+ \frac{\varepsilon}{\mu} \nabla \cdot \left(\left(\int_V (v - \mathbb{E}_q)(v \cdot \nabla q) dv \right) \bar{p} \right). \qquad (25)$$

We define the diffusion tensor D as before, i.e. as a multiple of the variance-covariance matrix of q:

$$D(x) := \frac{1}{\mu} \mathbb{V}_q = \frac{1}{\mu} \int_V (v - \mathbb{E}_q)(v - \mathbb{E}_q)^T q(x, v) dv. \qquad (26)$$

We collect two properties of D:

$$\int_V v(v - \mathbb{E}_q)^T q(\xi, v) dv = \int_V (v - \mathbb{E}_q)(v - \mathbb{E}_q)^T q(\xi, v) dv = \mu D(\xi),$$

and

$$\nabla \nabla (D \bar{p}) = \partial_i \partial_j (D^{ij} \bar{p})$$

$$= \partial_i \partial_j \left(\frac{1}{\mu} \int_V v^i (v^j - \mathbb{E}_q^j) q dv \bar{p} \right)$$

$$= \frac{1}{\mu} \partial_i \left(-\int v^i \partial_j \mathbb{E}_q^j \, q dv \bar{p} + \int v^i (v^j - \mathbb{E}_q^j) \partial_j q \, dv \bar{p} \right.$$

$$\left. + \int v^i (v^j - \mathbb{E}_q^j) q dv \partial_j \bar{p} \right)$$

$$= \frac{1}{\mu} \nabla \cdot \left(-\int v \, \mathrm{div} \mathbb{E}_q q dv \bar{p} + \int v(v - \mathbb{E}_q) \cdot \nabla q dv \bar{p} \right.$$

$$\left. + \int v(v - \mathbb{E}_q) q dv \cdot \nabla p \right)$$

Then, with (25), we arrive at the limit equation with correction term

$$\bar{p}_\sigma + \nabla \cdot (\mathbb{E}_q \bar{p}) = \varepsilon \nabla \left(\nabla (D(\xi) \bar{p}) + \frac{1}{\mu} \mathbb{E}_q (\nabla \cdot \mathbb{E}_q) \bar{p} \right). \qquad (27)$$

Equivalently, we can use the moments of q to write the limit equation as

$$\bar{p}_\sigma + \nabla \cdot (\mathbb{E}_q \bar{p}) = \frac{\varepsilon}{\mu} \nabla \big(\nabla (\mathbb{V}_q(\xi) \bar{p}) + \mathbb{E}_q(\nabla \cdot \mathbb{E}_q) \bar{p} \big). \qquad (28)$$

Critically, if $\mathbb{E}_q \approx 0$ (as in the parabolic case) we obtain the same diffusion term as for the parabolic scaling in (14). In fact, for $\mathbb{E}_q = 0$ we can simply rescale the hyperbolic limit equation (28) by $\tau = \varepsilon\sigma$ to obtain an identical limit to (14).

The Hyperbolic Scaling. Further to (4) we make the following assumptions:

(B1)
$$\sigma = \varepsilon t, \qquad \xi = \varepsilon x,$$

where σ, ξ are of order one.

Let $p(\sigma, \xi, v)$ be a solution of the scaled kinetic equation

$$\varepsilon p_\sigma + \varepsilon v \cdot \nabla_\xi p = \mu(q\bar{p} - p). \qquad (29)$$

Then the solution p can be split into $p = \bar{p}q + \varepsilon p^\perp$, where the leading order term $\bar{p}(\sigma, \xi)$ is approximated by the solution of the drift-diffusion equation

$$\bar{p}_\sigma + \nabla \cdot (\mathbb{E}_q \bar{p}) = \frac{\varepsilon}{\mu} \nabla \big(\nabla (\mathbb{V}_q(\xi) \bar{p}) + [\mathbb{E}_q(\nabla \cdot \mathbb{E}_q)] \bar{p} \big). \qquad (30)$$

From the construction it is expected that the approximation should be second order in ε, although to our knowledge this has not yet been shown.

5   The Moment Approach

Moment closure provides a third way to derive macroscopic equations from the transport model (1). As in the previous cases, it was first developed in a physical context to describe the dynamics of fluids and gases and we will therefore adopt the physical definitions within the present biological context. The principle players are mass, momentum and energy, with the goal of defining model equations for these quantities.[2]

---

[2]This section is an adaptation from [10].

Given a particle distribution $p(t, x, v)$, the mass is defined as

$$\bar{p}(t, x) = \int_V p(t, x, v) dv,$$

the momentum as

$$\bar{p}(t, x) U(t, x) := \int_V v p(t, x, v) dv,$$

and the internal energy by

$$E(t, x) = \int_V |v - U(t, x)|^2 p(t, x, v) dv.$$

The momentum implicitly defines the ensemble (or macroscopic) velocity

$$U(t, x) = \frac{1}{\bar{p}(t, x)} \int_V v p(t, x, v) dv.$$

The energy is the trace of the pressure tensor

$$\mathbb{P}(t, x) = \int_V (v - U(t, x))(v - U(t, x))^T p(t, x, v) dv,$$

in the sense that

$$E(t, x) = \operatorname{tr} \mathbb{P}(t, x).$$

In a physical context mass, momentum and energy have vey precise meanings yet applied to biology we must consider carefully their appropriate biological reinterpretation. The total mass, $\bar{p}$, and ensemble velocity, U, correspond directly to their physical quantities, describing respectively the total density of individuals and their average velocity. The momentum $\bar{p}U$ is somewhat different, since cells and animals generally cannot be regarded as hard spheres and hence $\bar{p}U$ is not the physical momentum an ensemble of cells would generate if it hits an object, for example. The biological momentum can simply be regarded as the average particle flux, i.e. the total density, $\bar{p}$, multiplied by the mean velocity, U. The energy is the trace of the full pressure tensor and direct interpretations of either pressure or tensor are hard to find. We can, instead, consider these from a statistical perspective. The ratio $p/\bar{p}$ is a probability density with respect to the velocity, with $U/\bar{p}$ the expectation and $\mathbb{P}/\bar{p}$ the variance-covariance matrix. Consequently, $U/\bar{p}$ gives the mean velocity and $\mathbb{P}/\bar{p}$ gives information on the breadth of the distribution $p/\bar{p}$. The variance-covariance tensor $\mathbb{P}/\bar{p}$ is symmetric, but can be anisotropic and allowing greater spread in one direction than others. The energy $E/\bar{p}$ is the (magnitude of the) variance with $\sqrt{E/\bar{p}}$ the standard deviation.

We need one more variable which, in the physical context, corresponds to the energy flux:

$$Q(t, x) = \int_V |v - U(t, x)|^2 (v - U(t, x)) p(t, x, v) dv.$$

The vector Q is a trace of a full third order moment, with magnitude dominated by cells not moving with the mean velocity and direction given by the mean direction of the outliers, relative to the ensemble velocity U.

In a similar way, we can also define the ensemble pressure tensor of the system

$$\mathbb{P}_0(t, x) = \int_V U(t, x) U(t, x)^T p(t, x, v) dv = \bar{p}(t, x) U(t, x) U(t, x)^T$$

and the ensemble energy flux

$$Q_0(t, x) = \int_V U^2(t, x) U(t, x) p(t, x, v) dv = \bar{p}(t, x) U^2(t, x) U(t, x).$$

Next, we will derive differential equations for the macroscopic quantities mass, $\bar{p}$, momentum, $\bar{p}U$, and energy, E. To obtain the mass conservation equation, we simply integrate (3) over V to obtain

$$\bar{p}_t(x, t) + \nabla \cdot (\bar{p}(t, x) U(t, x)) = 0. \tag{31}$$

The momentum equation is derived through multiplication of (3) by v and integrating (omitting space, time and v dependencies for clarity):

$$\int_V v p_t \, dv + \int_V v(v \cdot \nabla) p \, dv = \mu \int_V v q \, dv \, \bar{p} - \mu \int_V v p \, dv,$$

$$(\bar{p}U)_t + \nabla \cdot \int_V v v^T p dv = \mu \bar{p} \mathbb{E}_q - \mu \bar{p} U. \tag{32}$$

The pressure tensor can be written as

$$\mathbb{P} = \int_V (v - U)(v - U)^T p \, dv,$$

$$= \int v v^T p \, dv - \int U v^T p \, dv - \int v U^T p \, dv + \int U U^T p \, dv,$$

$$= \int v v^T p \, dv - \bar{p} U U^T. \tag{33}$$

We use this expression in (32) and obtain the momentum equation

$$(\bar{p}U)_t + \nabla \cdot (\bar{p} U U^T) = -\nabla \mathbb{P} + \mu(\bar{p} \mathbb{E}_q - \bar{p} U). \tag{34}$$

For the energy equation, we multiply (3) by v^2 and integrate:

$$\int v^2 p_t dv + \int v^2 (v \cdot \nabla) p \, dv = \mu \int v^2 q \, dv \, \bar{p} - \mu \int v^2 p \, dv,$$

$$E_t + \nabla \cdot \int v v^2 p \, dv = \mu \int v^2 q \, dv \, \bar{p} - \mu E. \tag{35}$$

We study the two integral terms in (35) separately. To obtain an expression for $\int v v^2 p$, we study the heat flux Q:

$$Q = \int |v - U|^2 (v - U) p \, dv,$$

$$= \int v v^2 p dv - \int v^2 U p dv - \int 2(vU) v p dv$$

$$+ \int 2(vU) U p dv + \int U^2 v p dv - \int U^2 U p dv,$$

$$= \int v v^2 p dv - UE - 2U \cdot \int v v^T p dv + 2U \cdot (\bar{p} U U^T),$$

$$= \int v v^2 p dv - UE - 2U \cdot \mathbb{P}, \tag{36}$$

where we used (33) in the last equality.

To obtain the second order q term, we compute

$$\text{tr} \, \mathbb{V}_q = \int_V (v^i - \mathbb{E}_q^i)(v_i - \mathbb{E}_{qi}) q \, dv,$$

$$= \int v^i v_i q dv - \int v^i \mathbb{E}_{qi} q dv - \int \mathbb{E}_q^i v_i q dv + \int \mathbb{E}_q^i \mathbb{E}_{qi} q dv,$$

$$= \int v^2 q dv - \mathbb{E}_q^2. \tag{37}$$

Hence the energy equation (35) becomes

$$E_t + \nabla \cdot (EU) = -\nabla Q - 2\nabla \cdot (U \cdot \mathbb{P}) + \mu(\text{tr} \, \mathbb{V}_q + \mathbb{E}_q^2 - E). \tag{38}$$

The equations for mass, $\bar{p}$, momentum, $\bar{p} U$, and energy, E, are given by (31), (34), (38) respectively. However, this system is not closed, due to the inclusion of the higher order moments $\mathbb{P}$ and Q. To resolve this, we can attempt a derivation of differential equations for these higher moments, although in doing so even higher order moments will appear: if fact, the sequence of moment equations is unending and we face a *moment closure problem*. Thus, we must find a mechanism for estimating the higher order moments in order to close the system of equations (31), (34), (38). Two standard ways of finding a moment closure are through (1) the equilibrium distribution and (2) entropy maximisation. Here we focus on the first method, noting that details of the entropy method can be found elsewhere (e.g. [16]).

5.1 Moment Closure

The principal assumption here is that the system is close to equilibrium and that the higher order moments are dominated by this equilibrium. Earlier, we computed $\ker \mathscr{L} = \langle q \rangle$. Hence the equilibrium distribution has the form

$$p_e(t, x, v) = \bar{p}(t, x)q(x, v).$$

For this distribution, we can explicitly compute the moments:

- Mass,

$$\bar{p}_e(t, x) = \int \bar{p}(t, x)q(x, v) = \bar{p}(t, x);$$

- Momentum,

$$\bar{p}_e(t, x)U_e(t, x) = \int v\bar{p}(t, x)q(x, v)dv = \bar{p}(t, x)\mathbb{E}_q(x);$$

- Pressure tensor,

$$\mathbb{P}_e(t, x) = \int (v - \mathbb{E}_q)(v - \mathbb{E}_q)^T \bar{p}(t, x)q(x, v)dv = \bar{p}(t, x)\mathbb{V}_q(x); \qquad (39)$$

- Energy flow,

$$Q_e(t, x) = \int |v|^2 v\bar{p}(t, x)q(x, v)dv = \bar{p}(t, x)\mathbb{T}_q(x), \qquad (40)$$

where we introduce the third order moment of q

$$\mathbb{T}_q(x) = \int v^2 v q(x, v)dv.$$

These formulae reveal that at equilibrium all momentum is carried by the ensemble, which is moving in the mean network direction $\mathbb{E}_q$, and that all energy and pressure is produced by the variance-covariance matrix of the underlying distribution. The above expressions for the pressure tensor and energy flux are employed to close system (31), (34), (38) for mass, momentum, and energy. We should stress that here *we are making an approximation* and that even though we retain the equality sign $\bar{p}, U, E$ are *approximations* to the exact $\bar{p}, U, E$ values.

Moment Closure. In addition to (4) we assume that

(C1) the macroscopic quantities $\mathbb{P}$ and Q are given by their equilibrium distributions (39), (40).

Then the mass $\bar{p}$, the momentum $\bar{p}U$ and the energy E are approximated by the solution of the closed system:

$$\bar{p}_t + \nabla(\bar{p}U) = 0 \tag{41}$$

$$(\bar{p}U)_t + \nabla \cdot (\bar{p}UU^T) = -\nabla(\bar{p}\mathbb{V}_q) + \mu(\bar{p}\,\mathbb{E}_q - \bar{p}U) \tag{42}$$

$$E_t + \nabla \cdot (EU) = -\nabla(\bar{p}\mathbb{T}_q) - 2\nabla \cdot (U \cdot (\bar{p}\mathbb{V}_q))$$

$$+\mu(\operatorname{tr} \mathbb{V}_q + \mathbb{E}_q^2 - E) \tag{43}$$

We note that for the closed system (41)–(43), the first two equations are independent of the energy E. Hence, (43) decouples and we can study the first two equations (41) and (42) independently.

5.2 Fast Flux Relaxation

The derivatives on the left hand side of (41)–(43) all have characteristic form $\partial_t \phi + \nabla \cdot (U\phi)$, termed the directional derivative of ϕ in the direction of the flow U (also known as the material derivative or characteristic derivative). As a special case we assume that the flux relaxes quickly to its equilibrium, i.e. we set

$$0 = -\nabla(\bar{p}\mathbb{V}_q) + \mu(\bar{p}\,\mathbb{E}_q - \bar{p}U),$$

which we can solve for $\bar{p}U$ to give

$$\bar{p}U = -\frac{1}{\mu}\nabla(\bar{p}\mathbb{V}_q) + \bar{p}\mathbb{E}_q.$$

Using this expression in (41) yields the drift-diffusion equation

$$\bar{p}_t + \nabla(\bar{p}\mathbb{E}_q) = \frac{1}{\mu}\nabla\nabla(\bar{p}\mathbb{V}_q). \tag{44}$$

Fast Flux Relaxation. In addition to (4) we assume that:

(C1) The macroscopic quantities $\mathbb{P}$ and Q are given by the equilibrium distributions as in (39), (40);

(C2) The momentum $\bar{p}U$ relaxes fast to its equilibrium.

Then the total mass $\bar{p}(t, x)$ is approximated by the solution of the drift-diffusion limit equation

$$\bar{p}_t + \nabla(\bar{p}\mathbb{E}_q) = \frac{1}{\mu}\nabla\nabla(\bar{p}\mathbb{V}_q). \tag{45}$$

6   Comparison Between Scalings

In this section we will summarise the various scaling methods and compare and contrast our findings. First we will focus on the forms of the limit equations themselves, with an explanation of the relationships between them, before proceeding to examine their underlying assumptions. For convenience of comparison, we unify the notation by setting $u = \bar{p} = \bar{p}_0$ and specifying a generic time coordinate t (noting that t had been rescaled to τ for the derivation of the parabolic and hyperbolic limits).

6.1   Relationships Between Limit Equations

The three scaling approaches resulted in the following four limit equations:

- Parabolic scaling (PS),

$$u_t = \frac{1}{\mu}\nabla\nabla(\mathbb{V}_q u); \tag{PS}$$

- Hyperbolic scaling (HS),

$$u_t + \nabla \cdot (\mathbb{E}_q u) = 0; \tag{HS}$$

- Hyperbolic scaling with correction terms (HC),

$$u_t + \nabla \cdot (\mathbb{E}_q u) = \frac{\varepsilon}{\mu}\nabla\nabla(\mathbb{V}_q u) + \frac{\varepsilon}{\mu}\nabla \cdot (\mathbb{E}_q(\nabla \cdot \mathbb{E}_q)u); \tag{HC}$$

- Moment closure (MC),

$$u_t + \nabla(\mathbb{E}_q u) = \frac{1}{\mu}\nabla\nabla(\mathbb{V}_q u). \tag{MC}$$

Clearly the above equations reveal significant overlap. For example, moment closure (*MC*) is a combination of the parabolic (*PS*) and hyperbolic scaling (*HS*), containing both diffusion and drift terms. Consequently, we refer to the parabolic scaling as the *diffusion-dominated* case, with the hyperbolic scaling the *drift-dominated* case. More formally, the relationships between the limiting equations can be grouped into the following lemma.

Lemma 1. *We summarise the relationships into five scenarios.*

1. **(Diffusion-dominated)** *In the case $\mathbb{E}_q = 0$ all three approaches (PS), (HC), (MC) lead to the parabolic limit (PS), while (HS) is trivial.*
2. **(Diffusion-dominated)** *If $\mathbb{E}_q \approx O(\varepsilon^2)$, then equations (HC) and (MC) coincide with the parabolic limit (PS) to order ε.*
3. **(Drift-diffusion limit)** *If $\mathbb{E}_q \approx O(\varepsilon)$ equation (HC) is identical to (MC) to leading order (assuming a suitable scaling of time in (HC)).*
4. **(Drift-dominated)** *If $\mathbb{V}_q \approx O(\varepsilon)$, then (MC) coincides with the hyperbolic scaling (HS) to leading order.*
5. **(Drift-dominated)** *If $\mu \approx O(\varepsilon^{-1})$, then (MC) once again coincides with (HS) to leading order.*

6.2 Assumptions Behind Limit Equations

Having explored the relationships behind the limit equations, we next consider their underlying assumptions.

(Parabolic) Here the expectation $\mathbb{E}_q = 0$ and there exists a small parameter $\varepsilon > 0$ such that either $\tau = \varepsilon^2 t, \xi = \varepsilon x$, where τ and ξ are both of order one, or $\mu = \varepsilon^{-2}\tilde{\mu}, s = \varepsilon^{-1}\tilde{s}$, where $\tilde{\mu}$ and $\tilde{s}$ are both of order one.

(Hyperbolic) There exists a small parameter $\varepsilon > 0$ such that $\sigma = \varepsilon t, \xi = \varepsilon x$, where σ and ξ are both of order one.

(Moments) The higher moments $\mathbb{P}$ and Q are given by the equilibrium distribution and the momentum ρU relaxes quickly.

While an all-encompassing interpretation of these assumptions is somewhat difficult, we provide the following intuitive scenarios. In the following section, these distinctions will be illuminated further through specific applications.

(Parabolic) The time scale is one in which particles are fast and turn frequently, with movement close to a Brownian random movement. The environment provides no specific directional cue (or, at least, a relatively weak directional cue) and hence $\mathbb{E}_q \simeq 0$ (i.e. movement up or down a given direction is effectively equal). Directional bias could be included through possible anisotropy of the variance-covariance tensor $\mathbb{V}_q$ of the underlying medium.

(Hyperbolic) Once again, time and space scales are chosen such that particles are fast and turn often. But now the movement has a very clear directional component, $\mathbb{E}_q \neq 0$ and the drift component dominates.

(Moments) Here it is assumed that the pressure tensor is close to the pressure tensor of the equilibrium. Effectively, the system as a whole is near to equilibrium with subsequently small differential pressure terms. This implies that the population density p is "somewhat" closely aligned with the underlying tissue.

We note that all three methods lead to an anisotropic diffusion equation of the form

$$U_t = \nabla\nabla(DU) \tag{46}$$

i.e. the diffusion tensor lies inside the two derivatives. In the literature, anisotropic diffusion is usually associated with an equation in divergence form,

$$V_t = \nabla(D\nabla V). \tag{47}$$

This second form is derived from material physics, where the material flux is taken to be proportional to the gradient ∇V with proportionality factor D. As we also discuss in Sect. 7.1.3 below, the above two models are quite different. If D is positive definite, (47) obeys the maximum principle and solutions converge to homogeneous steady states (on bounded domains with zero-flux boundary conditions, for example). In contrast, equation (46) does not have a maximum principle and, as we see later, spatial patterns can evolve.

When deriving diffusion equations from stochastic processes, both of the above versions (46) and (47) can be generated. For example, Othmer and Stevens [32] present a careful analysis that reveals how different assumptions for an individual's local response to the environment results in distinct macroscopic models, including the above two forms. Here we have shown how a model of type (46) arises very naturally. It is certainly possible that a distinct set of assumptions to those used in this paper could also give rise to a model of type (47), however we do not take this further at present.

7   Examples and Applications

During the last few sections we have established a toolkit for generating distinct macroscopic equations, originating from the same transport model for movement of an individual (whether cell or organism) in an oriented environment. In this section we demonstrate these findings through a combination of examples and some specific applications.

7.1   Bidirectional and Nondirectional Environments

Here we consider environments in which the orientational cues do not provide a single direction to the biased movement. Examples range from the movement of wolves along seismic lines, hikers along footpaths, animals along roads or cells along collagen fibres: i.e., while there is a tendency to move with the alignment of the environment, there is no specific "up" or "down". As previously specified, we model this by assuming symmetry in q:

$$q(x, -v) = q(x, v),$$

with the direct consequence

$$\mathbb{E}_q = 0 \quad \text{and} \quad \mathbb{V}_q(x) = \int_V v v^T q(x, v) dv.$$

In relation to the above scaling methods, item (1) of Lemma 1 applies: we have no drift term and all methods lead (eventually) to the diffusion limit

$$\bar{p}_t = \nabla(\nabla D(x)\bar{p}), \tag{48}$$

where $D(x) = \frac{1}{\mu}\mathbb{V}_q(x)$ is an anisotropic diffusion tensor.

7.1.1   Isotropic Diffusion: The Pearson Walk

We illustrate the above with the simplest version of a transport process as expressed by (3) in a completely uniform directional field (i.e. we have a nondirectional environment): the Pearson walk. Individuals are assumed to move with a constant speed s ($V = s\mathbb{S}^{n-1}$) and the underlying directional field is uniform:

$$q(x, v) = \frac{1}{|V|} = \frac{s^{1-n}}{|\mathbb{S}^{n-1}|}.$$

Again, q is symmetric and hence $\mathbb{E}_q = 0$. The variance is computed as

$$\mathbb{V}_q = \int v v^T q(v) dv = \frac{s^{1-n}}{|\mathbb{S}^{n-1}|} s^2 s^{n-1} \int_{\mathbb{S}^{n-1}} \gamma \gamma^T d\sigma = \frac{s^2}{|\mathbb{S}^{n-1}|} \frac{|\mathbb{S}^{n-1}|}{n} \mathbb{I}_n = \frac{s^2}{n} \mathbb{I}_n,$$

where $\mathbb{I}_n$ denotes the identity matrix.

Hence, the drift component will be zero and the diffusion is isotropic with diffusion constant[3]

$$d = \frac{s^2}{\mu n}.$$

7.1.2 Anisotropic Diffusion Example

We present a specific example together with some simulations of the transport model and its diffusive limit. Specifically, we consider a migrating population within a simple rectangular landscape (set to be of dimensions $[-10, 10] \times [-10, 10]$) with an oriented section centring on the origin. The orientational field strength is assumed to reduce with distance, effectively becoming isotropic in the periphery. See Fig. 2a–d for a representation of this environment.

For the directional distribution q we consider the bimodal von Mises distribution:

$$q(x, \theta) = \frac{1}{4\pi I_0(k)} \left(e^{k\theta \cdot \gamma} + e^{-k\theta \cdot \gamma} \right), \tag{49}$$

where $\theta \in \mathbb{S}^1$ defines the movement direction of the population and $\gamma \in \mathbb{S}^1$ defines the dominating alignment of the local environment. I_n denotes the modified Bessel function of first kind of order n. Note that the von Mises distribution is the analogue of a normal distribution on a circle. The parameter k defines the strength of anisotropy and is termed the *parameter of concentration*. The above bimodal von Mises distribution clearly has two local maxima, one for $\theta = \gamma$ and one for $\theta = -\gamma$ [3]. For $k \to 0$ it converges to a uniform distribution (i.e. isotropic), while for $k \to \infty$ it converges to a sum of two point measures in directions γ and $-\gamma$.

To represent an environment in which anisotropy varies in the manner described, we assume $k(x)$ decays exponentially with distance from the origin

$$k(x) = k_0 e^{-r|x|^2},$$

where, in this example, we set $k_0 = 10$ and $r = 0.25$. This leads to high anisotropy in the centre of the domain and almost no directional bias in the periphery. Generally, γ could vary in space (for example, as in a curving road) however here we set it constant and in the direction of the diagonal, $\gamma(x) = (1/\sqrt{2}, 1/\sqrt{2})^T$. Figure 2a represents the environmental anisotropy for the central portion of the field, with the orientation and size of k represented by the direction and length of the individual line segments. For the three field positions indicated we plot the corresponding bimodal von Mises distributions in Fig. 2b–d.

---

[3]A general formula for directional moments, such as $\int \gamma \gamma^T d\gamma = |\mathbb{S}^{n-1}|/n \; \mathbb{I}_n$ can be found in [16].

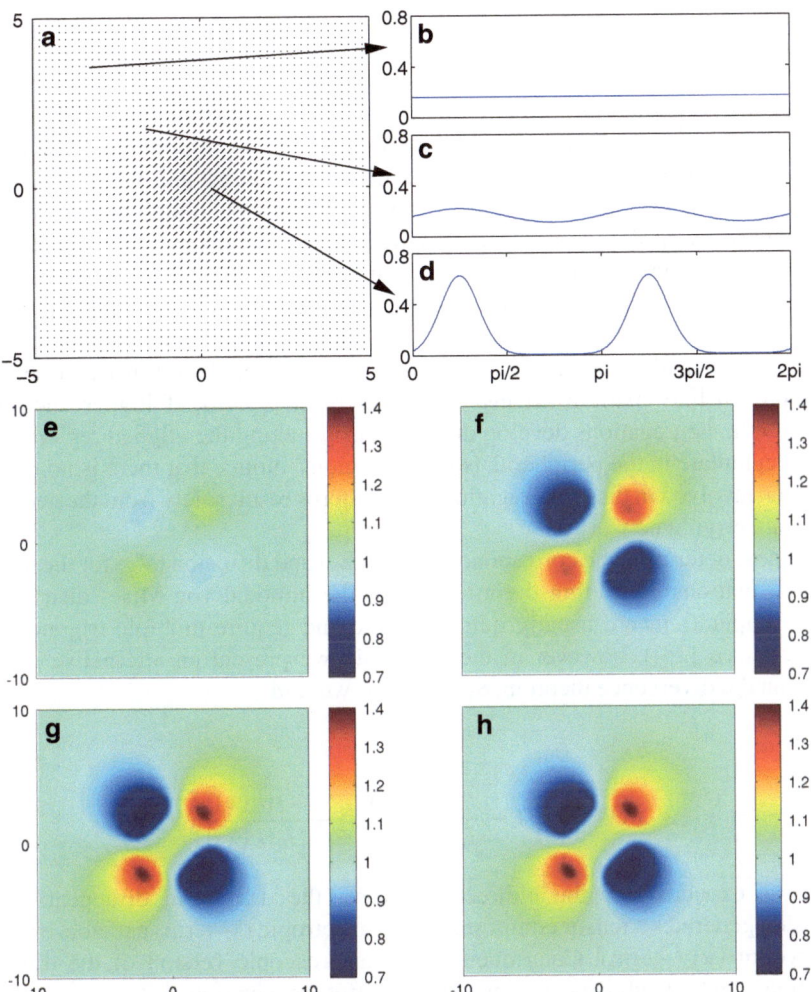

Fig. 2 Population heterogeneity arising due to bidirectional orientation of the environment. (a)–(d) representation of the imposed anisotropy, with (a) representing strength of anisotropy k (length of line segments) and alignment in the field (figure truncated at ± 5 to aid clarity of presentation) and (b)–(d) plotting the corresponding distribution (49) at each point indicated, as a function of $\theta = (\cos\phi, \sin\phi)$ for $\phi \in [0, 2\pi)$. Note that two dominating and equal orientations arise corresponding to $\gamma = \pm(\sqrt{2}/2, \sqrt{2}/2)$. (e)–(g) Simulation of the transport model (3) under the imposed q, showing the predicted macroscopic cell density $\bar{p}$ at time $t = 50$ for (e) $s = 0.1, \mu = 0.01$; (f) $s = 1, \mu = 1$; (g) $s = 10, \mu = 100$. (h) Simulation of the parabolic limit (51) at the same time $t = 50$ with $s^2/\mu = 1$ and the diffusion tensor as computed from (50). For details of the numerical implementations we refer to the Appendix

We first simulate the original transport model by substituting the above k and γ into (49) and solving (3). For details of the numerical methods used throughout this section, we refer to the Appendix. We assume the population is initially homogeneous and unaligned, with $p(x, v, 0) = \text{constant}$ and $\bar{p}(x, 0) = 1$. To limit the impact from boundaries we impose periodic boundary conditions along edges. In Fig. 2e–g we plot the macroscopic cell density $\bar{p}(x, t)$ at $t = 50$ for three distinct speeds, s, and turning rate, μ: (e) $s = 0.1, \mu = 0.01$; (f) $s = 1, \mu = 1$; (g) $s = 10, \mu = 100$. Note that the parabolic limit corresponds to the limiting scenario in which $s \to \infty$, $\mu \to \infty$ with s^2/μ constant and we can therefore expect (g) to most accurately reflect the solution to the parabolic model. The simulations reveal the impact of the environmental anisotropy on the population. Far from the origin the population is almost uniformly distributed. Nearer the centre a heterogeneous population distribution arises due to movement into the aligned region with subsequent transport in the direction of alignment. The bidirectional movement in this region results in symmetry in the population distribution, with a "dumbbell-like" pattern arising composed from regions of higher and lower density. The aggregations develop due to transport along the aligned region where they accumulate in the peripheral, isotropic regions. Notice that there is no taxis or adhesion involved in these aggregations; the patterns result solely from the geometry of the underlying network.

We next determine the corresponding drift ($\mathbb{E}_q$) and diffusion ($\mathbb{V}_q$) for the macroscopic equations by finding the moments of the bimodal von Mises distribution. Such computations are usually quite involved and require multiple trigonometric integrals (see [28]), however in the Appendix we present an alternative method based on the divergence theorem. Specifically, we find

$$\mathbb{E}_q(x) = 0\,,$$

$$\mathbb{V}_q(x) = \frac{1}{2}\left(1 - \frac{I_2(k(x))}{I_0(k(x))}\right)\mathbb{I}_2 + \frac{I_2(k(x))}{I_0(k(x))}\gamma\gamma^T\,. \tag{50}$$

Thus, as expected for the bidirectional case, the drift term disappears while diffusion generates a tensor composed from an isotropic ($\mathbb{I}_2$-term) and non-isotropic component ($\gamma\gamma^T$-term). Consequently, the macroscopic version of the transport equation simulated above is the anisotropic diffusion equation

$$\bar{p}_t = \frac{s^2}{\mu}\nabla(\nabla\mathbb{V}_q(x)\bar{p})\,, \tag{51}$$

where the heterogeneous and anisotropic diffusion tensor is given by (50) using the choices for γ and $k(x)$ above that define our direction distribution. Simulations are shown in Fig. 2h for a simulation of (51) with $s^2/\mu = 1$, with $\bar{p}(x, t)$ plotted at $t = 50$. Notably, the population distribution quantitatively matches the output from the transport model under the simulated parabolic limit scaling of s and μ.

7.1.3 Steady States

The above simulations suggest a capacity of the model to generate inhomogeneous steady states, at first a little surprising for a pure diffusion model. Closer scrutiny of (51) reveals how these patterns could arise as we demonstrate through the one-dimensional example. Consider the following distinct models for movement of a population within an interval:

$$u_t = (D(x)u_x)_x \tag{52}$$

and

$$u_t = (D(x)u)_{xx} \tag{53}$$

with homogeneous Neumann conditions assumed at the boundaries. Equation (53) can be expanded into $u_t = (D'(x)u + D(x)u_x)_x$, revealing an additional advective term with advective velocity D' in comparison to (52). To determine the impact of this extra term we examine steady states for (52) and (53).

At steady state, (52) leads to $(D(x)u_x)_x = 0$ which, after integrating and applying the boundary conditions, yields $D(x)u_x = 0$. This implies $u_x = 0$ and $u(x)$ is constant at steady state. This is what we expect for a pure diffusion process. Steady states for (53), on the other hand, satisfy $(D(x)u)_{xx} = 0$ and hence we find $(D(x)u)_x = 0$. Thus, $D(x)u = c$ (constant) and

$$u(x) = \frac{c}{D(x)}.$$

For spatially varying $D(x)$, (53) clearly allows nonuniform steady states, with the corresponding $u(x)$ being high or low in small or large diffusion regions, respectively. The additional advective term lies at the heart of this nontrivial steady state.

7.1.4 Application to Seismic Line Following

Having confirmed that the diffusion model (51) can accurately capture predicted behaviour of the original transport model, at least under relevant scalings, we now apply the method to tackle a specific ecological problem: wolf movement in certain habitats. The model as discussed is particularly useful for describing the movements of populations in environments containing linear features such as roads, rivers, valleys, or seismic lines. Work by McKenzie and others [28, 29] determined the movement patterns of wolves in a typical Western Canadian habitat, consisting of boreal forest cut by seismic lines. Seismic lines are clear-cut straight lines (with a width of about 5 m) used by oil exploration companies for testing of oil reservoirs. Typical densities are approximately 3.8 km of lines on 1 km^2 and both wolves and ungulates (such as caribou) use these lines to move and forage, leading to significant impact on predator prey-interactions.

To describe the movement of wolves in such a habitat, McKenzie used GPS data generated from four individual wolves and estimated parameters for a diffusion-advection model, dividing the habitat into three areas: (i) seismic lines, (ii) near seismic lines (less or equal 50 m), and (iii) far from seismic lines (larger than 50 m). Wolves demonstrated preferred movement along lines, while occasionally leaving lines to reenter forest. In particular, wolf movement data on seismic lines supported a fit to the directional distribution given by the bimodal von Mises distribution (49), where $\gamma(x) \in \mathbb{S}^1$ now describes the direction of the seismic line and $\theta \in \mathbb{S}^1$ the movement direction of the wolves.

To model this scenario we consider the parabolic limit of an underlying transport model in which wolf direction varies according to being on or off a seismic line. With no up or down information provided by the seismic line, we therefore have a bidirectional local environment and can expect the density of wolves, $w(x,t)$, to follow the anisotropic diffusion equation

$$w_t = \frac{s^2}{\mu} \nabla(\nabla \mathbb{V}_q w) , \tag{54}$$

where the anisotropic diffusion tensor $\mathbb{V}_q$ is given by (50), $\gamma(x)$ will correspond to the direction of a seismic line while $k(x)$ varies according to a position on or off a seismic line.

To illustrate the applicability, consider for the moment a coordinate system aligned with a seismic line, i.e. $\gamma = e_1$. Here we can directly compute the diffusion tensor:

$$\mathbb{V}_{\tilde{q}} = \begin{pmatrix} \frac{1}{2}\left(1 + \frac{I_2(k)}{I_0(k)}\right) & 0 \\ 0 & \frac{1}{2}\left(1 - \frac{I_2(k)}{I_0(k)}\right) \end{pmatrix} .$$

The term $I_2(k)/I_0(k)$ enhances the mobility along a seismic line and reduces mobility in perpendicular direction. Moreover, for $k \to \infty$ (corresponding to an increasing strength of anisotropy), $I_2(k)/I_0(k) \to 1$ and the above diffusion tensor collapses to one-dimensional diffusion along the seismic line.

Away from the seismic lines wolves show no clear tendency to migrate towards or away from seismic lines [28]. Effectively, away from the lines we set $k(x) = 0$ in the bimodal von Mises distribution (49) and we obtain the isotropic diffusion tensor:

$$\mathbb{V}_{\tilde{q}} = \frac{1}{2}\mathbb{I}_2. \tag{55}$$

Using these ideas, we next simulate the expected population distribution for wolves in a typical habitat containing seismic lines. The aerial photograph in Fig. 3a is of a Northern Alberta landscape in winter, demonstrating a woodland habitat criss-crossed with a combination of roads (thicker lines) and seismic lines (thinner lines). This image was digitised into a binary map, Fig. 3b, showing areas of seismic lines (or roads) (white) and away from seismic lines (black).

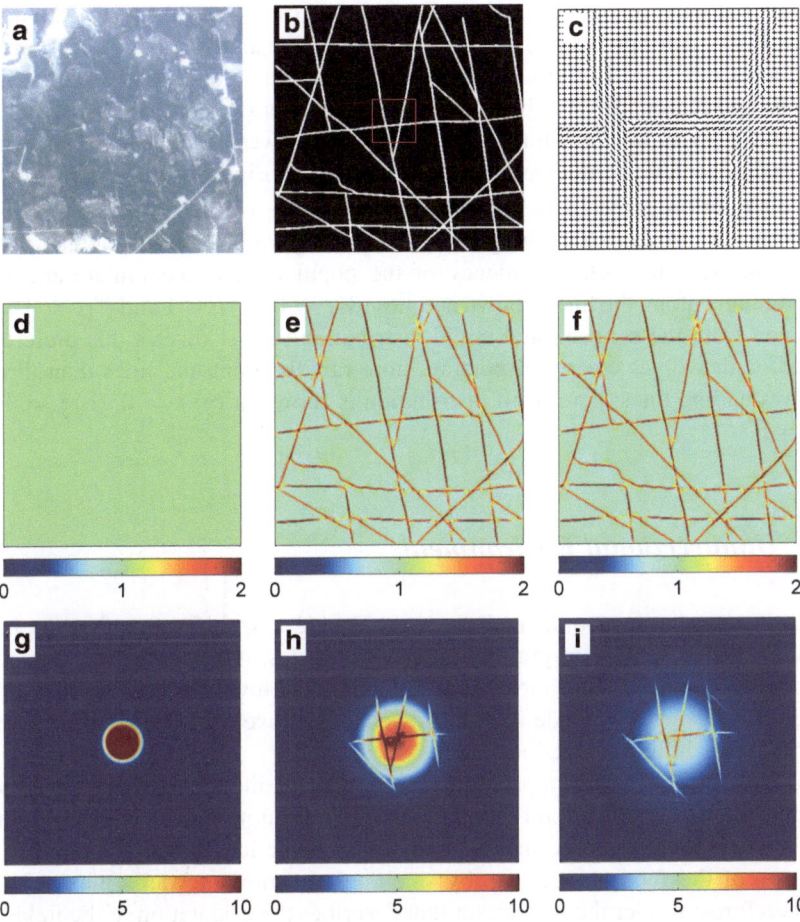

Fig. 3 Wolf distribution in anisotropic environments. (**a**) Aerial photograph of a Northern Alberta (Canada) landscape, showing criss-crossing seismic lines and roads. (**b**) Binary map created from (**a**) with *lines* marked as *white*. (**c**) Blow-up of boxed region in (**b**), showing detail of the anisotropic diffusion tensor automatically generated from the image in (**b**). (**d**–**f**) Numerical simulation of (54) for a uniform distribution $w(x, 0) = 1$, using the computed diffusion tensor generated from (**b**) and setting $s^2/\mu = 1$. Wolf density $w(x, t)$ is plotted at times (**d**) $t = 0$, (**e**) $t = 1$ and (**f**) $t = 10$. (**g**–**i**). Numerical simulation for $w(x, 0) = 100e^{-|x-x_c|^2}$ (where x_c marks the domain centre), showing $w(x, t)$ at times (**g**) $t = 0$, (**h**) $t = 1$ and (**i**) $t = 5$. Note that the simulated domain is a little larger than that plotted, with the surrounding zone assumed isotropic and implemented to reduce the impact of boundary conditions (note that this has negligible impact on the qualitative results presented). For details of the numerical implementation we refer to the Appendix

An automated processing of this image was applied to calculate the orientation at a point specified as seismic line, with this orientation determining the vector field $\gamma(x)$ used to compute the anisotropic diffusion tensor (50). In Fig. 3c this anisotropy is represented for a small square section indicated by the boxed area in 3b, with the long axes at each point representing the direction (and strength) of the alignment. We set $k = 2.5$ for points marked as on a seismic line and $k = 0$ for points marked as off a seismic line. To limit the impact from boundary conditions we remark that the digitised region in B was buffered with a perimeter of isotropic diffusion.

Preliminary simulations for the distribution of wolves, w, are shown for two initial conditions: a uniform distribution $w(x, 0) = 1$ in Fig. 3d–f and a 2D Gaussian-type distribution centered in the field for Fig. 3g–i, $w(x, 0) = 100e^{-|x-x_c|^2}$. In the former we observe the emergence of a spatially variable wolf population from homogeneity, with a clear tendency of the population to accumulate and move preferentially along the lines, shown at times (d) $t = 0$, (e) $t = 1$ and (f) $t = 10$. The diffusion from the concentrated initial distribution further reveals this preferential spread, with wolves clearly dispersing more rapidly along the lines than through the surrounding lines; here, wolf distribution is shown at (g) $t = 0$, (h) $t = 1$ and (i) $t = 5$.

7.2   Unidirectional Environments

In many cases an environmental cue can provide a specific direction, as in the magnetic fields used by migrating turtles and whales, the slope of the ground for hilltopping butterflies, the movement of organisms towards food sources or the current of a river. To include such cues we can remove the symmetry assumption for q imposed in the bidirectional case.

To examine how this impacts on the scaling limit we consider the specific example of attraction to a food supply. We let $F(x)$ denote a given food distribution, with $x \in \mathbb{R}^2$, and assume that individuals more or less accurately identify the direction of the food source (e.g. by smelling) and move towards maxima of F. We therefore consider the unit vector that describes the orientation of the field to be given by

$$\gamma(x) = \frac{\nabla F(x)}{\|\nabla F(x)\|}.$$

Since orientation of individuals is rarely perfect (i.e. movement will not be directly in the direction of the food) we take a (unimodal) von Mises distribution about the gradient of F:

$$\tilde{q}(x, \theta) = \frac{1}{2\pi I_0(k)} e^{k\theta \cdot \gamma}. \tag{56}$$

The above defines a direction distribution in which individuals align and migrate in the direction of the source. Note that varying degree of alignment could also be incorporated, for example through allowing k to depend on the size of F or $\|\nabla F(x)\|$. To determine the macroscopic terms we again compute the moments of the distribution (see Appendix):

$$\mathbb{E}_q(x) = \frac{I_1(k)}{I_0(k)}\gamma\,; \tag{57}$$

$$\mathbb{V}_q = \frac{1}{2}\left(1 - \frac{I_2(k)}{I_0(k)}\right)\mathbb{I}_2 + \left(\frac{I_2(k)}{I_0(k)} - \left(\frac{I_1(k)}{I_0(k)}\right)^2\right)\gamma\gamma^T\,. \tag{58}$$

Notably, the drift term $\mathbb{E}_q$ is now nonzero and in the direction of $\nabla F(x)$, whereas the diffusion term has two components: an isotropic part and an oriented nonisotropic part, which is proportional to $\nabla F(x)\nabla F(x)^T$. The resulting macroscopic equation is therefore of the form of an anisotropic drift-diffusion equation

$$p_t + s\nabla(\mathbb{E}_q p) = \frac{s^2}{\mu}\nabla(\nabla\mathbb{V}_q p)\,. \tag{59}$$

It is worth noting two limiting scenarios. For the parameter of concentration k becoming small (i.e. the food source provides a weak orientational cue), then

$$\lim_{k\to 0}\mathbb{E}_q = 0 \qquad \lim_{k\to 0}\mathbb{V}_q = \frac{1}{2}\mathbb{I}_2\,,$$

and we obtain uniform isotropic diffusion and no accumulation at the food source. For the parameter of concentration k becoming large (i.e. the food source provides a strong orientational cue), then

$$\lim_{k\to\infty}\mathbb{E}_q = \gamma \qquad \lim_{k\to\infty}\mathbb{V}_q = 0\,,$$

and hence we obtain the pure drift equation in which cells move directly towards the food source with speed s.

7.2.1 Anisotropic Diffusion-Drift Example

To illustrate how unidirectional environments impact on patterning, we present a scenario analogous to the example of Sect. 7.1.2. Specifically, we consider a population in a landscape with a unidirectional patch in the centre of the domain. We assume the above von Mises distribution (56) with the main orientation along the diagonal $\gamma = (1/\sqrt{2}, 1/\sqrt{2})^T$,

$$q(x, \theta) = \frac{1}{2\pi I_0(k)}e^{k(x)\theta\cdot\gamma}\,.$$

Once again $k(x)$ is assumed to decay exponentially from the centre to the periphery of the domain, with

$$k(x) = k_0 e^{-r|x|^2}\,.$$

Here we set $k_0 = 5$ and $r = 1.0$.

We again perform a direct simulation of the original transport model (3) with the above choice for q and solving subject to the same initial and boundary conditions as for the example of Sect. 7.1.2. As we observe in Fig. 4e, the directed patch significantly impacts on the subsequent distribution of the population. Rapid transportation through the oriented region results in a markedly decreased population density within this region. This generates a large "plume"-like structure adjacent to this region.

We simulate the corresponding anisotropic diffusion-drift equation. For the above von Mises distribution we compute the heterogeneous drift and diffusion terms from (57) and (58) respectively and substitute these into (59). Simulations show an excellent quantitative match with the transport model, Fig. 4f, once again confirming the validity of the macroscopic scaling process.

7.2.2 Relation to Haptotaxis and Chemotaxis

As a brief remark we note that unidirectional environments can be reinterpreted in terms of modelling haptotaxis (directed migration of cells in response to regions of high adhesivity in the ECM), chemotaxis (directed movement in response to chemical gradients) and other forms of gradient following. Haptotaxis and chemotaxis are typically modelled by an advective type term in PDE models (e.g. see [2, 19, 22, 27, 31]), with cell velocity proportional to the adhesion/chemical gradient.

The present work provides new motivation for such models. For example, we assume $F(x)$ describes the ECM adhesivity field surrounding a cell and take the von Mises distribution (56) to describe oriented movement towards higher adhesion, i.e. we take q to be given by

$$q(x, v) = \delta_{s(\|\nabla F\|)}(\|v\|) \frac{1}{2\pi I_0(k)} \exp\left(k \frac{v \cdot \nabla F(x)}{\|v\| \, \|\nabla F(x)\|}\right).$$

Furthermore, we let the speed s depend on the strength of the gradient, $s(\|\nabla F(x)\|)$. Since $\mathbb{E}_q \neq 0$, the parabolic limit does not apply and we employ instead the hyperbolic scaling. Drift subsequently dominates with diffusion of lower order and the corresponding macroscopic model becomes (to leading order)

$$u_t + 2\pi I_1(k) \nabla \cdot \left(\frac{s(\|\nabla F\|)}{\|\nabla F\|} \nabla F u\right) = 0.$$

The field F could also be reinterpreted to describe other forms of tactic migration.

7.3 Singular Distributions

The theories above have been derived for regular measures $q \in L^2$ only and, while it is possible to extend some of the results to singular measures (see for

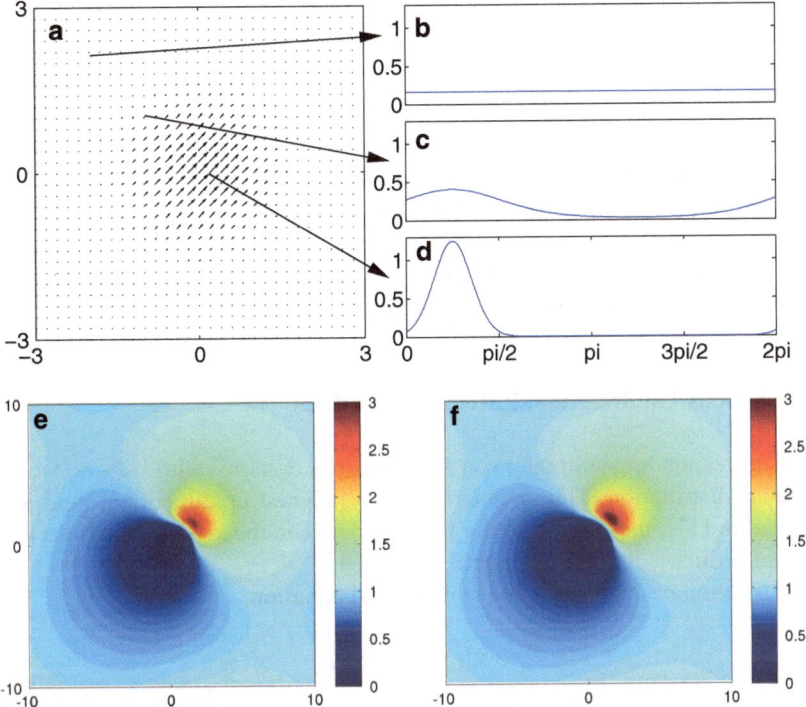

Fig. 4 Population heterogeneity arising due to unidirectional orientation of the environment. (a)–(d) representation of the imposed anisotropy, with (a) representing strength of anisotropy k (length of line segments) and the directional alignment of the field (figure truncated at ± 3 to aid clarity of presentation) and (b)–(d) plotting the corresponding distribution (56) at each point indicated, plotted as a function of $\theta = (\cos\phi, \sin\phi)$ for $\phi \in [0, 2\pi)$. Note that the dominating orientation corresponds to $\gamma = (\sqrt{2}/2, \sqrt{2}/2)$. (e) Simulation of the transport model (3) under the imposed q, showing the predicted macroscopic cell density $\bar{p}$ at time $t = 50$ for $s = 10$ and $\mu = 100$. (f) Simulation of the diffusion-drift limit (59), using $s = 10$ and $s^2/\mu = 1$ and plotted at $t = 50$, with the diffusion tensor computed from (58) and the drift term calculated according to (57). For details of the numerical implementation we refer to the Appendix

example [7, 20]), the mathematical overhead becomes enormous; here we simply apply the formal limit equations in good faith. Singular measures, however can play an important role either in describing certain oriented fields or representing a limit scenario for previously considered cases.

7.3.1 Strictly Bidirectional: Degenerate Diffusion

If we consider the earlier bimodal von Mises distribution (49) and let $k \to \infty$ we converge to two point measures in directions γ and $-\gamma$. Such distributions could be considered as completely aligned and bidirectional networks. Specifically, we let

$$q(x) := \frac{1}{2}(\delta_{\gamma(x)}(v) + \delta_{-\gamma(x)}(v)) ,$$

and find

$$\mathbb{E}_q = 0 \quad \text{and} \quad \mathbb{V}_q = \gamma\gamma^T . \tag{60}$$

Thus, there is zero drift and diffusion is given by a rank-one tensor $\mathbb{V}_q$, i.e. diffusion occurs only along the $\gamma/ - \gamma$ axis. The corresponding diffusion tensor $D = \frac{s^2}{\mu}\mathbb{V}_q$ is degenerate and not elliptic, hence the general solution theory for parabolic equations does not apply. In a forthcoming paper we develop methods that allows us to describe *very weak* solutions for such degenerate problems [21].

7.3.2 Strictly Unidirectional: Relation to ODEs

For the corresponding unimodal von Mises distribution (56) with $k \to \infty$ we obtain a singular distribution. This defines a strictly aligned unidirectional field and, as described in [17], there is a striking relation between these limit equations and the theory of ordinary differential equations (ODE).

The solution of the autonomous differential equation

$$\dot{x}(t) = f(x(t)) \tag{61}$$

in the domain $\mathbb{R}^n$ is given by the solution semigroup $\Phi(t, x_0)$ which describes orbits that are tangential to the vector field $f(x)$. In our notation here, we assume that this vector field $f(x) \in V$ defines a given direction at each point in $\mathbb{R}^n$ and define

$$q(x, v) = \delta_{f(x)}(v), \tag{62}$$

where δ_f denotes the point measure with mass in $f \in V$. In this case we find

$$\mathbb{E}_q(x) = f(x), \quad \text{and} \quad \mathbb{V}_q = 0.$$

This is a clearly drift-dominated situation and the hyperbolic scaling is appropriate. Item (4) of Lemma 1 applies and we obtain the limit equation

$$u_t + \nabla(f(x)u) = 0.$$

This hyperbolic PDE has the characteristics

$$\dot{x}(t) = f(x(t)),$$

which is the ODE from above. Hence typical movement paths of particles in an environment given by a singular measure (62) are orbits of the corresponding ODE.

7.4 Life in a Stream

An example that amalgamates various cases above (nondirectional, unidirectional and singular) is the movement of living organisms in a stream (which, for convenience, is assumed to be two dimensional).

Movement can be split into two principal contributions: (i) transport due to the current, and (ii) active movement by the individuals. For transport due to the current we let $\gamma(x)$ denote the direction of the stream (assumed quasi-constant over the timescale of interest), and let $q_1(\theta) = \delta_{\gamma(x)}(\theta)$ define the stream current. We augment this transport with a degree of turbulence, expressed via the random movement contribution $q_2(\theta) = |\mathbb{S}^{n-1}|^{-1}$.

For the active movement we assume individuals are biased towards a given food source

$$q_3(\theta) = \frac{1}{2\pi I_0(k)} e^{k\,\theta\cdot\Gamma(x)}, \quad \text{with} \quad \Gamma(x) = \frac{\nabla F(x)}{\|\nabla F(x)\|},$$

where $F(x)$ describes the distribution of food inside the stream. To simplify computations, we assume individuals have a preferred speed s, i.e. $V = s\mathbb{S}^{n-1}$. Hence, q is a convex combination of the above effects:

$$q(x, v) = s^{1-n} \left(\alpha_1 q_1(\hat{v}) + \alpha_2 q_2(\hat{v}) + \alpha_3 q_3(\hat{v}) \right),$$

where $\alpha_1 + \alpha_2 + \alpha_3 = 1$, and $\alpha_i \geq 0$ for $i = 1, \ldots, 3.$, and $\hat{v} = v/\|v\|$ denotes the unit vector in direction of v.

In this case, the macroscopic drift component is given by

$$\mathbb{E}_q = \alpha_1 \gamma(x) + s\alpha_3 2\pi I_1(k)\Gamma(x).$$

Drift arises as the interplay between transport due to the stream $\gamma(x)$ and movement towards the food source $\Gamma(x)$. The diffusion term is given by

$$D(x) = \frac{s^2}{\mu} \left[\frac{\alpha_2}{2}\mathbb{I}_2 + \frac{\alpha_3}{2} \left(1 - \frac{I_2(k)}{I_0(k)} \right) \mathbb{I}_2 + \alpha_3 \left(\frac{I_2(k)}{I_0(k)} - \left(\frac{I_1(k)}{I_0(k)} \right)^2 \right) \Gamma(x)\Gamma(x)^T \right],$$

derived from a combination of random movement and the imperfect response to the food source. We note that more detailed modelling of river ecosystems and species survival has been undertaken by Lutscher et al. [26].

8 Discussion

The principal aims of this paper have been to demonstrate the effectiveness of transport equations as a method for modelling cell or animal movement, to explain and summarise the various scaling limits that allow their approximation to distinct macroscopic models, and to consider a few pertinent ecological applications, such as wolf movement on seismic lines, attraction to a food source and movement in rivers.

The transport model is a natural model for movement, relying as it does on experimentally measurable data such as speeds and turning rates for its key inputs. While it is certainly possible to study the transport model directly, both the analytical and numerical overheads can be costly. For example, the numerical solution of the simple (and assumed 2D) transport model given by (3) requires discretisation not only over space, but also orientation; extensions to relevant scenarios such as 3D, variable speeds or more intricate turning functions would significantly add to the computational time. Simplifying to the relatively straightforward macroscopic model, which still possesses details of the underlying microscopic processes in its macroscopic parameters, allows far faster numerical computation while opening the vault to a wealth of analytical tools.

Typically the scaling methods considered here (parabolic scaling, hyperbolic scaling, and moment closure) are studied separately and it can be difficult for unfamiliar readers to determine why one method is chosen over another. By focussing on a specific formulation of a transport model, originally developed to describe cell movement in network tissues, we could transparently derive the various limiting equations and expose the assumptions that underlie them.

Responding to a question posed during the introduction, it would be bold to categorically state a "best" method and instead models must be treated on a case by case basis. Succinctly, it comes down to the relative size of drift and diffusion terms: when the model is drift-dominated, as occurs for environments with a strong cue in a specific direction, the hyperbolic approximation applies; when the model is diffusion-dominated, as for environments with either nondirectional or bidirectional orientation, the parabolic limit is appropriate; if the two effects are of a similar order then either the moment closure or the hyperbolic model with corrections provide the most appropriate approximation.

It is worth noting that the clarity of the analysis here is a direct product of the simplicity of our transport model. Full analyses for more general kinetic equations can become highly technical and fill entire textbooks (for example, see [8] for diluted gases or [5, 37] for biological applications). With the aim of illuminating the various scaling limits we have made a number of convenient assumptions and it is worth describing some of the limiting factors here, and their potential importance for biological applications.

- We have not considered time-varying habitats. In many instances, the environment can change considerably on the timescale of movement, either independently (for example, the changing position of the sun or alterations in wind strength) or through direct modification by the migrating population (e.g. formation of pheromone trails by ants or restructuring the ECM by cells). The addition of t-dependence in the orientation function q adds a significant level of complexity and, while the scaling limits do apply, they require detailed analysis and consideration on a case by case basis. For details of such analyses in the context of mesenchymal cell migration we refer to [17].

- In this paper, the environment has been assumed to only impact on the turning of individuals, not on their speed. While it is trivial to extend the original transport model to incorporate more general speed dependencies, the subsequent computations to calculate the scaling limits are often complex and obscure their basic features. We note that in the context of taxes above, we have given one simple example on how to perform scaling for nonconstant speeds.

- Appropriate boundary conditions on bounded domains require special attention. For example in the case of the seismic lines above, what would be meaningful boundary conditions on and off the seismic lines for both the original transport model and the subsequent macroscopic limits? We circumnavigated this issue in the simulations by buffering the simulated region with a surrounding isotropic region and using periodic boundary conditions, however other conditions could certainly be considered. For example, zero-flux boundary conditions could be one relevent choice, as assumed in [29].

- More complicated formulations for the turning kernel $T(x, v, v')$ and non-constant turning rates $\mu(t, x, p, v)$ arise naturally in many applications. Obviously, any such choice should be tailored according to the application under analysis, however the ensuing calculations can become highly intricate. One important yet complicated case is the incorporation of interactions between individuals. For example, the patterns formed by many migrating populations, from bird flocks to wildebeest, are highly structured through the response of an individual to the movement of a neighbour.

- The simple model here has neglected aspects such as a resting phase (individuals are assumed to move continuously) or population kinetics. For example, modelling the impact of seismic lines on the predator-prey dynamics of wolves and caribou would require an extension of the model to include a separate caribou population and appropriate predator-prey interactions. Again, while tailoring the original transport model to include such extensions is relatively straightforward, the subsequent calculation of scaling limits would require treatment on a case by case basis.

- On a technical side, in our theorems we have typically used the notion "is approximated by" to denote the formal limit considerations. Rigorously, to refer to an approximation property would require proof of convergence in an appropriate function space and we have completely omitted these issues from these studies.

Migration, whether cellular or animal, clearly is immensely relevant to a plethora of crucial biological and ecological processes. Distinct methods offer different advantages, allowing multiple windows through which the underlying mechanisms can be observed. In this paper, our aim has been to concentrate on the transport (and associated macroscopic) equations, with the key aim of shedding illumination on this useful modelling approach.

Appendix

9 Moments of von Mises Distributions

The appendix is used to present an alternative method for computing moments of a von Mises distribution. Usually, moments are computed through explicit trigonometric integrations (see e.g. [3, 28, 30]) however here we instead apply the divergence theorem. While this method is easily generalised to arbitrary space dimensions, explicit integration becomes increasingly cumbersome with increases in the space dimension.

Given a unit vector $\gamma \in \mathbb{S}^{n-1}$, we first study the (unimodal) von Mises distribution

$$q(\theta) = \frac{1}{2\pi I_0(k)} e^{k\theta \cdot \gamma} \tag{63}$$

In the main text it is noted that the moments employ Bessel functions and we begin by collecting a few of their properties. If $J_n(x)$ denote the Bessel functions of first kind, then

$$I_n(x) := (-i)^{-n} J_n(ix)$$

denotes the Bessel function of first kind with purely imaginary argument, or the *modified Bessel functions*. For these we have the relation

$$I_n(k) = \frac{1}{2\pi} \int_0^{2\pi} \cos(n\phi) e^{k\cos\phi} d\phi . \tag{64}$$

Two further important relations include the differential recurrence

$$\frac{d}{dx}(x^n J_n(x)) = x^n J_{n-1}(x) \tag{65}$$

for $n \geq 0$, and the recurrence relation

$$J_{n+1}(x) = \frac{2n}{x} J_n(x) - J_{n-1}(x). \tag{66}$$

9.1 Unimodal von Mises Distribution

To compute the total mass of the (unimodal) von Mises distribution (63) we denote the angle between θ and γ by ϕ:

$$\int_{\mathbb{S}^1} q(\theta) d\theta = \frac{1}{2\pi I_0(k)} \int_0^{2\pi} e^{k\cos\phi} d\phi = 1,$$

where we used (64).

To compute the expectation, we note

$$2\pi I_0(k)\mathbb{E}_q = \int_{\mathbb{S}^1} \theta e^{k\theta \cdot \gamma} ,$$

$$= \int_{\mathbb{B}_1(0)} \mathrm{div}_v \, e^{kv \cdot \gamma} dv ,$$

$$= \int_{\mathbb{B}_1(0)} k\gamma e^{kv \cdot \gamma} dv ,$$

$$= k\gamma \int_0^1 \int_0^{2\pi} e^{rk\cos\phi} r\, dr\, d\phi ,$$

$$= k\gamma \int_0^1 2\pi r I_0(rk) dr ,$$

$$= 2\pi k\gamma \int_0^1 r I_0(rk) dr .$$

To solve the last integral, we use (65) and write

$$r I_0(rk) = \frac{irk J_0(irk)}{ik} = \frac{1}{ik}\frac{d}{dx}(x J_1(x))|_{x=irk} = \frac{1}{ik}\frac{d}{dr}(r J_1(irk)).$$

Then

$$\int_0^1 r I_0(rk) dr = \frac{1}{ik} r J_1(ik) = \frac{1}{ik} i I_1(k) = \frac{I_1(k)}{k}. \qquad (67)$$

and we find

$$\mathbb{E}_q = \frac{I_1(k)}{I_0(k)} \gamma. \qquad (68)$$

The variance-covariance matrix is given by

$$\mathbb{V}_q = \int_{\mathbb{S}^1} (\theta - \mathbb{E}_q)(v - \mathbb{E}_q)^T q(\theta) d\theta = \int_{\mathbb{S}^1} \theta\theta^T q(\theta) d\theta - \mathbb{E}_q \mathbb{E}_q^T .$$

To find the second moment of q we consider two test vectors $a, b \in \mathbb{R}^2$ and employ index notation for automatic summation over repeated indices

$$2\pi I_0(k) a \int_{\mathbb{S}^1} \theta\theta^T q(\theta) d\theta \, b = \int_{\mathbb{S}^1} a_i \theta^i b_j \theta^j e^{k\theta^l \gamma_l} d\theta$$

$$= \int_{\mathbb{S}^1} \theta^i (a_i b_j \theta^j e^{k\theta^l \gamma_l}) d\theta$$

$$= \int_{\mathbb{B}_1(0)} \frac{\partial}{\partial v^i} (a_i b_j v^j e^{kv^l \gamma_l}) dv$$

$$= \int_{\mathbb{B}_1(0)} a_i b_i e^{kv \cdot \gamma} dv + \int_{\mathbb{B}_1(0)} a_i (v \cdot b) k \gamma_i e^{kv \cdot \gamma} dv$$

$$= a \cdot b \int_{\mathbb{B}_1(0)} e^{kv \cdot \gamma} dv + ka \cdot \gamma \, b \cdot \int_{\mathbb{B}_1(0)} v e^{kv \cdot \gamma} dv \quad (69)$$

The first integral in (69) can be solved directly

$$\int_{\mathbb{B}_1(0)} e^{kv \cdot \gamma} dv = \int_0^1 \int_{\mathbb{S}^1} e^{rk\theta \cdot \gamma} r \, dr \, d\theta = \int_0^1 2\pi r I_0(rk) dr = 2\pi \frac{I_1(k)}{k},$$

where we used (64) and (67) in the penultimate and ultimate step respectively. Using (64) we can transform the second integral from (69) as follows:

$$\int_{\mathbb{B}_1(0)} v e^{kv \cdot \gamma} dv = \int_0^1 \int_{\mathbb{S}^1} r\theta e^{rk\theta \cdot \gamma} r \, dr \, d\theta = \int_0^1 r^2 \int_{\mathbb{S}^1} \theta e^{rk\theta \cdot \gamma} d\theta$$

$$= 2\pi \gamma \int_0^1 r^2 I_1(rk) dr, \quad (70)$$

where we used (68) in the last step.

Now we use the differential recurrence relation (65) to write

$$r^2 I_1(rk) = -\frac{1}{ik^2} (irk)^2 J_1(irk) = -\frac{1}{ik^2} \frac{d}{dx} (x^2 J_1(x))|_{x=irk} = -\frac{1}{k} \frac{d}{dr} (r^2 J_1(irk)).$$

Continuing from (70) we find

$$\int_{\mathbb{B}_1(0)} v e^{kv \cdot \gamma} dv = -2\pi \gamma \int_0^1 \frac{1}{k} \frac{d}{dr} (r^2 J_1(irk)) dr = -2\pi \gamma J_2(ik) = 2\pi \gamma \frac{I_2(k)}{k}.$$
$$(71)$$

Substituting all the integrals back into (69)

$$a \int_{\mathbb{S}^1} \theta \theta^T q(\theta) d\theta \, b = a \cdot b \frac{2\pi \frac{I_1(k)}{k}}{2\pi I_0(k)} + ka \cdot \gamma \frac{2\pi \gamma \cdot b \frac{I_2(k)}{k}}{2\pi I_0(k)}$$

$$= a \left(\frac{1}{k} \frac{I_1(k)}{I_0(k)} \mathbb{I}_2 + \gamma \gamma^T \frac{I_2(k)}{I_0(k)} \right) b.$$

Finally, we use the identity (66) for $n = 1$ to replace

$$\frac{1}{k} \frac{I_1(k)}{I_0(k)} = \frac{1}{2} \left(1 - \frac{I_2(k)}{I_0(k)} \right)$$

and the second moment is given by

$$\int_{\mathbb{S}^1} \theta\theta^T q(\theta)d\theta = \frac{1}{2}\mathbb{I}_2 + \frac{I_2(k)}{I_0(k)}\left(\gamma\gamma^T - \frac{1}{2}\mathbb{I}_2\right). \qquad (72)$$

Together with the formula for the expectation (68) we find

$$\mathbb{V}_q = \int_{\mathbb{S}^1} \theta\theta^T q(\theta)d\theta - \mathbb{E}_q\mathbb{E}_q^T$$

$$= \frac{1}{2}\mathbb{I}_2 + \frac{I_2(k)}{I_0(k)}\left(\gamma\gamma^T - \frac{1}{2}\mathbb{I}_2\right) - \left(\frac{I_1(k)}{I_0(k)}\right)^2\gamma\gamma^T \qquad (73)$$

$$= \frac{1}{2}\left(1 - \frac{I_2(k)}{I_0(k)}\right)\mathbb{I}_2 + \left(\frac{I_2(k)}{I_0(k)} - \left(\frac{I_1(k)}{I_0(k)}\right)^2\right)\gamma\gamma^T. \qquad (74)$$

Clearly, if the parameter of concentration k becomes small (i.e. $k \to 0$) then $\mathbb{E}_q \to 0$ and $\mathbb{V}_q \to \frac{1}{2}\mathbb{I}_2$.

9.2 Bimodal von Mises Distribution

Computations for the bimodal von Mises distribution

$$q(\theta) = \frac{1}{4\pi I_0(k)}\left(e^{k\theta\cdot\gamma} + e^{-k\theta\cdot\gamma}\right)$$

are very similar. Since the bimodal von Mises distribution is symmetric (or undirected) we have $\mathbb{E}_q = 0$ and $\mathbb{V}_q = \int \theta\theta^T q(\theta)d\theta$. We apply formula (72) for each of the components $e^{k\theta\cdot\gamma}$ and $e^{-k\theta\cdot\gamma}$ separately and sum. We find

$$\mathbb{V}_q = \frac{1}{2}\left(1 - \frac{I_2(k)}{I_0(k)}\right)\mathbb{I}_2 + \frac{I_2(k)}{I_0(k)}\gamma\gamma^T.$$

10 Numerical Methods

10.1 Simulations of Transport Model

Simulations of the transport model (3) were performed with a Method of Lines (MOL) approach, in which space and velocity are discretised into a high-dimensional system of time-dependent ODEs (the MOL-ODEs). For the transport equations presented, the rectangular spatial domain (of dimensions $L_x \times L_y$) was discretised into a uniform mesh of 201 by 201 points, while velocity $v = s(\cos\alpha, \sin\alpha)$ (for $\alpha \in [0, 2\pi)$) was discretised into 100 uniformly spaced orientations with a fixed speed s. Spatial terms for particle movement were approximated in conservative form using a third-order upwinding scheme, augmented by flux-limiting to maintain positivity. The resulting MOL-ODEs were integrated in time using the ROWMAP stiff systems integrator [40], with a fixed absolute and relative error tolerance of 10^{-7}. Similar approaches to those above were employed in [34].

10.2 Simulations of Macroscopic Models

Simulations of both the anisotropic diffusion (51) and anisotropic drift-diffusion (59) model were performed with a similar MOL approach. The anisotropic diffusion term was factored into diffusive and convective terms and solved in conservative form, applying a central difference scheme for the former and first order upwinding for the latter. The additional drift terms in the drift-diffusion model were also solved with first order upwinding and the resulting MOL-ODEs were integrated in time using ROWMAP with error tolerances of 10^{-7}. For the two simulations in Figs. 2 and 4 we used 201 by 201 mesh points for the spatial discretisation, while for the simulations in Fig. 3 we use 500 by 500 mesh points. We note that simulations with finer spatial discretisations and smaller tolerances demonstrated no appreciable quantitative difference.

Acknowledgements Work of Thomas Hillen was supported by NSERC and work of Kevin J. Painter was supported by BBSRC.

References

1. W. Alt, Biased random walk model for chemotaxis and related diffusion approximation. J. Math. Biol. **9**, 147–177 (1980)
2. A.R.A. Anderson, M.A.J. Chaplain, E.L. Newman, R.J.C. Steele, A.M. Thompson, Mathematical modelling of tumour invasion and metastasis. J. Theor. Med. **2**, 129–154 (2000)
3. E. Batschelet, *Circular Statistics in Biology* (Academic, London, 1981)
4. N. Bellomo, *Modeling Complex Living Systems - Kinetic Theory and Stochastic Game Approach* (Birkhauser, Basel, 2008)
5. N. Bellomo, M.L. Schiavo, *Lecture Notes on the Mathematical Theory of Generalized Boltzmann Methods* (World Scientific, Singapore, 2000)
6. G.P. Brown, B.L. Phillips, J.K. Webb, R. Shine, Toad on the road: use of roads as dispersal corridors by cane toads (bufo marinus) at an invasion front in tropical australia. Biol. Cons. **133**(1), 88–94 (2006)
7. J.A. Carrillo, R.M. Colombo, P. Gwiazda, A. Ulikowska, Structured populations, cell growth and measure valued balance laws. J. Diff. Equ. **252**, 3245–3277 (2012)
8. C. Cercignani, R. Illner, M. Pulvirenti, *The Mathematical Theory of Diluted Gases* (Springer, New York, 1994)
9. F.A.C.C. Chalub, P.A. Markovich, B. Perthame, C. Schmeiser, Kinetic models for chemotaxis and their drift-diffusion limits. Monatsh. Math. **142**, 123–141 (2004)

10. A. Chauviere, T. Hillen, L. Preziosi, Modeling cell movement in anisotropic and heterogeneous network tissues. Networks and Heterogeneous Media **2**, 333–357 (2007)
11. Y. Dolak, C. Schmeiser, Kinetic models for chemotaxis: hydrodynamic limits and spatio-temporal mechanics. J. Math. Biol. **51**, 595–615 (2005)
12. G.A. Dunn, J.P. Heath, A new hypothesis of contact guidance in tissue cells. Exp. Cell Res. **101**, 1–14 (1976)
13. P. Friedl, E.B. Bröcker, The biology of cell locomotion within three dimensional extracellular matrix. Cell Motil. Life Sci. **57**, 41–64 (2000)
14. P. Friedl, K. Wolf, Tumour-cell invasion and migration: diversity and escape mechanisms. Nat. Rev. **3**, 362–374 (2003)
15. S. Guido, R.T. Tranquillo, A methodology for the systematic and quantitative study of cell contact guidance in oriented collagen gels. Correlation of fibroblast orientation and gel birefringence. J. Cell. Sci. **105**, 317–331 (1993)
16. T. Hillen, On the L^2-closure of transport equations: the general case. Discrete Contin. Dynam. Syst. Ser. B **5**(2), 299–318 (2005)
17. T. Hillen, M^5 mesoscopic and macroscopic models for mesenchymal motion. J. Math. Biol. **53**(4), 585–616 (2006)
18. T. Hillen, H.G. Othmer, The diffusion limit of transport equations derived from velocity jump processes. SIAM J. Appl. Math. **61**(3), 751–775 (2000)
19. T. Hillen, K.J. Painter, A user's guide to PDE models for chemotaxis. J. Math. Biol. **58**, 183–217 (2009)
20. T. Hillen, P. Hinow, Z.A. Wang, Mathematical analysis of a kinetic model for cell movement in network tissues. Discrete Contin. Dynam. Syst. - B **14**(3), 1055–1080 (2010)
21. T. Hillen, K.J. Painter, M. Winkler, Anisotropic diffusion in oriented environments can lead to singularity formation. Eur. J. Appl. Math. doi:10.1017/S0956792512000447
22. E.F. Keller, L.A. Segel, Initiation of slime mold aggregation viewed as an instability. J. Theor. Biol. **26**, 399–415 (1970)
23. M. Lachowicz, Microscopic, mesoscopic and macroscopic descriptions of complex systems. Prob. Eng. Mech. **26**, 54–60 (2011)
24. H.P. Lipp, A.L. Vyssotski, D.P. Wolfer, S. Renaudineau, M. Savini, G. Troster, G. Dell'Omo, Pigeon homing along highways and exits. Curr. Biol. **14**(14), 1239–1249 (2004)
25. K.J. Lohmann, C.M. Lohmann, C.S. Endres, The sensory ecology of ocean navigation. J. Exp. Biol. **211**, 1719–1728 (2008)
26. F. Lustcher, E. Pachepsky, M.A. Lewis, The effect of dispersal patterns on stream populations. SIAM Rev. **478**, 749–7725 (2005)
27. P.K. Maini, Spatial and spatio-temporal patterns in a cell-haptotaxis model. J. Math. Biol. **27**, 507–522 (1989)
28. H.W. McKenzie, Linear Features Impact Predator-Prey Encounters: Analysis and First Passage Time, MSc thesis, University of Alberta, 2006
29. H.W. McKenzie, E.H. Merrill, R.J. Spiteri, M.A. Lewis, How linear features alter predator movement and the functional response. Interface Focus **2**(2), 205–216 (2012)
30. P. Moorcroft, M.A. Lewis, *Mechanistic Home Range Analysis* (Princeton University Press, Princeton, 2006)
31. G. Oster, J.D. Murray, A. Harris, Mechanical aspects of mesenchymal morphogenesis. J. Embryol. Exp. Morphol. **78**, 83–125 (1983)
32. H.G. Othmer, A. Stevens, Aggregation, blowup and collapse: the ABC's of taxis in reinforced random walks. SIAM J. Appl. Math. **57**, 1044–1081 (1997)
33. H.G. Othmer, S.R. Dunbar, W. Alt, Models of dispersal in biological systems. J. Math. Biol. **26**, 263–298 (1988)
34. K.J. Painter, Modelling migration strategies in the extracellular matrix. J. Math. Biol. **58**, 511–543 (2009)
35. K.J. Painter, T. Hillen, Mathematical modelling of glioma growth: the use of Diffusion Tensor Imaging (DTI) data to predict the anisotropic pathways of cancer invasion. J. Theor. Biol. doi:10.1016/j.jtbi.2013.01.014

36. G. Pe'er, D. Saltz, H. Thulke, U. Motro, Response to topography in a hilltopping butterfly and implications for modelling nonrandom dispersal. Anim. Behav. **68**, 825–839 (2004)
37. B. Perthame, *Transport Equations in Biology* (Birkhäuser, Basel, 2007)
38. A.M. Reynolds, T.K. Dutta, R.H. Curtis, S.J. Powers, H.S. Gaur, B.R. Kerry, Chemotaxis can take plant-parasitic nematodes to the source of a chemo-attractant via the shortest possible routes. J. R. Soc. Interface **8**, 568–577 (2011)
39. S.M. Tomkiewicz, M.R. Fuller, J.G. Kie, K.K. Bates, Global positioning system and associated technologies in animal behaviour and ecological research. Phil. Trans. R. Soc. B **365**, 2163–2176 (2010)
40. R. Weiner, B.A. Schmitt, H. Podhaisky, Rowmap–a row-code with Krylov techniques for large stiff ODEs. Appl. Num. Math. **25**, 303–319 (1997)
41. P.C. Wilkinson, J.M. Lackie, The influence of contact guidance on chemotaxis of human neutrophil leukocytes. Exp. Cell Res. **145**, 255–264 (1983)
42. K. Wolf, R. Muller, S. Borgmann, E.B. Brocker, P. Friedl, Amoeboid shape change and contact guidance: T-lymphocyte crawling through fibrillar collagen is independent of matrix remodeling by MMPs and other proteases. Blood **102**, 3262–3269 (2003)
43. A. Wood, P. Thorogood, An analysis of in vivo cell migration during teleost fin morphogenesis. J. Cell Sci. **66**, 205–222 (1984)

Incorporating Complex Foraging of Zooplankton in Models: Role of Micro- and Mesoscale Processes in Macroscale Patterns

Andrew Yu. Morozov

Abstract There is a growing understanding that population models describing trophic interactions should benefit from the increasing knowledge of the complex foraging behavior of individuals constituting those populations. A notable example is the modelling of planktonic food chains where the foraging behavior of herbivorous zooplankton is often complicated and involves active vertical displacement (migration) in the water column with the aim of optimizing the fitness under constantly varying environmental conditions such as distribution of predators, location of food, temperature gradient, oxygen concentration, etc. Vertical migration of zooplankton takes place on different time and space scales ranging from seconds and centimeters to months and the size of the whole euphotic zone. Taking into account active foraging behavior of zooplankton would alter theoretical predictions obtained with earlier plankton models where such behavior has often been ignored—especially in the mean-field models which operate with integrated species biomasses/densities. In this paper, I revisit two important aspects of incorporating patterns of active zooplankton feeding in models, based on recent progress in field observations and experiments. Firstly, I investigate how complex foraging movement of herbivores in the column can alter the shape of the zooplankton functional response on different spatial and temporal scales—in particular, I scale up the local functional response to macroscales (the whole euphotic zone) and show the emergence of a sigmoid functional response (Holling type III) on the macroscale based on a non-sigmoid local response on microscales. Secondly, I theoretically investigate the role of intra-population variability of the feeding behavior of grazers (implying physiological and behavioral structuring of a population) in the persistence of the whole population under predation pressure. I show that structuring

A.Yu. Morozov (✉)
Department of Mathematics, University of Leicester, LE1 7RH Leicester, UK

Shirshov Institute of Oceanology, Moscow, Russia
e-mail: am379@leicester.ac.uk

M.A. Lewis et al. (eds.), *Dispersal, Individual Movement and Spatial Ecology*,
Lecture Notes in Mathematics 2071, DOI 10.1007/978-3-642-35497-7_8,
© Springer-Verlag Berlin Heidelberg 2013

of the population according to feeding behavior would enhance the population persistence in a eutrophic environment thus preventing species extinction.

1 Introduction

It has been well recognized in ecology that spatial heterogeneity is a crucial factor shaping population dynamics and affecting species persistence ([8, 51, 66, 88]). The growth of a population often takes place in a highly heterogeneous environment characterized by a pronounced variation in the species fitness. In the case organisms have the ability to actively move within a large part of the habitat they can adjust their spatial location to improve their living conditions by acquiring more food, escaping from natural enemies, etc. An important ecological example of such behaviour is the active vertical migration of herbivorous zooplankton in the water column in lakes and the ocean. Although in the horizontal direction the active displacement of plankters is seriously impeded by a pronounced turbulence ([1, 36, 68]), mesozoopolankton such as copepods can quickly adjust their vertical location and find the optimal depth within the entire euphotic zone (i.e. the zone where the light intensity is enough to make possible photosynthesis of phytoplankton) depending on the given distribution of predators and food conditions, as well as abiotic factors as temperature, salinity, etc. [10, 37, 53, 60, 65, 90]. Since copepods constitute the main source of food for small pelagic fish (the upper trophic level) and can also control the primary production via intensive grazing, their correct description in models is becoming of crucial importance when simulating the biochemical cycles, sustainable fishery management, toxic plankton blooms, marine biodiversity, etc. Moreover, excluding patterns of active foraging behavior of grazers can be somewhat of a bottleneck in improving the predictive power of plankton models [14, 65].

Active vertical displacement of herbivorous zooplankton in the column takes place on different time and space scales (see Fig. 1). On microscales (seconds and dozens of centimeters, up to 1–2 m) zooplankton show active foraging behaviour by performing small foraging jumps and accumulating in micropatches of high food density [30, 70, 113]. On the intermediate time and space scales (1–3 h and dozens of meters), organisms perform short-term exchanges between surface layers which are rich in food (phytoplankton) and deeper layers, which contain less food but are safer from predators [65, 77, 95]. On a daily time scale zooplankton can show regular diel vertical migration where the organisms ascend to upper (surface) layers for feeding at night and stay in deep layers during the day time. It is believed that this strategy allows herbivores to escape from visual predators [10, 60, 90] and/or because of the energy gain in deeper waters due to low temperature [49, 72]. Finally, zooplankton exhibit variations in movement behavior on a longer time scale (varying from several weeks to months) which is related to the ontogenetic plankton cycles where zooplankters can even leave the limits of the euphotic zone and descend to deeper layers [91, 117].

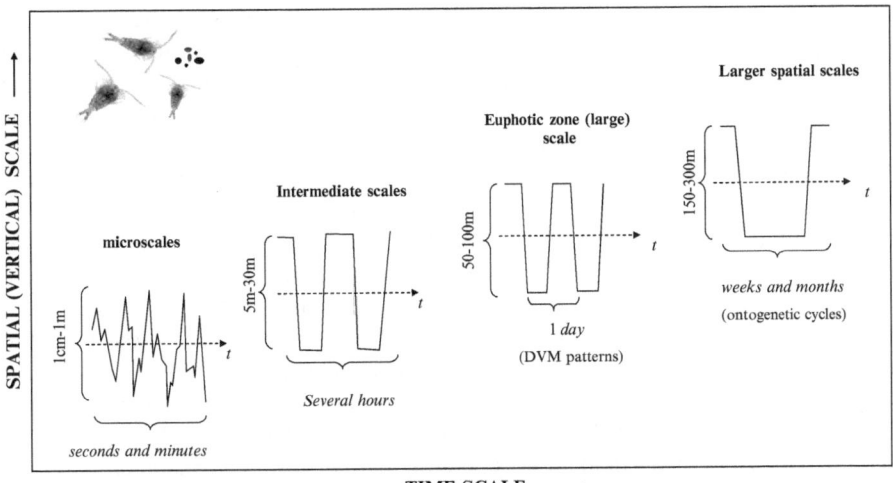

Fig. 1 Active foraging behavior of herbivorous zooplankton in the ocean and deep lakes over different spatial and temporal scales. For details and the literature references see the text

There exist a large number of publications concerning the modelling of active vertical migration of zooplankton. Most of these publications, however, provide models of the regular diel vertical migration (DVMs) taking place on the scale of the whole euphotic zone. In particular, it has been shown that such migrations can be an optimal strategy for the persistence of a population under the threat of predation by visual predators [39, 49, 57, 67, 107]. On the other hand, there also exist a high number of theoretical works on zooplankton movement on microscales ([3, 33, 45, 105, 106, 114]). Such works usually model the movement of zooplankton on microscales as a fractional random walk ([9, 23, 105, 106]) and even as a Levy flight ([3, 120]) and are justified by experimental material on zooplankton movement recorded by cameras [30, 70, 113]. Regrettably though, foraging behavior of zooplankton herbivores on the intermediate time and space scales is studied much less both regarding observation/experimental and modelling works.

An important reason for the lack of studies on intermediate time and space scales is that the active vertical movement of grazers at intermediate scales is often non-synchronized: in other words, exchange of the individuals between the horizontal layers in the column can take places without alteration of the profile of the population as a whole [21, 65, 77]. Such non-synchronized vertical migration is rather hard to investigate in vivo since this would require labelling and tracking a large number of small-size organisms in a highly turbulent environment. Another important reason for the mentioned lack of knowledge is rather coarse methods of sampling. As a result, the existence of any fine structure of plankton layers is often overlooked. This concerns, for example, the so-called thin plankton layers with a characteristic vertical width up to few meters but having plankton densities several orders higher than outside the layers [22, 52]. There is an opinion that foraging of zooplankton in these high density food patches can be crucial for the survival of grazers [13, 64, 82] since the food density outside the patches is often below the feeding threshold of zooplankton (i.e. the minimal concentration of food required below which grazing does not occur). Interestingly, the thin layers of zooplankton and these of algae sometimes do not overlap, resulting in complex foraging jumps of zooplankters into the food layer and back [54, 64, 65].

There is a growing body of evidence that the active foraging of zooplankton on different scales should be incorporated into plankton models (e.g. [14, 65]). However, each ecological model is a simplification of reality and it is impossible for it to encompass all details on the movement of individuals. On the other hand, quite often we simply do not need to describe the individual behavior on a microscale, when, for example, we are interested in the functioning of the whole plankton community. As such, the problem of transition between the modelling scales arises: processes taking place on a finer scale should be implicitly incorporated into a model operating on a larger scale [2, 17, 32, 59]. As a result, the model on a larger scale can be considered as a mean-field model operating with the average characteristics (e.g. the mean species densities, food concentration, etc.). An interesting and practically relevant problem is how to implicitly include active foraging behavior of individual zooplankters on a smaller time and spatial scale (i.e. without using a fine spatial and time resolution as well as a detailed description of interaction between organisms) into a coarse-scale plankton model.

In this paper, I shall address two issues related to including active foraging behavior of zooplankton in models and scaling them up. Firstly, I will consider the zooplankton functional response on different spatial and temporal scales and I will show that the shape of the emerging global functional response of a community can be substantially altered from to the local response of a single individual. Secondly, I shall model the role of intra-population variability in the feeding behavior of grazers in the persistence of the whole population under predation pressure. I find that structuring of the population according to feeding behavior can enhance the population persistence in eutrophic environments (characterized by a high nutrient load) thus preventing species extinction.

The paper is organized as follows. In Sect. 2, I compare the Eulerian and the Lagrangian approaches in the modelling of herbivorous zooplankton. In Sect. 3, I provide two general definitions of the zooplankton functional response based on the Eulerian and the Lagrangian frameworks, and discuss their applicability. Then I demonstrate the emergence of Holling type III (sigmoid) global functional response from a non-sigmoid local response. Section 4 is devoted to the modelling of the role of behavioral structuring on the survival of a population of grazers. Finally, in Sect. 5, I provide a general discussion on the incorporation of foraging behavior in plankton models and consider possible applications of our results for some other (non-planktonic) ecosystems.

2   The Lagrangian vs. the Eulerian Approach
in the Modelling of Zooplankton Dynamics

When modelling zooplankton dynamics in the water column, a critical issue is to choose an adequate modelling framework. In ecological modelling there exist the two main approaches: the Eulerian and the Lagrangian frameworks. According to the Eulerian approach, the distribution of organisms in space is regarded as continuous and is described in terms of the population density. The Lagrangian models are known as well as individual-based models (IBMs) where each individual (or a homogeneous group of individuals or super-individual) is explicitly modelled as a discrete entity [46, 47, 104]. Thus each individual/group is described by a set of variables (e.g. age, filtration rate, size, nutrition condition, etc.), and the behavior of an organism/group is governed by a set of prescribed rules. The dynamics at the population level emerges as a result of interactions of a huge number of individuals and their environment [4, 15, 63, 65]. Note that currently there is a tendency in the literature to implement the IBM framework when modelling zooplankton.

Each of the two modelling approaches has its advantages and disadvantages. A general discussion and comparison between the two approaches in theoretical ecology and, in particular, in plankton modelling, should be a matter of separate discussion (e.g. [46, 121]). An advantage of IBMs is the possibility of a more detailed description of the behavioral aspects of organisms as well as heterogeneity of physiological traits within populations. Thus, the central idea is obtaining the population dynamics from the first principals, i.e., by describing the life and feeding cycles in all possible mechanical details. The Lagrangian approach allows us to include complex movement of animals more easily than the Eulerian approach, especially when the movement of each individual is not synchronized in space and time. An important example is the unsynchronized vertical migration of zooplankton characterized by a constant short-term exchange of organisms between the surface and deeper layers, with only little change in the vertical profile of zooplankton as a whole [21, 65, 77]—the Lagrangian-based framework allows us to model the situation when the grazing of a zooplankter is not just a function of the ambient food any more, but is a reflection of the physiological condition of the organism [65].

Implementation of the Lagrangian approach has some disadvantages, however. An important shortcoming of IBMs is that we are not able to describe the behavior of a zooplankter on the individual level in full detail—this behaviour is still poorly understood. A typical IBM depends on a large number of un-measurable parameters, and in such a situation, including or omitting some features in feeding strategy on a microscale (individual level) can result in a large error on a macroscale (population level). As a result, the central idea of IBMs—to obtain emergent population dynamics from first principles—becomes seriously undermined. It is to be noted that the number of herbivorous zooplankters in the water column is usually rather large ($>10^3$–10^4 inds. per square meter) and this would require a large number of state variables describing all the organisms, incoming a large computational cost.

The problem becomes practically unsolvable when we are interested in modelling the dynamics of a planktonic metapopulation inhabiting an area with a horizontal dimension of dozens of kilometers (or considering the regional scale). In this case, the classical density-based approach can be more natural.

Interestingly, as it has been shown in theoretical ecology, the complex behaviour of animals on an individual level can be included on the population level via density dependant models based on the Fokker-Planck formalism [18, 40, 44, 119]. Note however, that the resultant equations can differ from the classical reaction-diffusion-advection type equations (e.g. [119]). On the other hand, there also exist standard techniques for incorporating a non-heterogeneous life trait distribution within a population of grazers, as well as the age structure of the population in density-based models ([69], see also Sect. 4 of this paper). In particular, complex interactions between Daphnia spp. and phytoplankton can be successfully described based on physiologically structured models ([29] and the references therein). Finally, the feeding cycles of zooplankton, including periods of active grazing and digestion, can be incorporated into simple density-based models ([76,77]). In this paper, I shall use the density-based (Eulerian) approach when modelling interactions in planktonic communities; however, when suggesting a general definition of the zooplankton functional response (Sect. 3.1), I shall discuss implementation of both the Eulerian-based and the Lagrangian-based frameworks.

3   Modelling and Scaling the Zooplankton Functional Response

In theoretical ecology the functional response of a predator/grazer was initially defined as the specific consumption rate of food by an individual per unit of time [56, 110]. Later on, it was well recognized that such a definition depends on the time and space scales under consideration [20, 32, 75, 99]. In plankton ecology the importance of spatial and temporal scales in feeding is less well recognized, for instance, than in terrestrial ecology. Conventionally, a zooplankton functional response is determined based on experimental feeding of organisms in microcosms. Tremendous amounts of literature exist on this topic showing that the feeding rate of a zooplankter in laboratory settings can be well described by a certain function of food which is referred to as a functional response ([38, 101] for a review). However, the direct interpretation of microcosm plankton experiments in ecosystem models on a larger scale is tricky and not always possible (e.g. [79]). This is mostly related to the two following aspects. Firstly, the environment in which species interactions take place is highly heterogeneous, thus the question of correct averaging arises. Secondly, the foraging cycles of grazers imply periods of active consumption and periods of rest (digestion) and those periods are often characterized by different food densities [21,65]. As a result, grazing and digestion can be separated in space. Thus, the conventional definition of the functional response, based on the assumption of a homogeneous small-sized (laboratory) environment needs to be refined.

3.1 Defining the Zooplankton Functional Response in Real Ecosystems

The existence of the zooplankton functional response on different temporal and spatial scales is a fundamental issue for modelling and to address this issue one should provide a rigorous definition of such a response. Below I suggest two definitions based on the Eulerian and Lagrangian frameworks.

(i) The Eulerian-based definition. Consider a certain domain Ψ which is a part of an n-dimensional habitat ($n = 1, 2, 3$). We are interested in the amount of food $E_{T,\Psi}$ that individuals belonging to the given species consumed within this domain during the observation time T. The $E_{T,\Psi}$ quantity can be re-written as

$$E_{T,\Psi} = \langle E(t) \rangle_{T,\Psi} = \frac{\langle E(t) \rangle_{T,\Psi}}{\langle Z \rangle_{T,\Psi}} \langle Z \rangle_{T,\Psi} = F \langle Z \rangle_{T,\Psi} , \qquad (1)$$

where $E(t)$ is the instantaneous rate of food consumption and $Z(t)$ is the instantaneous biomass of predators in the domain Ψ. Thus, to compute the total consumption of food in Ψ over time T by the predators one needs to multiply the biomass $\langle Z_{T,\Psi} \rangle$ the predators and the quantity F, which is mathematically a functional (i.e. a function of functions) since its value depends on the spatial distributions of species. The $\langle \rangle$ symbol denotes averaging

We shall define F as a functional response of predators in the case where the consumption of food can be described as

$$E_{T,\Psi} = F \left(\langle P \rangle_{T,\Psi}, \langle Z \rangle_{T,\Psi} \right) \langle Z \rangle_{T,\Psi} , \qquad (2)$$

up to the necessary degree of accuracy required for a given model. In other words, we require that F should be a function of the total amount of food $\langle P \rangle_{T,\Psi}, \langle$ in the domain. The size of the domain Ψ and the period of time T in the above definitions depend on the modelling purposes. In the limiting case, when the volume of Ψ and T tend to zero, we obtain the "local" functional response $F = F(P(\vec{r}), \vec{r})$, i.e. consumption of grazers in a given space point at a given moment time. The concept of the local functional response is implemented in most PDE-based models in oceanography [48,89]. In the other limiting case Ψ represents the whole euphotic water column and T is approximately equal to one day and the conventional modelling framework at those scales is the "classical" mean field plankton models ([31,34]).

(ii) The Lagrangian-based definition. According to the Lagrangian framework, we do not consider a fixed spatial domain. Instead, we follow the trajectories of the individuals though their paths. Thus, for the consumption rate of N individuals we obtain

$$E_{N,T} = \sum_{i=1}^{N} \langle e_i(t) \rangle_T = \frac{NB}{NB} \sum_{i=1}^{N} \langle e_i(t) \rangle_T = F_1 \cdot Z , \qquad (3)$$

where B is the average biomass of an individual; Z is the total biomass of N zooplankters; ei is the instantaneous consumption rate of individual i and the symbol $\langle \rangle$ now denotes averaging of the consumption rate along the path of a zooplankter.

We can define the functional response F_1 in the case the consumption rate of the whole population can be computed (up to the necessary degree of accuracy) as the product between zooplankton biomass and the average food concentration in the habitat, i.e.

$$E_{N,T} = F_1 (\langle P \rangle_T, \langle Z \rangle_T) \cdot Z , \qquad (4)$$

Unlike the Eulerian approach, the Lagrangian-based functional response is a function of the food density averaged over the layers where organisms mostly graze. The use of (4) requires the knowledge of individual paths of zooplankters and their grazing rates along those paths. One of the techniques for computing foraging paths uses individual-based modelling (IBM).

The question of applicability of the above definitions (2) and (4) is a matter of much discussion in the literature. For instance, it has been frequently observed that the local Eulerian-based functional response does not exist at all in natural plankton communities. In other words, very often there is no apparent correlation between the ambient food density and the ingestion rate of copepods [11, 25, 77, 112, 116]. This is not only the result of pronounced environmental noise but is due to the fact that the locations of the active food consumption and those of the rest can be different. For instance, a large density of zooplankton in layers with poor nutrition conditions can be explained by the fact that organisms migrate to those layers for digestion or to avoid predators [25, 65].

Note that the Lagrangian-based definition (4) can provide a better fit to the field data than the Euler-based definition, which can be seen from the following illustrative example based on feeding data on Calanus spp. in situ (the Central Barents Sea, 2003–2005). All details regarding the collection of material and methods can be found in [79]. Figure 2a shows the local functional response based on the Eulerian framework (the ingestion rates are plotted against the ambient food density), whereas Fig. 2b represents the functional response based on the Lagrangian definition constructed using the same data set.

To construct the Lagrangian-based response, we need to know the exact depths where the organisms are grazing for food before collection. Although we normally ignore those depths, we can try to reconstruct Lagrangian-based response by proceeding in the following way. Zooplankton samples were collected in three separate layers (0–20; 20–50; 50–100 m) and at each depth we considered the ingestion rate averaged over all individuals. I used the assumption that the ingestion rate of an individual is an increasing function of food density. Based on this assumption, I compared ingestion rates I_i and I_{i+1} ($i = 1, 2$) in each pair of adjacent layers with the average chlorophyll densities P_i and P_{i+1}, respectively. In the case where $I_i \leq I_{i+1}$, but $P_i \geq P_{i+1}$ I suggested that organisms caught in

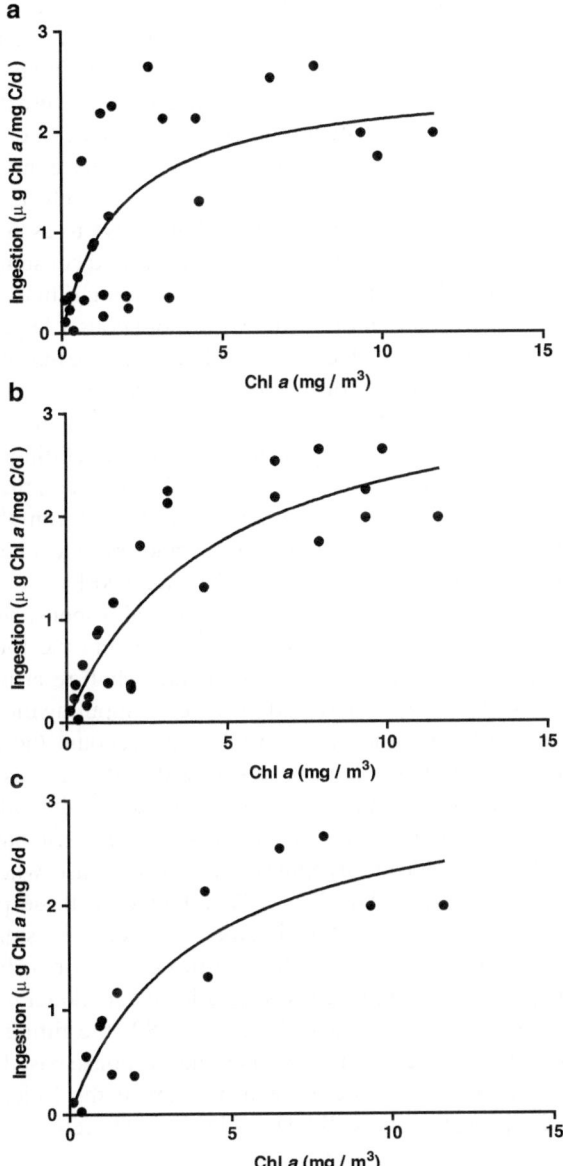

Fig. 2 Functional responses of herbivorous copepods (*Calanus Finmarchicus*, CIV, Central Barents Sea, 2003–2005) measured in situ. (**a**) Local functional response constructed based on the Eulerian framework, i.e. ingestion rates are plotted against the ambient food density. (**b**) Local functional response constructed based on the Lagrangian framework, i.e., ingestion rates are plotted against the food densities, where organisms were feeding the last time before capture. (**c**) Local functional response constructed considering only the actively feeding zooplankton. The fitting curves are obtained based on nonlinear regression (LSM), using the Monod curve as fitting functions. For details on constructing the functional responses see the text

layer $i + 1$, in fact, consumed their food in layer i. In this case, I considered that I_{i+1} corresponds to the density P_i. Alternatively, for $I_i \geq I_{i+1}$, but $P_i \leq P_{i+1}$ I suggested that the actual consumption by organisms caught in layer i was in layer $i + 1$, i.e. that I_i corresponds to P_{i+1}. Moreover, in the case where the ratio I_{i+1}/I_i was close to unity, but P_i was substantially larger than P_{i+1}, I considered that the actual grazing of the organisms caught in layer $i + 1$ took place in layer i. The field observation shows (see [79] for details) that $P_2 > P_3$ for each station and, thus it is easy to prove that the above described algorithm allows to assign I_i to P_j in a unique way. In other words, we assumed that the rate of food consumption was close to linear at low chlorophyll densities (up to $P = 4$–5 mg/m^3 Chl a). The biological justification for the above assumptions is to avoid anomalously large ingestion rates in layers with small food density. Overall, I should emphasize that such a simplified method can give us only estimates of the actual 12 locations (stations) of grazing. We performed the statistical treatment of both functional responses in Fig. 2 based on the least square method (LSM) using the Monod curve as a fitting function, which gives $R^2 = 0.46$ for Fig. 1a and $R^2 = 0.78$ for Fig. 2b. Based on the comparison of R^2 as well on the fact Fig. 2b shows less scattering of points from the fitting curve, one can conclude that the Lagragian framework would provide a better description of zooplankton functional response than the Euler framework.

Finally, when constructing the zooplankton functional response one can take into account only those consumers which are currently grazing the food and exclude those ones which are digesting food at the moment. I shall refer to those grazing individual as the actively feeding zooplankton. When computing the grazing impact of a zooplankton population, one needs to take into account the contribution of only those feeders. A major problem, however, is that it is almost impossible to distinguish between actively feeding and resting animals when collecting samples [21]. Despite this fact, we can try to reconstruct such a response based on the data set from Fig. 2a. Here I used the hypothesis that in the case where $I_i \leq I_{i+1}$, but $P_i \geq P_{i+1}$ or $I_i \geq I_{i+1}$, but $P_i \leq P_{i+1}$ I ignored those points. In other words, I ignored anomalously large ingestion rates in layers with small food density suggesting that those organisms do not feed in those layers but digest food. The resultant graph is presented in Fig. 2c showing a local functional response of with less scattering of points than Fig. 2a with $R^2 = 0.80$ (the fitting curve was the Monod function). The approach using the functional response based on the actively feeding zooplankton can be considered as a mixture of the Euler-based and the Lagrangian-based approaches and will be used in the next section.

3.2 Emergence of a Sigmoid (Holling Type III) Overall Zooplankton Functional Response

A number of plankton models ignore the explicit vertical resolution and consider the species densities averaged over the column. To describe the grazing of herbivorous, one needs to consider an overall/global functional response in the entire euphotic zone, i.e. to scale up the local/microscale functional response. Interestingly, the overall functional response can be of different Holling classification type compared to the local response and this is a result of the active foraging behavior of zooplankton (I show this below). In particular, an accelerating overall functional response (Holling type III) can emerge from a non-sigmoid (Holling type I or II) local responses.

The overall functional response of zooplankton in the column can be constructed based on the definition (2), where the domain Ψ includes the whole euphotic zone. However, to avoid the situation shown in Fig. 2a, I shall take into account only the actively feeding zooplankton and denote the vertical distribution of such zooplankters by $Z_a(h)$, where h is the depth. Note that the profile of $Z_a(h)$ can be rather different from that of $Z(h)$ which is the total (bulk) zooplankton density since the latter includes also individuals which are currently not feeding (e.g. digesting) [25, 76, 77]. Let us suppose that the instantaneous consumption of the actively feeding zooplankton can be described via the local functional response $f(P)$, where P is the local density of food (phytoplankton). The overall functional response will be defined as

$$F = \frac{1}{Z_0 H} \int_0^H f(P(h), h) Z_a(h) dh \, , \tag{5}$$

where Z_0 is the total amount of zooplankton, H is the total depth of the euphotic zone where phytoplankton can grow.

The actual distribution of the actively grazing feeders in the column is a matter of much discussion in the literature [11, 25, 77, 116]. In this work, I shall assume that the distribution of the actively feeding zooplankton in the water column is an ideal free distribution. Some field evidence of an ideal free distribution of grazing zooplankton can be found in [43, 61, 80]. In the simplest case, one can suggest that the distribution of feeders follows the distribution of food

$$Z_a(h) = \frac{P(h)}{\langle P \rangle} \cdot Z_0 a \, , \tag{6}$$

where $\langle P \rangle$ is the spatial mean density of the phytoplankton; $Z_0 a$ gives the total amount of actively foraging animals. Note that one can also take into account possible interference between grazers in the column which can be parameterized by

$$Z_a(h) = \frac{P^\mu(h)}{\langle P^\mu \rangle} \cdot Z_0 a \, , \tag{7}$$

where μ is a parameter describing the strength of interference of grazers. Some theoretical background for parameterization (7) can be found in [83] (see also [62, 111] for other possible parameterizations). In particular, $\mu > 1$ means a larger degree of interference of predators compared to the "classical" ideal free distribution; the situation with $\mu < 1$ would signify a lesser degree of competition among the foragers in patches with high food density. Note that some field observations and experimental studies in plankton towers show that the ideal free distribution provides a suitable approximation of real profile patterns of the actively feeding zooplankton [26, 53, 61, 80]. I do not take into account a possible time lag between the changes in profile of chlorophyll and the response of zooplankton to such changes. A large delay in response of zooplankton to changes of chlorophyll profiles would not be realistic in real ecosystems since changing in the vertical profile of phytoplankton takes from some days to a week while the active vertical displacement of zooplankton within a 100 m layer takes 6–10 h [10, 21, 90].

The dynamics of phytoplankton in the water column is described by the following partial differential equation

$$\frac{\partial P}{\partial t} = D\frac{\partial^2 P}{\partial t^2} + r_0 \cdot \exp\left(-\lambda h - \gamma \int_0^h P(h)dh\right) P\left(1 - \frac{P}{K}\right) - Z_a \cdot f(P) \,, \quad (8)$$

where the first term in (8) gives the random vertical displacement of phytoplankton due to turbulent diffusion in the column; the second describes the algal growth and the last term stands for the local grazing. The coefficient r_0 is the maximal per capita algal growth, which depends on the availability of nutrients; the exponential multiplier describes the light attenuation due to absorption by water and because of self-shading; $f(P)$ is the local functional response of herbivores, K is the carrying capacity taking into account mutual interference of algae. To parameterize the local functional response, I use the "classical" hyperbolic (Monod) parametrization [42]

$$f(P) = \frac{aP}{1 + bP} \,, \quad (9)$$

where a and b are the coefficients with an obvious meaning. Note that this type of response has been found for most herbivorous zooplankton in laboratory experiments ([27, 50, 55, 58] see also Fig. 2b,c). I assume that the total amount of zooplankton in the water column $Z_0 = const$ on the considered time scale. I also neglect the diel regular vertical migrations which would highly affect the ideal free distributions (6) and (7). I consider (8) with the zero-flux boundary conditions. To obtain a continuous range of $\langle P \rangle$, one needs to vary a control model parameter. I have chosen r_0 as the control parameter. This would allow the modelling of the occurrence of an algal bloom arising as a response to an increase of water temperature, light intensity, etc. [115].

Figure 3 shows the overall functional responses as functions of the average amount of phytoplankton $\langle P \rangle$ calculated for the local response of Holling type II with a large half-saturation constant ($1/b \gg 1$) (a) and with a small half-saturation constant (b). The functional responses are constructed for different intensities of interference of grazers μ. I consider realistic model parameters from the literature [7, 31, 50, 101] giving $0.5 < r < 21/d$; $0.01 < a < 0.3\ d/(\mu g\,Cl^{-1})$; $0.005 < b < 0.2\ \mu g\,Cl^{-1}$; $0.005 < \lambda < 0.15\ 1/m$; $0.0005 < \gamma < 0.0051/(m\,\mu g\,Cl^{-1})$; $D = 1m^2/d$. The local responses are shown by dashed lines. One can clearly see from the graphs the emergence of a sigmoid overall

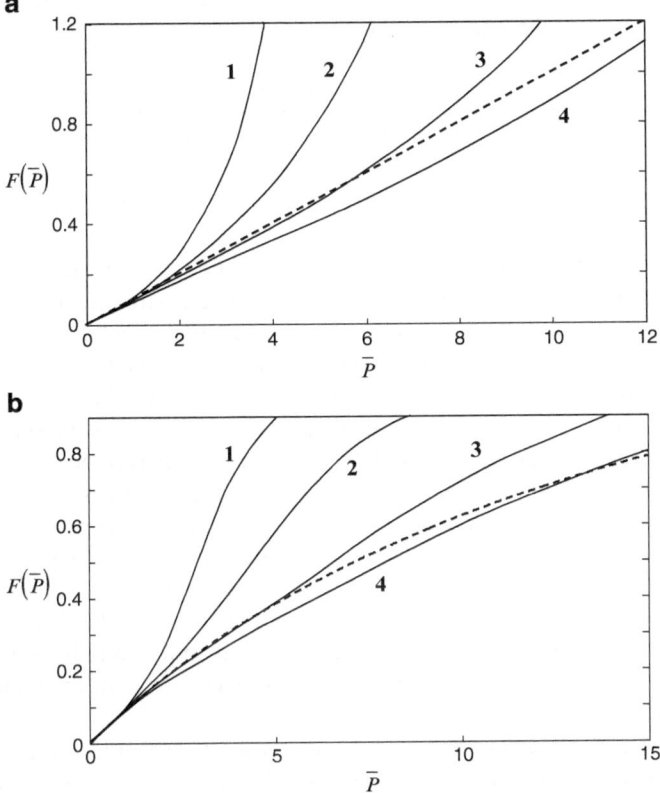

Fig. 3 Overall/global zooplankton functional responses of the zooplankton population in the entire euphotic zone constructed for varying degree of strength μ of grazer interference. The curves 1–4 correspond to $\mu = 0.8$; $\mu = 1$; $\mu = 1.2$; $\mu = 1.35$, respectively. (**a**) Overall functional responses obtained for Holling type I (linear) local response; (**b**) Overall functional responses obtained for Holling type II local response ($b = 0.06$). In both figures the overall response is shown by *bold curves*; local functional responses are depicted by *dashed lines*. The other parameters are $D = 1\ \mathrm{m^2}/d, a = 0.1\ d/\mu\mathrm{g\,Cl^{-1}}; H = 100\ \mathrm{m}; Z_{0a} = 1\ \mu\mathrm{g\,Cl^{-1}}$

functional response (Holling type III) from the local non-sigmoid (Holling type II or I) response having a concave downward part. Such an alteration of types of responses requires a small saturation in the grazing rate and a small degree of grazers' interference ($1 \geq \mu$).

The self-accelerating behavior of the overall functional response shown in Fig. 3 can be proven analytically as well. In Appendix A I demonstrate that in the case the diffusion term is small compared to the local growth rate and the grazing term, the overall functional response can be approximated by (21), which, however, results in a rather cumbersome explicit expression (21) is obtained for the a linear local functional response). This expression can be simplified depending on the magnitudes of λ, μ (see Appendix A). In the simplest case (no interference of grazers, $\mu = 1$) the overall functional response is given by (26)–(28). By taking into account the first three terms in the Taylor expansion for $F(\langle P \rangle)$ and obtain

$$F(\langle P \rangle) \approx \alpha\lambda \left(\frac{1 + \exp(\lambda H)}{2(\exp(\lambda H) - 1)} \langle P \rangle + \frac{\gamma H^2 \langle P \rangle^2}{6} + \frac{1 + \exp(\lambda H)}{24 \exp(\lambda H) - 1} H^3 \gamma^2 \langle P \rangle^3 \right),$$

(10)

Based on (10) one can prove that $F'(P) - PF(P) > 0$, which is the stability condition for predator-prey interactions in a eutrophic environment [87].

It is possible to come up with a simple (but not mathematically strict) explanation of the observed emergence of Holling type III functional response. Figure 4a shows the vertical distribution profiles of distribution of actively feeding zooplankton (plotting the ratio $Z_a(h)/Z_{0a}$; the vertical distribution of phytoplankton is the same) constructed, for the sake of simplicity, for $\mu = 1$ (no interference of grazers). An increase in the total amount of phytoplankton $\langle P \rangle$ leads to a sharper gradient of algal distribution (because of algal self-shading). The distribution of grazers $Z_a(h)$ follows that of the food and it results in a larger proportion of zooplankton feeding in food-rich layers, thus increasing the total consumption rate. Note that the emergence of an overall sigmoid functional response due to the above mechanism is possible in the case of a pronounced depth (deep waters), but is impossible for small H since the distribution of plankton becomes more homogeneous and, thus, closer to the local response.

The interference between the grazers (increase in μ) would impede the above alteration between the types of responses. This can be obtained directly from expression (27) for $F(\langle P \rangle)$ found in Appendix A for $\lambda = 0$. Differentiation of those expressions shows that $F'(P) - PF(P) < 0$ for large μ, thus not satisfying the stability condition [87]. However, the impact of interference of the grazers can be better understood directly from Fig. 4b where the distribution of actively feeding grazers (the ratio $Z_a(h)/Z_{0a}$ is shown for the same total amount of phytoplankton in the system ($\langle P \rangle = 4 \ \mu g C l^{-1}$). One can see that the competition between the grazers results in homogenization of the vertical distribution of active grazers, thus the local functional response is approached. Interestingly, the overall functional response can be even slightly smaller than the local one (see Fig. 3b) despite the fact that a substantial part of the active feeders are located in food-rich surface layers.

It is important to stress here that the emergence of a Holling type III overall response due to active food searching behaviour of grazers has observational background. In particular, it was found that zooplankton species which exhibit non-sigmoid functional response under laboratory conditions show a different overall functional response in real ecosystems [76, 79]. In particular, increasing the total amount of phytoplankton in the system can result in displacement of zooplankton towards surface layers of high food concentration, with feeding taking place mostly in those layers (see [5, 79].

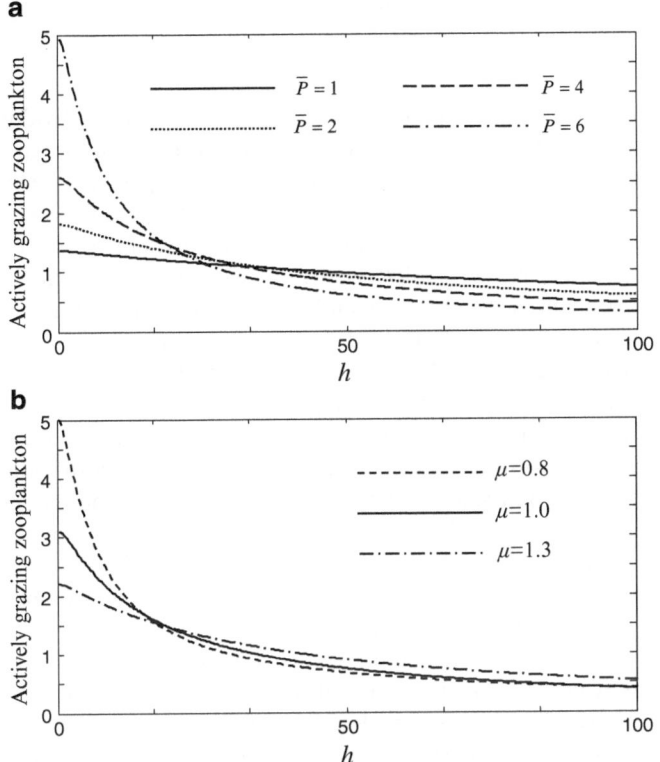

Fig. 4 (**a**) Mechanism of emergence of a sigmoid zooplankton functional response. Stationary vertical distributions of actively feeding grazers ($Z_a(h)/Z_{0a}$) are shown for different total amounts of food $\langle P \rangle$ ($\mu = 1$, no interference of grazers). An increase in $\langle P \rangle$ results in a sharper gradient of food distribution, the feeders follow the distribution of food and move for feeding to food-rich surface layers. The spatial distribution of phytoplankton is proportional to that of zooplankton and is not shown in the figure. (**b**) The influence of interference of grazers on the consumption rate. Vertical profiles of actively feeding grazers are shown for the same total amount of phytoplankton ($\langle P \rangle = 4\,\mu g\,Cl^{-1}$) for different μ. Enhancing the competition of grazing (increasing μ) results in a more pronounced homogeneity in vertical distribution

4 The Role of Intra-population Variability of Zooplankton in Population Persistence

In this section I address another important issue related to the active feeding behavior of zooplankton in the water column: the within-population variability of grazers and its role in the population survival and persistence.

4.1   Describing the Intra-population Variability
of Zooplankton Grazers

An important intrinsic property of any real population is that individuals forming
this population often differ from each other: organisms have various sizes, different
ability to move, and, finally, they can have different personal behaviour (e.g.
preference for staying in risky or safe environment, aggressiveness, etc.). There
exist a large number of theoretical works considering dynamics of such structured
populations ([24, 29, 69, 73, 74, 97, 118]). It has been shown that taking into
account intra-population difference would seriously alter modelling outcomes. A
proper review on models of structured populations and a comparison with their
unstructured analogues should be done elsewhere. In most previous publications,
however, the authors have considered population structuring with respect to the
age or size of individuals or due to some physiological traits. Less studied are
population which are structured according to different behavior of individuals (but
see [96]). In this section I shall construct a simple model combining physiological
and behavioural structuring of a population of grazers regulated by top predation
(carnivorous zooplankton or planktivorous fish).

Zooplankters are known to show a large interindividual variability in their
feeding patterns [92, 93, 108]. Figure 5 demonstrates a large variability in the
consumption rates of individual zooplankters (*Calanus spp.*) obtained in laboratory
(data provided by prof. E. Arashkevich and colleagues). In the figure, the individual
consumption rates of copepods are plotted for different temperatures but for the
same food density. One can see a large deviation in the consumption rates of
the grazers which can be as large as one order of magnitude. Another important
observation is that most of the individuals conserve their consumption character-
istics/traits for different environmental conditions (various temperatures), i.e. their
ability of consuming food at high, intermediate or low rates (see Fig. 5b). Thus, the
whole population can be described as physiologically structured. Interestingly,
the pronounced difference in food consumption does not seem to be related to
the variation in the individual sizes of organisms which were close to each others
(not shown result). Note also that zooplankters also show a large interindividual
variability in their swimming rates which could result in a large variation in the rate
of food consumption [108].

The observed individual differences in the consumption and swimming rates
of grazers could eventually result in an intra-population difference in foraging
behaviour. Indeed, zooplankton herbivores often migrate to the upper layers with
higher food abundance despite the predation risk [10, 65, 90]. Often the grazers
implement the eat and run strategy which consists in quickly filling the gut and
leaving the risky environment [95]. Since the ability of filling guts can substantially
vary from individual to individual, the individuals with high ingestion rate can leave
the surface layers faster than the others, thus spending more time outside the risky
environment [26, 107]. As such, variation in physiological traits within a zooplankton population can translate itself into different foraging strategies/behaviour and,

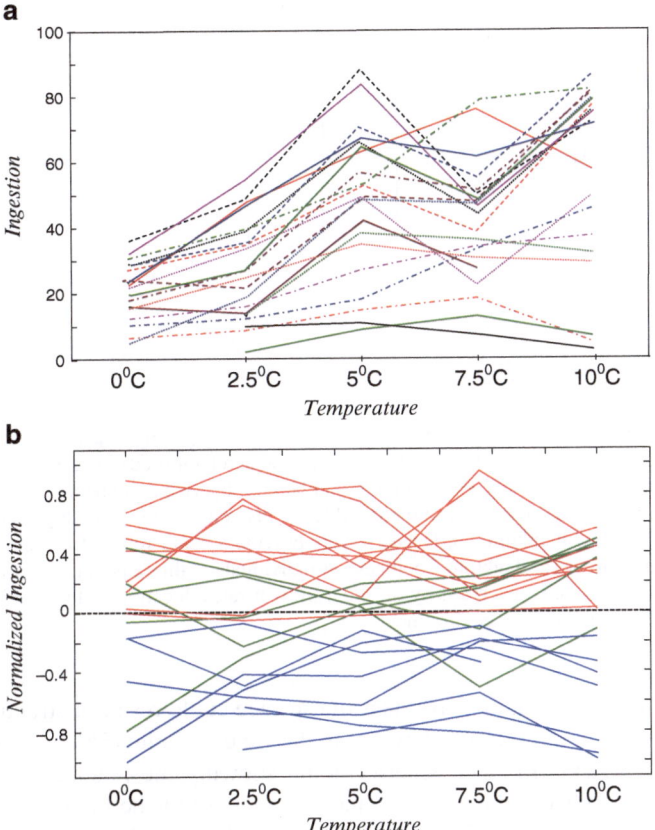

Fig. 5 Individual ingestion rates (mg Chl *a*/ind./day) of copepods Calanus spp. measured in the laboratory under different feeding conditions (different temperatures). Each *curve* describes the ingestion rate of a single individual. (**a**) The absolute values of ingestion rates. (**b**) Normalized ingestion rates compared to the mean value for the given temperature. *Red* and *blue curves* correspond to individuals with ingestion rates lying, respectively, above and below the population mean value. *Green curves* describe individuals with highly variable ingestion rates which can be both above and below the mean value. One *curve* crossing zero is depicted in *red* since it shows a persistent behavior very close to the mean values. Absence of points for certain individuals can be explained by the fact that the organisms are suggested to be depressed in those experiments and simply did not consume food. The data have been obtained by Prof. E. Arashkevich and colleagues

as a result, into segregation of organisms in space. A similar scenario of segregation of grazers within the same population has been found by Fossheim and Primicerio [37], where different copepodite developmental stages were separated in the column in the presence of top predators (fish). Note that similar differentiation in behaviour due to differences in physiological traits has been found for other non-planktonic species, such as fish [12,19], octopuses [71] and some mammals [100]. In particular, it was reported in a population of salmon that the interindividual variability in

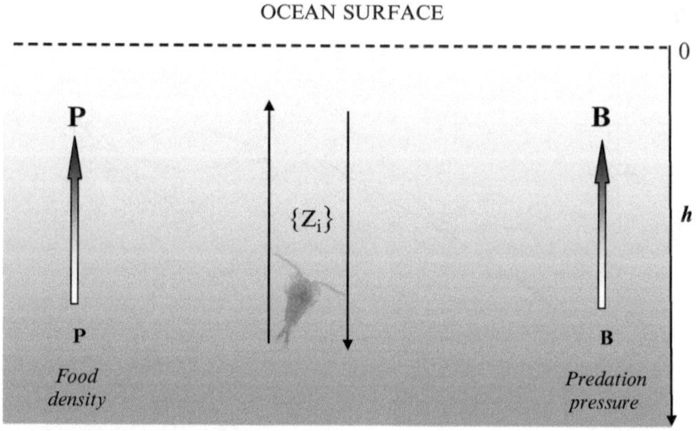

Fig. 6 Schematic diagram explaining construction of physiologically structured model (11)–(12) of trophic interactions between herbivorous zooplankton and their predator (carnivorous zooplankton and/or planktivourous fish). Predation on zooplankton by visual predators takes place mostly in the surface layers (the risky environment) which is also characterized by high food abundance (high density of phytoplankton P). Deeper layers provide a better refuge from the predators but are less abundant in phytoplankton. Zooplankton individuals within a population are divided into cohorts Z_i, which are characterized by different growth rates, location of feeding in the column and the time spent in the risky environment

willingness to take predation risk near the surface could result in structuring of the patterns of vertical migration behaviour in the water column [35].

In this paper, I suggest a generic model showing the potential role of intra-population variability in the life traits and behaviour of the herbivorous zooplankton in persistence and stability. Schematically, the model is depicted in Fig. 6. The food density (phytoplankton) increases in the layers towards the surface. At the same time, the efficiency of visual predators is higher near the surface, thus there is a trade-off between food density and the mortality due to predation risk. In the model, the zooplankton population Z is divided into n cohorts/groups (Z_i) with different behaviour. In particular, cohorts Z_i vary with respect to the amount time spent feeding in surface layers with high predation risk as well as in the depth of feeding. Thus, the mortality rate of zooplankton, which in the model is due to predation, becomes cohort-dependant. I consider that different cohorts can exhibit different growth rates due to the variation of time spent in food-rich layers.

The trophic interactions between the grazers and their predator (carnivorous zooplankton or planktivorous fish) are described via the following differential equations

$$\frac{dZ_i}{dt} = \sum_j w_{ij} R_j Z_j - g_i(Z_i)B , \qquad (11)$$

$$\frac{dB}{dt} = B\left(\sum_i \omega g_i(Z_i) - \delta\right) , \qquad (12)$$

where Z_i and B are biomass of the zooplankton in cohort i ($i = 1, n$) and the predator, respectively. The sum in (11) describes the growth rate of Z_i due to the reproduction of all cohorts; the contribution of cohort j to the growth rate of cohort i is described by the weight w_{ij}. I further call coefficients w_{ij} the demographic factors which I consider to be density independent. I require that the sum of the demographic factors w_{ij} over all cohorts should be equal to unity.

One would expect that $w_{ii} \gg w_{ij}$, i.e. the offspring of each cohort mostly belong to the same cohort. However, I do not formally impose such a restriction by considering as well the possibility of $w_{ii} \approx w_{ij}$. In this paper, I assume that the demographic factors are at genetic equilibrium, i.e. the demographic factors do not change in time (cf. [16]). The coefficient R_j describes the overall per capita growth rate of cohort j. For the sake of simplicity I consider that all R_j are constant, i.e. no intraspecific competition. Such an assumption allows to model plankton dynamics in euphotic environments.

The parameters describing the predators of zooplankton are: the functional response $g_i(Z_i)$, which is different for different cohorts; ω is the food utilization coefficient and Δ is the mortality rate of the predator. For the sake of simplicity, I consider the functional response g_i with saturation of Holling type II given by the Monod parametrization

$$g_i = a_i \frac{Z_i}{1 + b \sum_j \frac{a_j Z_j}{a}} , \qquad (13)$$

where b is the coefficient characterizing the saturation of predation at high densities of Z_i. The coefficients a_i, which are proportional to the attack rates, are different for different zooplankton cohorts. I suggest that ai are larger for those cohorts where the individuals intentionally stay longer in more risky part of the habitat (surface layers) with higher predation pressure. When including the effects of saturation, I take into account the fact that the actual amount of zooplankton which is available for predation in surface layers should be multiplied by certain weights. Those weights would model the relative duration of the zooplankton cohorts stay in the more risky environment, thus they should be a function of the predator attack rate. I suggest that such weights are proportional to the attack rates, since the relative difference in a_i characterizes the relative time spent in surface layers; a denotes in (13) the average values of a_i. I should note that our findings remain qualitatively the same in the case of the 'classical' Holling type II response with the same weights, i.e. when removing a_j and a in the denominator of (13).

4.2 Analysis of the Model of the Intra-population Variability of Zooplankton

Model (12)–(13) has been intensively investigated both numerically and partially analytically. In this paper, I focus mostly on the case where there are only two zooplankton cohorts $n = 2$ and only briefly discuss how the patterns obtained for

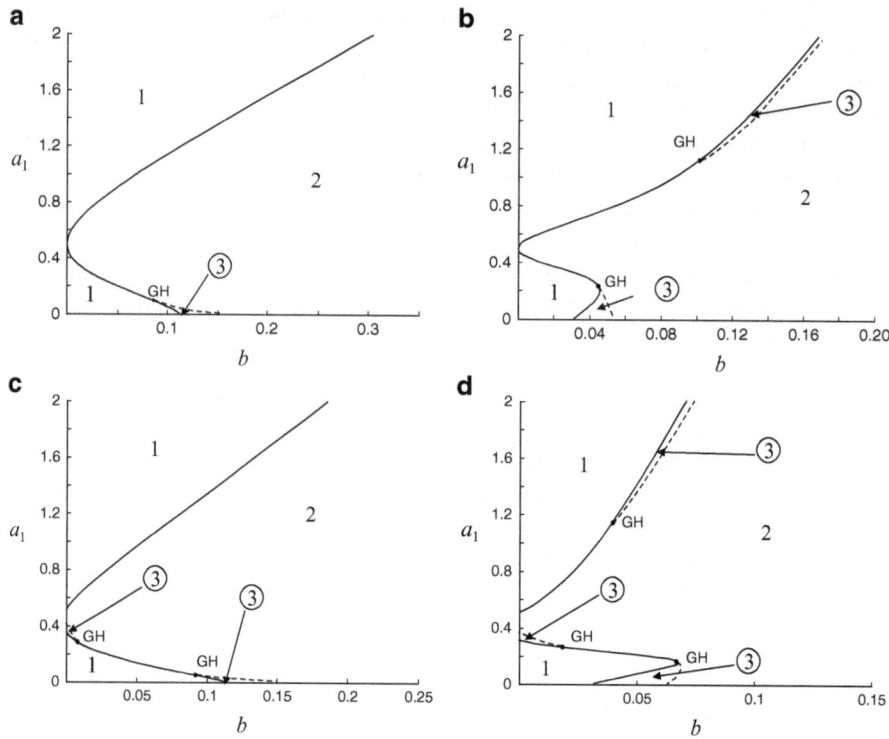

Fig. 7 Bifurcation diagrams in the $a_1 - b$ plane constructed for system (11)–(12) for the number of zooplankton cohorts $n = 2$. Dynamical regimes are described in the text. The system can be regulated in domains 1 and 3 (the solutions are bounded). The *solid curve* is the Hopf bifurcation curve. The *dot-dashed curves* represent limit cycle bifurcation curves. The points GH denote general bifurcation points. (**a**), (**b**) Equal per capita growth rates $R_1 = R_2 = 1$; (**a**) $w_{11} = w_{22} = 0.5$; (**b**) $w_{11} = w_{22} = 0.8$ (**c**), (**d**) Different per capita growth rates $R_1 = 1; R_2 = 2$. (**c**) $w_{11} = w_{22} = 0.5$; (**d**) $w_{11} = w_{22} = 0.8$. The other parameters are $a_2 = 0.5, \omega = 0.25, \delta = 0.1$

$n = 2$ will change with an increase in n. The stability of the stationary states of the system for $n = 2$ is addressed in Appendix B (for small b). In particular, I show that the system has a unique nontrivial stationary state which, depending on model parameters, can be either stable or unstable. Note that the system (11)–(12) becomes the classical predator-prey model in the case where all cohorts are equal. This system is always globally unstable for $b > 0$ [6, 87] and is neutrally stable (Lotka-Volterra model) for $b = 0$.

Figure 7 provides an insight into parametric portraits for $n = 2$. The diagrams are obtained with the help of the software MATCONT [28]. The diagrams are constructed in the (a_1, b) plane, the other parameters being fixed. Figure 7a,b describe the scenario when the two cohorts have equal per capita rates $R_1 = R_2$ but different attack rates $a_1 \neq a_2$. Ecologically, it signifies that the larger mortality due to predation (consequently, a larger time spent in surface layers) has no influence on the reproduction rate. In domain 1 the unique coexistence state (Z_1, Z_2, B) is locally stable, thus small perturbations of the state will eventually vanish. This stationary state, however, is not globally stable. It is surrounded by an unstable limit cycle: for initial species densities located far away from the state, the trajectories would go to infinity and other factors (lack of resources, competition) should limit the population growth. In domain 2 the nontrivial stationary state is globally unstable: all trajectories starting nearby will eventually go to infinity. In the rather narrow domains 3, trajectories unwind from the unstable stationary state to a stable limit cycle. This cycle is enclosed by an unstable outer cycle.

One can see from the Fig. 7a,b that the fact that the attack rates of predators on different cohorts are different can provide the stability of the whole population thus preventing it from extinction. Indeed, for equal attack rates by predators, the population will exhibit oscillations with gradually increasing amplitude, with the minimal species densities approaching zero. Thus producing individuals which are subjected to more predation can be beneficial for the whole population. Interestingly, even in the absence of saturation ($b = 0$), the Lotka-Volterra system can become stable. This fact is analytically proven in Appendix B. Figure 7a is constructed for the equal demographic factors ($w_{ij} = 0.5$). Taking into account a more realistic scenario when $w_{ii} > w_{ij}$—i.e. the offspring of each cohort mostly belong to the same cohort- results in shrinking of the stability domains 1 (see Fig. 7b, mind the difference between the scales on the b-axis).

I shall now consider a more realistic scenario, where increase/decrease in the attack rate a results into a relative increase/decrease in the growth rate R. This is shown in the diagrams in Fig. 7c,d constructed for $R_2 > R_1$. One can see that compared to the case with $R_2 = R_1$, the size of domains of stabilization (1, 3) have shrunk. Moreover, the diagrams predict that the stabilization will take place when the cohorts with the larger growth rates should have a smaller mortality rate, i.e. for $R_i > R_j$ there should be $a_i < a_j$. This conclusion is analytically justified in Appendix B ($b = 0$). On the contrary, for $R_i > R_j$ and $a_i > a_j$ and a small absolute difference between a_i and a_j the system will be globally unstable. Thus, the existence of a cohort with individuals dwelling more time in food rich (and predation risky) layers and having larger reproduction rates would be destabilizing. Finally, the stability of the system is restored for $R_i > R_j$ and $a_i > a_j$ in the case where the difference between a_i and a_j is sufficiently large. This corresponds to the lower stability domain 1.

It is natural to suggest that the parameters R_i and a_i are not independent but related via a certain trade-off function $a_i = a_i(R_i)$ which has the same functional from $a = a(R)$ for all cohorts. The parameters R and a are potentially related since they both depend on the amount of time spent in food-rich but predator risky environment. For the sake of simplicity consider first that $b = 0$. In the case $a(R)$ is a decreasing function, the coexistence stationary state in the model will always be stable regardless of the shape of this function. On the other hand, an increasing function a(R) may have a destabilizing effect. Interestingly enough, even if $a(R)$ is a decreasing function, the system's stabilization can be still possible provided the rate of decrease $a(R)$ is sufficiently large. This is related to the fact that for $a_i \rightarrow a_j$ and $R_i \rightarrow R_j$ the size of the instability interval on the a-axis vanishes. This can be seen from the comparison of the upper and the lower figures in Fig. 7 (see also Appendix B). Let us vary the parameter a_1 for a fixed a_2. I shall decrease R_1 starting from $R_1 = R_2$; it will result in an increase of a_1. It is shown that (Appendix B) in the case where the gradient of the trade-off function is larger than that of (45), the stability of the stationary state will be guaranteed. Taking into account the saturation in predation $b > 0$ will result in some changes. In particular, the stability of the stationary state will require a larger degree of scattering between a_1 and a_2 (larger absolute values of $a_i \rightarrow a_j$). Thus, even the existence of a fast decreasing trade-off function $a(R)$ will not automatically signify the stabilization of the system (see Fig. 7): a threshold value ofjiaa.should be exceeded in this case.

Note that the stabilization in the structured population of grazers requires that the physiological traits and behavioral patterns within a cohort remain constant. I shall refer to such structuring as the genetic structuring of a population. In contrast, there can be a temporal structuring within a population where the traits of each cohort vary in time due to some stochastic processes. As a result, the splitting of a population into cohorts occurs for a short time period (which is smaller than the individual lifespan time) after which the individuals swap between different cohorts. Model (11)–(12) predicts that stabilization takes place only for a genetically structured population. Figure 8a shows damped oscillations of a genetically structured population of grazers consisting of two cohorts with different attack rates of predator ($a_1 = 1$, $a_2 = 0.6$). The trajectory will tend to a stable stationary state. The situation is different when the population is temporally structured (Fig. 8b). In this case, instantaneously the population still consists of two cohorts with the same vulnerabilities of predation ($a_1 = 1$, $a_2 = 0.6$), however, every Δt time units there is a probability $1/2$ of exchange between the cohorts: individuals swap between the two cohorts. As a result, the oscillations of density become constantly increasing (Fig. 8b) which would result in a population collapse. In this case, the dynamics is equivalent to a non-structured predator-prey system with a Holling type II functional response, which is globally unstable.

Considering a larger number n of cohorts does not qualitatively alter the previous results on stabilization. In particular, bifurcation diagrams constructed for $n = 3, 4$ are qualitatively similar to those from Fig. 7. In general, stability of the coexistence state requires the existence of a trade-off function $a(R)$ with a supercritical slope at $a_i = a_j$ and $R_i = R_j$. Finally, when n is large, the discrete framework (11)–(12) based on the use of a system of ODEs should be replace by a continuous distribution of the life traits in the population suggesting an infinitely large number of cohorts and the model becomes transformed into a system of two integro-differential equations. In this case, the shape of $a(R)$ will play a critical role in the system stability along with the distribution of the demographic factors w_{ij}. Investigation of such a model will be a part of future research.

One important question concerning this section is which trade-off functions relating the predation risk and the growth rate of zooplankton will be realistic. Although there exists a large amount literature on this topic, this issue is a matter of

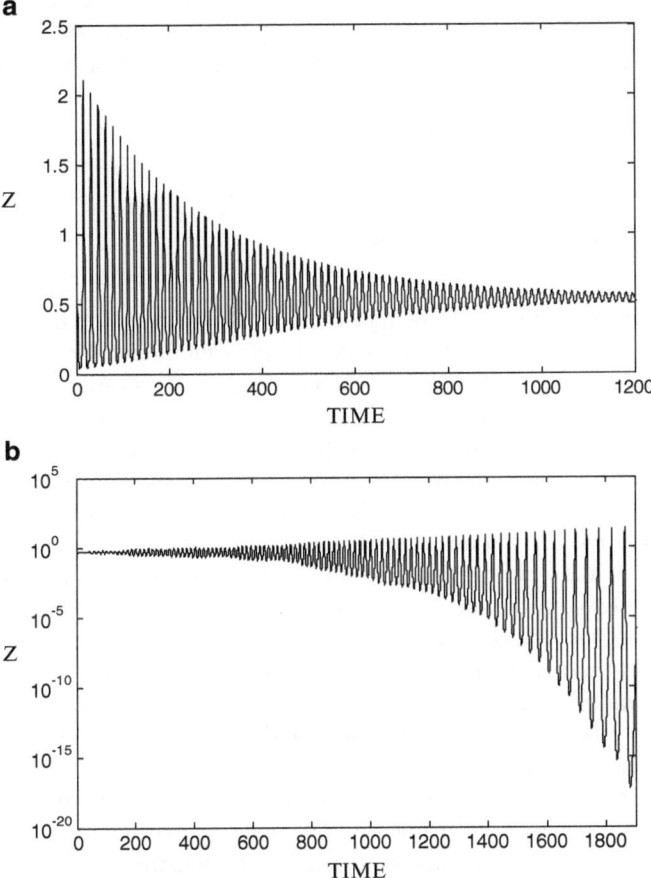

Fig. 8 Explaining the difference between the genetic structuring and the temporal structuring of a population ($n = 2$). (**a**) Genetic structuring: the vulnerability to predation for each subpopulation remain constant $a_1 = 1, a_2 = 0.6$. The total biomass of grazers Z exhibits damped oscillations and the trajectories tend to a stable stationary state. (**b**) Temporal structuring: at each moment of time the values a_i are different and equal to 1 or 0.6; however, every $\Delta = 0.5$ time units there is a probability of 0.5 of swapping between a_i and a_j. The other parameters are $R_1 = R_2 = 1$; $w_{11} = w_{22} = 0.5, a_2 = 0.5, \omega = 0.25, \delta = 0.1; b = 0.02$. The population of grazers will eventually attain very low densities and will collapse

much discussion [10,37,65,67]. Most of the conclusions are derived from modelling results or simply based on common sense (!). For instance, there is an opinion that those zooplankters which spend more time feeding in the risky surface layers should be be compensated by an increase in growth rate [41,57,107]. Some grazers which quickly fill their gut in the surface layer and run away to digest the consumed food would need to spend large amounts of energy on vertical migration, thus this energy will not be available for reproduction [65,95]. This would imply that $a(R)$ should be an increasing function. An important point, however, is that the time spent by zooplankters in the food-rich layer can be a poor indicator of the growth rate. Indeed, individuals who need to stay in surface layers more may simply be poor feeders with low growth rates. The major problem is that there is still a lack of data on the individual variability of zooplankton behavior in situ.

5 Discussion and Conclusions

Taking into account complex foraging behavior of zooplankton in the water column on different spatial and temporal scales is of vital importance for improving plankton models. Very often, however, one needs to incorporate movements of grazers implicitly, especially in models operating on large scales (e.g. the scale of the whole euphotic zone). Scaling up the grazing rate of zooplankters from microscales has its own particular features as compared to some other non-planktonic ecosystems. Firstly, there is a pronounced heterogeneity of the aquatic environment in the vertical direction due to the light attenuation with depth, water stratification in the column resulting in sharp turbulence, temperature, salinity gradients, patchy distribution of predators, etc. Secondly, the zooplankton grazers are usually fast moving organisms, so they are able to cover the whole euphotic zone in a short time period (hours) which is much smaller than the generation time of the population varying from several months to years. Thirdly, the movement of grazers is much faster than the characteristic rate of change of the spatial food distribution (phytoplankton or microzooplankton). Finally, the behavior of grazers depends on the part of the habitat where they are currently dwelling (e.g. between the surface and deep layers). As such, the conventional techniques of theoretical ecology of extrapolation of small-scale dynamics to larger scales (e.g. the aggregation approach, the scale transition approach, the modified mean-field approach) become inappropriate in this case (cf. [2, 17, 32, 94]). For instance, the scale transition framework becomes inefficient in the case where the environmental properties are substantially different [17]. Some other mathematical tools might be needed in this case.

Interpretation of laboratory experiments on zooplankton foraging in modelling can be tricky and should be done with care. For example, it has been nicely demonstrated that in laboratory settings some copepods exhibit swimming behavior which can be described as a fractal random walk (e.g. [105, 106, 108]). The scaling exponent of the motion was estimated [108] which could, in principal, allow us to extrapolate those results to larger scales. However, the characteristic size of the laboratory settings, was rather small (up to 1–2 m) and potential extrapolation of the results even to intermediate scales (10–30 m) looks problematic. In particular, the fractal random walk behavior cannot describe the eat and run strategy of copepods which is observed in reality on intermediate scales [65, 77, 95]. On such scales movement of animals includes persistent ascending and descending of grazers, i.e. a ballistic motion. Our general understanding is that the complex (multi) fractal food searchery observed on microscales would be only a part of the zooplankton feeding cycle. This goes along with the current understanding of movement in theoretical ecology. Indeed, it is now well recognized that patterns of animal movement can be often considered as a sequence of different phases/modes of movements [81, 86, 109], with each phase corresponding to a different type of activity of the animal, and potentially having different statistical properties. The sequential order of different phases on larger spatiotemporal scales gives the lifetime patch of an individual [86]. Active foraging of zooplankton is a good example of phase-based movement of animals.

In this paper I have considered two examples of the implicit incorporation of active foraging of plankton grazers into models. First, I considered the implementation of the zooplankton functional response on different scales. I demonstrated that scaling up the functional response to the size of the whole euphotic zone can result in alteration of the type of response and, as a result, a sigmoid (Holling type III) overall functional response can emerge from the local non-sigmoid local response. This fact is of importance since investigation of trophic chain models reveals that a Holling type III response usually enhances stability of eutrophic ecosystems [6, 87]. Interestingly, there is a vivid discussion in the literature on the adequacy of implementation of a sigmoid type of response in plankton models [78, 84, 102, 103]. The point is that it has been observed in experimental studies that for most herbivores the functional response is non-sigmoid, i.e. it is either of type I or II ([27, 50, 55, 58]) and, as a result, a rather strong opinion in the literature is that implementation of Holling type III in plankton models is biologically meaningless [27, 84, 103]. The results of Sect. 3 somewhat challenge this opinion and consider this problem from a different angle.

I suggest that the emergence of a sigmoid functional response when scaling up to macroscopic level would be observed for the carnivorous zooplankton as well due to a similar scenario/mechanism. Moreover, a similar alteration of the type of functional response resulting in the emergence of a sigmoid overall response from a local non-sigmoid response due to prey patchiness and predator aggregation has been found in another predatorprey system [85]. The ecosystem under study was an acarine predatorprey system involving the aggregation of predators in food patches. I predict that such a scenario can be found in other non-planktonic systems where there is a strong feedback between the spatial distribution of prey and its total abundance in the system resulting in a sharper gradient of spatial prey distribution (see Fig. 4a).

Another important result we have seen is that the interindividual structuring of grazers according to their behavior can enhance the persistence of the whole population and prevent the species from extinction (Sect. 4). Thus, variation in the behavioral strategy, which translates itself into a variation of time spent in dangerous parts of the environment, would be beneficial for the whole population. Stabilization of the system consisting of grazers and their predators (fish) in an environment with an unlimited carrying capacity requires two major conditions. Firstly, some life traits (in particular, the vulnerability to predation) of individuals should be genetically different, i.e. the mean values for a fixed individual should remain constant.

Secondly, in the case of high variation in the growth rate R among the individuals, there should be a certain trade-off between R and the vulnerability a. Stabilization is guaranteed when a larger growth rate signifies less vulnerability. In the opposite case, the stabilization can be still possible, but requires a supercritical value of slope of the functional dependence $a(R)$. I should admit, however, that the reported mechanism of the enhancement of persistence in plankton communities is still to be tested. The major problem arising here is that the spatial structuring of zooplankton (if it exists in reality) would take place on intermediate spatial scales and it is rather hard or even impossible to follow the trajectory of each zooplankter *in vivo* or in plankton towers. A possible method of experimental justification of the suggested mechanism could be to search for correlation between the vertical location of individual grazers in the column and certain physiological traits of those grazers (e.g. swimming speed, feeding rate, etc.) which can be revealed after sampling.

Note that our results on the stability of the dynamics of structured populations can be applied as well to some other non-planktonic predator-prey systems. Indeed, the key-factor assuring the system stability is the genetic difference in the vulnerability of individuals to predators. As such, in any population where individuals are characterized by a substantially different vulnerability to predation, the above stabilization mechanism could be realized. For instance, such a situation can be possible in the case of the existence of refuges for prey and structuring of the prey population according to the mean time spent in those refuges. Different attack rates on different prey individuals can be due to high genetic variation in physiological characteristics such as mobility, the ability of detecting the predator, etc. Finally, I should note that our result that physiologically and/or behaviorally structured populations are less prone to extinction is in a good agreement with previous theoretical works [96, 98] providing a somewhat different mechanism of stabilization based on intra-population competition.

Among the important challenges for further progress in understanding and modelling patterns of active zooplankton foraging behaviour I would like to highlight the investigation of intermediate scale processes. New studies should include the collection of plankton samples on finer scales. On the other hand, a proper mathematical framework for modelling the movement of plankton on those scales is still lacking. I suggest that such a framework should be based on a density-dependant (Eulerian) approach, but include complex behavioral aspects of the animals. In the absence of such a framework, the current tendency in the literature is the implementation of IBMs, and while I agree full-heartedly with the need for development of such models in marine ecology, I do argue that the need for implementation of IBMs still needs to be justified for herbivorous zooplankton which are characterized by large population numbers and patchy spatial structure. I strongly believe that the density-dependant framework can provide powerful modelling tools capable of efficiently incorporating complex patterns of foraging behavior and the variability of physiological traits within populations.

Appendix 1

Here I derive the expression for the overall functional response of zooplankton. The equation determining the vertical profile of phytoplankton is given by

$$\frac{\partial P}{\partial t} = D\frac{\partial^2 P}{\partial t^2} + r_0 \cdot \exp\left(-\lambda h - \gamma \int_0^h P(h)dh\right) P - \frac{HP^\mu f(P)}{\int_0^H P^\mu(h)dh} Z_0 , \quad (14)$$

where D is the diffusion coefficient of vertical turbulence. I consider that the ecosystem is eutrophic, thus $r = const$ and $K \gg 1$. We are interested in computing the stationary vertical profile of zooplankton, which is determined by

$$0 = D\frac{\partial^2 P}{\partial t^2} + r_0 \cdot \exp\left(-\lambda h - \gamma \int_0^h P(h)dh\right) P - \alpha\frac{P}{1 + \beta P}\frac{HP^\mu}{\int_0^H P^\mu(h)dh} Z_0 , \quad (15)$$

To be able to provide an explicit analytical expression for the functional response we need to do some simplifications. As such, I neglect the diffusion process which are small terms compared to the local growth rate and the grazing. Also, I neglect saturation in the local functional response. We obtain

$$0 = r_0 \cdot \exp\left(-\lambda h - \gamma \int_0^h P(h)dh\right) P - \alpha\frac{HP^\mu}{\int_0^H P^\mu(h)dh} Z_0 , \quad (16)$$

One can simply re-write (16) in the following way

$$\lambda h + \gamma \int_0^h P(h)dh = -\ln\left[\frac{\alpha H}{r_0}\frac{P^\mu}{\int_0^H P^\mu(h)dh} Z_0\right] , \quad (17)$$

After differentiating (17) with respect to h, we obtain the following differential equation

$$\lambda P + \gamma = -\frac{\mu P'}{P} , \quad (18)$$

Integration of (18) which is a Bernoulli equation gives the stationary profile of phytoplankton:

$$P(h) = \frac{\lambda}{C \exp(\lambda h/\mu) - \gamma} , \quad (19)$$

where C is an integration constant. By integrating (19) over the whole water column, we can express C as a function of the spatial average phytoplankton density $\langle P \rangle$.

$$C = \gamma \frac{\exp\left(\frac{\lambda H + \gamma H \langle P \rangle}{\mu} - 1\right)}{\exp\left(\frac{\lambda H}{\mu}\right)\left(\exp\left(\frac{\gamma H \langle P \rangle}{\mu}\right) - 1\right)}, \tag{20}$$

Thus the vertical profile of phytoplankton becomes a function of $\langle P \rangle$, i.e. $P = P(h, \langle P \rangle)$. The overall functional response of zooplankton in the water column is obtained from (5), which gives

$$F(\langle P \rangle) = \frac{\alpha H \int_0^H P^{\mu+1}(h)dh}{\int_0^H P^\mu(h)dh}, \tag{21}$$

with $P = P(h, \langle P \rangle)$ is given by (19) and (20). Note that by integrating (21) one can obtain the explicit expression for the functional response which is rather cumbersome. However, one can obtain tractable analytical expressions for F for some limiting cases.

(i) $\mu = 1, \lambda = 0$. This signifies that there zooplankton is distributed according to the simplest ideal free distribution law and absorption of the light by water is small compared to the self-shading. Integration and simplification of (21) gives following the functional response

$$F(\langle P \rangle) = \frac{\exp(\gamma H \langle P \rangle) + \exp(-\gamma H \langle P \rangle) - 2}{(\gamma H)^2 \langle P \rangle}, \tag{22}$$

It is easy to prove that the overall functional response (22) is of Holling type III since the stability condition $F'(\langle P \rangle) - \langle P \rangle F(\langle P \rangle) < 0$ [87] is always satisfied for $\langle P \rangle > 0$. One can easily expand (22) into Taylor series:

$$F(\langle P \rangle) = \alpha \sum_{n=0}^{\infty} \frac{(\gamma H)^{2n} \langle P \rangle^{2n+1}}{(2n+2)!}, \tag{23}$$

One can use few first terms of expansion when modelling consumption rates at small and intermediate amounts of food.

(ii) $\mu = 1, \lambda \neq 0$. This signifies that there is no interference between the grazers; however there is absorption of light by water along with algal self-shading. Integration of (21) after some simplification gives the following Taylor expansion

$$F(\langle P \rangle) = \alpha \sum_{n=0}^{\infty} c_n \langle P \rangle^n, \tag{24}$$

where the coefficients c_n are determined from

$$c_{2n+1} = \frac{\alpha\lambda}{(2n+2)!} \frac{(1+\exp(\lambda H))) H^{2n+1}\gamma^{2n}}{\exp(\lambda H)-1} , \tag{25}$$

$$c_{2n} = \frac{\alpha\lambda H^{2n}\gamma^{2n-1}}{(2n+2)!} , \tag{26}$$

(iii) $\mu \neq 1$, $\lambda = 0$. This signifies that there is some interference between the grazers and the absorption of the light by water is still small compared to the self-shading. Integrating of (21) gives the following explicit expression for the functional response

$$F(\langle P \rangle) = \frac{(\mu-1)\,(\exp(\langle P \rangle \gamma H/\mu)-1)\,(\exp(\langle P \rangle \gamma H)-1)}{H\gamma\,(\exp(\langle P \rangle \gamma H/\mu)-\exp(\langle P \rangle \gamma H))} , \tag{27}$$

By considering the Taylor expansion of (27) we obtain

$$F(\langle P \rangle) = \alpha \sum_{n=0}^{\infty} s_n \langle P \rangle^{2n+1} , \tag{28}$$

It is impossible to obtain simple expressions for the coefficients s_n. However, the first three coefficients can be easily computed:

$$c_1 = 1; \; c_2 = \frac{(\gamma H)^2}{12\mu}; \; c_3 = \frac{(\gamma H)^4(\mu^2 - 4\mu + 1)}{720\mu^3} , \tag{29}$$

Note that for the most general case ($\mu \neq 1, \lambda \neq 0$), the expression for the overall functional response becomes rather untractable and only numerical methods can be used to reveal the shape of $F(\langle P \rangle)$.

An important question is about the stability of the profiles obtained. Numerical methods show that those profiles are stable; however, stability for an infinite carrying capacity $K \to \infty$ requires a certain threshold value of $\gamma > 0$.

Appendix 2

Here I analytically address the stability property of the system (11)–(12) in the case where there are two different zooplankton cohorts showing different behavior. For the sake of simplicity I shall also consider that the functional response of the predator is of Holling type I ($b = 0$). It is easy to prove the existence of a trivial stationary state (0,0,0) as well as semitrivial stationary states where the density of one of species is zero. Simple analysis shows that all those states are unstable. The nontrivial stationary states are determined from

$$0 = w_{11} R_1 Z_1 + w_{12} R_2 Z_2 - a_1 Z_1 B , \tag{30}$$

$$0 = w_{21} R_1 Z_1 + w_{22} R_2 Z_2 - a_2 Z_2 B , \tag{31}$$

$$0 = a_1 Z_1 + a_2 Z_2 - \delta/\omega , \tag{32}$$

From (30)–(32) one can easily find the stationary density B from the evident condition:

$$\Delta = \begin{vmatrix} w_{11} R_1 - a_1 B & w_{12} R_2 \\ w_{21} R_1 & w_{22} R_2 - a_2 B \end{vmatrix} = 0. \tag{33}$$

This gives the following "characteristic" equation for B

$$a_1 a_2 B^2 - (w_{11} R_1 a_2 + w_{22} R_2 a_1) B + (w_{11} w_{22} - w_{12} w_{21}) R_1 R_2 = 0 , \tag{34}$$

Note that a similar characteristic equation will provide the stationary B for in the case the number of zooplankton cohorts is n. Formally, (34) may have up to two positive solutions which are the roots of the quadratic equation. However, a rigorous analysis (resulting into rather cumbersome expressions) shows that in the case of two positive roots of (34), one of the roots always gives a negative stationary density Z_i. Thus, the nontrivial stationary state of (11)–(12) is unique provided it exists. The stability of the nontrivial stationary state is determined by the Jacobian matrix given by

$$J = \begin{pmatrix} R_1 w_{11} - a_1 B & R_2 w_{12} & -\delta a_1 Z_1 \\ R_1 w_{21} & R_2 w_{22} - a_2 B & -\delta a_2 Z_2 \\ \omega \delta a_1 B & \omega \delta a_2 B & 0 \end{pmatrix} . \tag{35}$$

However, a direct substitution of explicit expressions for the stationary states results in an analytically intractable formula. To have an analytical insight into the model properties, I shall consider the particular case, where the coefficients are related by $w_{11} w_{22} = {}_{21} w_{12}$, which is equivalent to $w_{11} + w_{22} = 1$. This happens, for example, when each cohort produces equal percentage of offspring belonging to it and to the other cohort $w_{11} = w_{22} = 0.5$. Under the above condition the stationary density of species are given by

$$Z_1 = \frac{\delta w_{11}}{\omega a_1 (w_{21} + w_{11})}; \; Z_1 = \frac{\delta w_{21}}{\omega a_2 (w_{21} + w_{11})}; B = \frac{w_{11} R_1 a_2 + w_{22} R_2 a_1}{a_1 a_2} , \tag{36}$$

The Jacobian matrix (35) computed at the point (36) becomes

$$J = \begin{pmatrix} -R_2 \frac{a_1}{a_2} \frac{w_{21} w_{12}}{w_{11}} & R_2 w_{12} & -\delta \frac{w_{11}}{\omega (w_{21} + w_{11})} \\ R_1 w_{21} & -R_1 \frac{a_2}{a_1} w_{11} & -\delta \frac{w_{21}}{\omega (w_{21} + w_{11})} \\ \omega \frac{R_1 w_{11}^2 a_2 + w_{12} w_{21} R_2 a_1}{a_2 w_{11}} & \omega \frac{R_1 w_{11}^2 a_2 + w_{12} w_{21} R_2 a_1}{a_2 w_{11}} & 0 \end{pmatrix} . \tag{37}$$

The characteristic equation for the eigenvalues of (35) is given by

$$\sigma^3 + \Omega_1\sigma^2 + \Omega_2\sigma + \Omega_3 = 0 , \tag{38}$$

where the coefficients Ω_i are determined by

$$\Omega_1 = Sp(J) = \frac{a_2^2 w_{11}^2 R_1 + a_1^2 w_{12} w_{21} R_2}{w_{11} a_1 a_2} , \tag{39}$$

$$\Omega_2 = \frac{\delta}{w_{11}} \frac{w_{11}^3 R_1 a_1 a_2 + w_{21}^2 w_{12} R_2 a_1 a_2 + w_{11}^2 w_{21} R_1 a_2^2 + w_{11} w_{21} w_{12} R_2 a_1^2}{a_1 a_2 (w_{11} + w_{22}))} , \tag{40}$$

$$\Omega_3 = det(J) = \delta \frac{\left(w_{11}^2 R_1 a_2 + w_{12} w_{21} R_2 a_1\right)^2}{a_2 a_1 w_{11}^2} , \tag{41}$$

The Routh-Hurwitz stability criterion requires that $\Omega_i > 0 \; i = 1, 3$ and $\Omega_1\Omega_2 - \Omega_3 > 0$, which can be rewritten in the following way

$$(a_2 - a_1)\left(a_2 + a_1\sqrt{\frac{R_2 w_{12}}{R_1 w_{11}}}\right)\left(a_2 - a_1\sqrt{\frac{R_2 w_{12}}{R_1 w_{11}}}\right) > 0 , \tag{42}$$

which is equivalent to the following conditions ($w_{12} = 1 - w_{22} = w_{11}$)

$$a_1 \in \left(0, a_2\sqrt{\frac{R_1}{R_2}}\right) \bigcup (a_2, +\infty) . \tag{43}$$

Here, for the sake of simplicity, I consider that $R_1 < R_2$ since one can easily derive the stability conditions for the opposite sign of this inequality. The stability would occur when the cohort of zooplankton having a larger per capita growth rate R_i has a smaller per capita mortality rate a_i due to predation i.e. for $R_2 > R_1$ we should have $a_2 < a_1$. On the contrary, for a_2 slightly smaller than a_1 (and $R_2 > R_1$) the stability conditions are not satisfied resulting in destabilization of the equilibrium. However, even in this case the stability can be still possible when the difference between a_i is supercritical, i.e. for $a_1 < a_2\sqrt{R_1/R_2}$.

Suppose that the coefficients a and R are not independent and related via a trade-off function, i.e. $a = a(R)$. In the case where such a function is a decreasing function of R, the stability conditions are satisfied (43) for any shape of $a = a(R)$. In the opposite case $a'(R) > 0$, one can easily derive a criterion which guarantees that the stability conditions are satisfied. One can fix a_2 and vary the value of $a1$ starting from $a_1 = a_2$ (corresponding to $R_1 = R_2$). The stability condition (43) requires that

$$a_1(R_1) < a_2\sqrt{\frac{R_1}{R2}} . \tag{44}$$

One can re-write (44) in terms of the difference between R_1 and R_2, i.e. $\Delta R = R_2 - R_1$

$$\Delta a = a_2 - a_1 < a_2 \left(1 - \sqrt{1 - \frac{\Delta R}{R_2}} \right). \tag{45}$$

For a small ΔR we have the condition

$$\Delta a < a_2 \left(1 - \sqrt{1 - \frac{\Delta R}{R_2}} \right) \approx a_2 \frac{\Delta R}{2 R_2} = A \Delta R. \tag{46}$$

Thus, in the case the slope of the trade-off relation $a = a(R)$ exceeds a certain constant A, the stationary state (36) will be always locally stable provided it exists. Note that one can analytically prove the same property for a more general case, where w_{ii} are arbitrary values ($i = 1, 2$). In other words, the unique non-trivial stationary state of the system is (locally) stable at least in the parametric region near $a_1 = a_2$ and $R_2 = R_1$ in the case the parameters a and R are related by a trade-off function which can be a linear function. Finally, since the system with $b = 0$ (no saturation in the functional response) is structurally stable, adding small saturation $b \ll 1$ will not violate the previously obtained results.

Acknowledgements I highly appreciated prof. S. V. Petrovskii (University of Leicester) for a careful reading and comments. Also I thank prof. Elena Arashkevich (Shirshov Institute of Oceanology) who kindly provided the data on *Calanus spp.* feeding in laboratory (Fig. 5).

References

1. E.R. Abraham, The generation of plankton patchiness by turbulent stirring. Nature **391**, 577–580 (1998)
2. P. Auger, S. Charles, M. Viala, J.C. Poggiale, Aggregation and emergence in ecological modelling: integration of ecological levels. Ecol. Model. **127**, 11–20 (2000)
3. F. Bartumeus, F. Peters, S. Pueyo, C. Marrassé, J. Catalan, Helical Lévy Walks: adjusting searching statistics to resource availability in microzooplankton. Proc. Natl Acad. Sci. **100**(22), 12771–12775 (2003)
4. H.P. Batchelder, C.A. Edwards, T.M. Powell, Individual-based models of zooplankton populations in coastal upwelling regions: implications of diel vertical migration on demographic success and near shore retention. Progr. Oceanogr. **53**, 307–333 (2002)
5. B. Bautista, R.P. Harris, Copepod gut contents, ingestion rates and grazing impact on phytoplankton in relation to size structure of zooplankton and phytoplankton during a spring bloom. Mar. Ecol. Prog. Ser. **82**, 41–50 (1992)
6. A.D. Bazykin, *Nonlinear Dynamics of Interacting Populations* (World Scientific, Singapore, 1998)
7. A. Beckmann, I. Hense, Beneath the surface: characteristics of oceanic ecosystems under weak mixing conditions - a theoretical investigation. Progr. Oceanogr. **75**, 771–796 (2007)
8. M. Begon, C.R. Townsend, J.L. Harper, *Ecology: From Individuals to Ecosystems*, 4th edn. (Blackwell Publishing, Oxford, 2005), p. 738
9. D.E. Boakes, E.A. Codling, G.J. Thorn, M. Steinke, Analysis and modelling of swimming behaviour in Oxyrrhis marina. J. Plankton Res. **33**, 641–649 (2011)

10. S.M. Bollens, B.W. Frost, Predator induced diel vertical migration in a marine planktonic copepod. J. Plankton Res. **11**, 1047–1065 (1989)

11. C.M. Boyd, S.M. Smith, T. Cowles, Grazing patterns of copepods in the upwelling system off Peru. Limnol. Oceanogr. **25**, 583–596 (1980)

12. S.V. Budaev, 'Personality' in the guppy (Poecilia reticulata): a correlation study of exploratory behavior and social tendency. J. Compar. Psychol. **111**, 399–411 (1997)

13. M.H. Daro, Migratory and grazing behavior of copepods and vertical distributions of phytoplankton. Bull. Mar. Sci. **43**, 710–729 (1988)

14. F. Carlotti, J.-C. Poggiale, Towards methodological approaches to implement the zooplankton component in "end to end" food-web models. Progr. Oceanogr. **84**, 20–38 (2010)

15. F. Carlotti, K.U. Wolf, A Lagrangian ensemble model of Calanus finmarchicus coupled with a1-D ecosystem model. Fisher. Oceanogr. **7**, 191–204 (1998)

16. B. Charlesworth, Selection in populations with overlapping generations. III. Conditions for genetic equilibrium. Theor. Popul. Biol. **3**, 377–395 (1972)

17. P. Chesson, M.J. Donahue, B.A. Melbourne, A.L. Sears, Scale transition theory for understanding mechanisms in metacomunities, in *Metacommunities: Spatial Dynamics and Ecological Communities*, ed. by M. Holyoak, A. Leibold, R.D. Holt (University of Chicago Press, Chicago, 2005), p. 513

18. M.G. Clerc, D. Escaff, V.M. Kenkre, Analytical studies of fronts, colonies, and patterns: combination of the Allee effect and nonlocal competition interactions. Phys. Rev. E **82**, 036210 (2010)

19. K. Coleman, D.S. Wilson, Shyness and boldness in pumpkinseed sunfish: individual differences are context-specific. Anim. Behav. **56**, 927–936 (1998)

20. C. Cosner, D.L. DeAngelis, J.S. Ault, D.B. Olson, Effects of spatial grouping on the functional response of predators. Theor. Popul. Biol. **56**, 65–75 (1999)

21. F.R. Cottier, G.A. Tarling, A. Wold, S. Falk-Petersen, Unsynchronised and synchronised vertical migration of zooplankton in a high Arctic fjord. Limnol. Oceanogr. **51**, 2586–2599 (2006)

22. T.J. Cowles, R.A. Desiderio, M.E. Carr, Small-scale planktonic structure: persistence and trophic consequences. Oceanography **11**, 4–9 (1998)

23. H.C. Crenshaw, L. Edelstein-Keshet, Orientation by helical motion. II. Changing the direction of the axis of motion. J. Math. Biol. **55**, 213–230 (1993)

24. J.M. Cushing, *An Introduction to Structured Population Dynamics* (SIAM, Philadelphia, 1998), p. 195

25. M.J. Dagg, K.D. Wyman, Natural ingestion rates of the copepods Neocalunus plumchrus and N. cristatus calculated from gut contents. Mar. Ecol. Prog. Ser. **13**, 37–46 (1983)

26. M.J. Dagg, B.W. Frost, J.A. Newton, Vertical migration and feeding behavior of Calanus pacificus females during a phytoplankton bloom in Dabob Bay, US. Limnol. Oceanogr. **42**, 974–980 (1997)

27. W.R. DeMott, Feeding selectivities and relative ingestion rates of Daphnia and Bosmina. Limnol. Oceanogr. **27**, 518–527 (1982)

28. A. Dhooge, W. Govaerts, Y. Kuznetsov, Matcont: a matlab package for numerical bifurcation analysis of ODEs. ACM TOMS **29**, 141–164 (2003). http://sourceforge.net/projects/matcont/

29. O. Diekmann, M. Gyllenberg, J.A. Metz, S. Nakaoka, A.M. de Roos, Daphnia revisited: local stability and bifurcation theory for physiologically structured population models explained by way of an example. J. Math. Biol. **61**, 277–318 (2010)

30. S.I. Dodson, S. Ryan, R. Tollrian, W. Lampert, Individual swimming behavior of Daphnia: effects of food, light and container size in four clones. J. Plankton Res. **19**, 1537–1552 (1997)

31. A.M. Edwards, J. Brindley, Zooplankton mortality and the dynamical behavior of plankton population models. Bull. Math. Biol. **61**, 202–339 (1999)

32. G. Englund, K. Leonardsson, Scaling up the functional response for spatially heterogeneous systems. Ecol. Lett. **11**, 440–449 (2008)

33. G.T. Evans, The encounter speed of moving predator and prey. J. Plankton Res. **11**, 415–417 (1989)

34. G.T. Evans, J.S. Parslow, A model of annual plankton cycles. Biol. Oceanogr. **3**, 327–347 (1985)
35. A. Ferno, I. Huse, J.-E. Juell, A. Bjordal, Vertical distribution of Atlantic salmon (Salmo salar L.) in net pens: trade-off between surface light avoidance and food attraction. Aquaculture **132**, 285–296 (1995)
36. C.L. Folt, C.W. Burns, Biological drivers of zooplankton patchiness. TREE **14**, 300–305 (1999)
37. M. Fossheim, R. Primicerio, Habitat choice by marine zooplankton in a high-latitude ecosystem. Mar. Ecol. Prog. Ser. **364**, 47–56 (2008)
38. B.W. Frost, A threshold feeding behavior in Calanus pacificus. Limnology and Oceanography **20**, 263–266 (1975)
39. W. Gabriel, B. Thomas, Vertical migration of zooplankton as an evolutionarily stable strategy. Am. Nat. **132**, 199–216 (1988)
40. C. Gardiner, *Stochastic Methods*, 4th edn. (Sringer, Berlin, 2009)
41. W. Geller, Diurnal vertical migration of zooplankton in a temperate great lake (L. Constance): a starvation avoidance mechanism? Archiv. Hydrobiol. **74**, 1–60 (1986)
42. W. Gentleman, A. Leising, B. Frost, S. Storm, J. Murray, Functional responses for zooplankton feeding on multiple resources: a review of assumptions and biological dynamics. Deep Sea Res. II **50**, 2847–2875 (2003)
43. J. Giske, R. Rosland, J. Berntsen, O. Fiksen, Ideal free distribution of copepods under predation risk. Ecol. Model. **95**, 45–59 (1997)
44. L. Giuggioli, F.J. Sevilla, V.M. Kenkre, A generalized master equation approach to modelling anomalous transport in animal movement. J. Phys. A **42**, 1–16 (2009)
45. T.C. Granata, T.D. Dickey, The fluid mechanics of copepod feeding in a turbulent flow: a theoretical approach. Progr. Oceanogr. **26**, 243–261 (1991)
46. V. Grimm, Ten years of individual-based modeling in ecology: what have we learned and what could we learn in the future? Ecol. Model. **115**, 129–148 (1999)
47. V. Grimm, S.F. Railsback, Agent-based models in ecology: patterns and alternative theories of adaptive behaviour, in *Agent-Based Computational Modelling: Contributions to Economics*, ed. by F.C. Billari, T. Fent, A. Prskawetz, J. Scheffran (Physica-Verlag, Heidelberg, 2006), pp. 139–152
48. N. Gruber, H. Frenzel, S.C. Doney, P. Marchesiello, J.C. McWilliams, J.R. Moisan, J. Oram, G.-K. Plattner, K.D. Stolzenbach, Eddy resolving simulation of plankton ecosystem dynamics in the California current system. Deep Sea Res. I **53**, 1483–1516 (2006)
49. B.P. Han, M. Straskraba, Modeling patterns of zooplankton diel vertical migration. J. Plankton Res. **20**, 1463–1487 (1998)
50. B. Hansen, K.S. Tande, U.C. Berggreen, On the trophic fate of Phaeocystis pouchetii (Hariot). III. Functional responses in grazing demonstrated on juvenile stages of Calanus finmarchicus (Copepoda) fed diatoms and Phaeocystis. J. Plankton Res. **12**, 1173–1187 (1990)
51. M.P. Hassell, R.M. May, Aggregation in predators and insect parasites and its effect on stability. J. Anim. Ecol. **43**, 567–594 (1974)
52. L.R. Haury, J.A. McGowan, P.H. Wiebe, Patterns and processes in the time- space scales of plankton distributions, in *Spatial Pattern in Plankton Communities*, ed. by J.H. Steele (Plenum Press, New York 1978), pp. 277–327
53. A.W. Herman, T. Platt, Numerical modelling of diel carbon production and zooplankton grazing on the scotian shelf based on observational data. Ecol. Model. **18**, 55–72 (1983)
54. A.W. Herman, Vertical patterns of copepods, chlorophyll, and production in Northeastern Baffin Bay. Limnol. Oceanogr. **28**, 709–719 (1983)
55. A.G. Hirst, A.J. Bunker, Growth of marine planktonic copepods: global rates and patterns in relation to chlorophyll a, temperature, and body weight. Limnol. Oceanogr. **48**, 1988–2010 (2003)
56. C.S. Holling, The components of predation as revealed by a study of small-mammal predation of the European pine sawfly. Can. Entomol. **91**, 293–320 (1959)

57. Y. Iwasa, Vertical migration of zooplankton: a game between predator and prey. Am. Nat. **120**, 171–180 (1982)
58. J.M. Jeschke, M. Kopp, R. Tollrian, Consumer-food systems: why type I functional responses are exclusive to filter feeders. Biol. Rev. **79**, 337–349 (2004)
59. P. Kareiva, Population dynamics in spatially complex environments: theory and data. Phil. Trans. R. Soc. B **330**, 175–190 (1990)
60. W. Lampert, Zooplankton vertical migrations: implications for phytoplanktonzooplankton interactions. Arch. Hydrobiol. Beih. Ergebn. Limnol. **35**, 69–78 (1992)
61. W. Lampert, Vertical distribution of zooplankton: density dependence and evidence for an ideal free distribution with costs. BMC Biol. **3**, 10 (electronic) (2005)
62. J. Latto, M.P. Hassell, Generalist predators and the importance of spatial density dependence. Oecologia **77**, 375–377 (1988)
63. A.W. Leising, Copepod foraging in patchy habitats and thin layers using a 2-D individual based model. Mar. Ecol. Prog. Ser. **216**, 167–179 (2001)
64. A.W. Leising, P.J.S. Franks, Copepod vertical distribution within a spatially variable food source: a foraging strategy model. J. Plankton Res. **22**, 999–1024 (2000)
65. A.W. Leising, J.J. Pierson, S. Cary, B.W. Frost, Copepod foraging and predation risk within the surface layer during night-time feeding forays. J. Plankton Res. **27**, 987–1001 (2005)
66. S.A. Levin, The problem of pattern and scale in ecology: the Robert H. MacArthur Award Lecture. Ecology **73**, 1943–1967 (1992)
67. S.H. Liu, S. Sun, B.P. Han, Diel vertical migration of zooplankton following optimal food intake under predation. J. Plankton Res. **25**, 1069–1077 (2003)
68. D.L. Mackas, C.M. Boyd, Spectral analysis of zooplankton spatial heterogeneity. Science **204**, 62–64 (1979)
69. P.S. Magal, S. Ruan (eds.), in *Structured Population Models in Biology and Epidemiology*. Lecture Notes in Mathematics, vol. 1936, Mathematical Biosciences Subseries (Springer, Berlin, 2008), p. 345
70. E. Malkiel, J. Sheng, J. Katz, J.R. Strickler, The three-dimensional flow field generated by a feeding calanoid copepod measured using digital holography. J. Exp. Biol. **206**, 3657–3666 (2003)
71. J.A. Mather, R.C. Anderson, Personalities of octopuses (Octopus rubescans). J. Compar. Psychol. **107**, 336–340 (1993)
72. J.A. McLaren, Effect of temperature on growth of zooplankton and the adaptive value of vertical migration. J. Fish. Res. Board Can. **20**, 685–727 (1963)
73. J.N. McNair, M.E. Boraas, D.B. Seale, Size-structure dynamics of the rotifer chemostat: a simple physiologically structured model. Hydrobiologia **387**, 469–476 (1998)
74. J.A.J. Metz, O. Diekmann, *The Dynamics of Physiologically Structured Populations* (Springer, Berlin, 1986), p. 511
75. J. Michalski, J.-C. Poggiale, R. Arditi, P. Auger, Macroscopic dynamic effects of migrations in patchy predatorprey systems. J. Theor. Biol. **185**, 459–474 (1997)
76. A.Y. Morozov, Emergence of Holling type III zooplankton functional response: bringing together field evidence and mathematical modelling. J. Theor. Biol. **265**, 45–54 (2010)
77. A.Y. Morozov, A.G. Arashkevich, Towards a correct description of zooplankton feeding in models: Taking into account food-mediated unsynchronized vertical migration. J. Theor. Biol. **262**, 346–360 (2010)
78. A.Y. Morozov, E. Arashkevich, Patterns of zooplankton functional response in communities with vertical heterogeneity: a model study. Math. Mod. Nat. Phen. **3**, 131–148 (2008)
79. A.Y. Morozov, E. Arashkevich, M. Reigstad, S. Falk-Petersen, Influence of spatial heterogeneity on the type of zooplankton functional response: a study based on field observations. Deep Sea Res. II **55**, 2285–2291 (2008)
80. A.Yu. Morozov, E.G. Arashkevich, A. Nikishina, K. Solovyev, Nutrient-rich plankton communities stabilized via predator-prey interactions: revisiting the role of vertical heterogeneity. Math. Med. Biol. **28**, 185–215 (2011)

81. J.M. Morales, P.R. Moorcroft, J. Matthiopoulos, J.L. Frair, J.G. Kie, R.A. Powell, E.H. Merrill, D.T. Haydon, Building the bridge between animal movement and population dynamics. Phil. Trans. R. Soc. B **365**, 2289–2301 (2010)
82. M.M. Mullin, E.R. Brooks, Some consequences of distributional heterogeneity of phytoplankton and zooplankton. Limnol. Oceanogr. **21**, 784–796 (1976)
83. W.W. Murdoch, C.J. Briggs, R.M. Nisbet, W.S.C. Gurney, A. Stewart-Oaten, Aggregation and stability in metapopulation models. Am. Nat. **140**, 41–58 (1992)
84. W.W. Murdoch, R.M. Nisbet, E. McCauley, A.M. Roos, W.S.C. De Gurney, Plankton abundance and dynamics across nutrient levels: tests of hypotheses. Ecology **79**, 1339–1356 (1998)
85. G. Nachman, A functional response model of a predator population foraging in a patchy habitat. J. Anim. Ecol. **75**, 948–958 (2006)
86. R. Nathan, W.M. Getz, E. Revilla, M. Holyoak, R. Kadmon, D. Saltz, P.E. Smouse, A movement ecology paradigm for unifying organismal movement research. Proc. Nat. Acad. Sci. **105**, 19052–19059 (2008)
87. A. Oaten, W.W. Murdoch, Functional response and stability in predatorprey systems. Am. Nat. **109**, 289–298 (1975)
88. E. Odum, G.W. Barrett, *Fundamentals of Ecology* (Thomson Brooks/Cole, Belmont, 2004), p. 598
89. T. Oguz, H. Ducklow, P. Malanotte-Rizzoli, J. Murray, E. Shushkina, V. Vedernikov, U. Unluata, A physical-biochemical model of plankton productivity and nitrogen cycling in the Black Sea. Deep Sea Res. **46**, 597–636 (1999)
90. M.D. Ohman, The demographic benefits of diel vertical migration by zooplankton. Ecol. Monogr. **60**, 257–281 (1990)
91. K.E. Osgood, D.M. Checkley, Seasonal variations in a deep aggregation of Calanus pacificus in the Santa Barbara Basin. Mar. Ecol. Prog. Ser. **148**, 59–69 (1997)
92. G.A. Paffenhöfer, Variability due to feeding activity of individual copepods. J. Plankton Res. **16**, 617–626 (1994)
93. G.A. Paffenhöfer, J.R. Strickler, K.D. Lewis, S. Richman, Motion behavior of nauplii and early copepodid stages of marine planktonic copepods. J. Plankton Res. **18**, 1699–1715 (1996)
94. M. Pascual, Computational ecology: From the complex to the simple and back. PLoS Comput. Biol. **1**, 2 (electronic) (2005)
95. S.J. Pearre, Eat and run? The hunger/satiation hypothesis in vertical migration: history, evidence and consequences. Biol. Rev. **78**, 1–79 (2003)
96. S.V. Petrovskii, R. Blackshaw, Behaviourally structured populations persist longer under harsh environmental conditions. Ecol. Lett. **6**, 455–462 (2003)
97. S.V. Petrovskii, A.Y. Morozov, Dispersal in a statistically structured population: Fat tails revisited. Am. Nat. **173**, 278–289 (2010)
98. S.V. Petrovskii, R.P. Blackshaw, B.-L. Li, Persistence of structured populations with and without the Allee effect under adverse environmental conditions. Bull. Math. Biol. **70**, 412–437 (2008)
99. J.C. Poggiale, Predator-prey models in heterogeneous environment: emergence of functional response. Math. Comput. Model. **27**, 63–71 (1998)
100. D. Reale, B.Y. Gallant, M. Leblanc, M. Festa-Bianchet, Consistency of temperament in bighorn ewes and correlates with behaviour and life history. Anim. Behav. **60**, 589–597 (2000)
101. E. Saiz, A. Calbet, Scaling of feeding in marine calanoid copepods. Limnol. Oceanogr. **52**, 668–675 (2007)
102. O. Sarnelle, A.E. Wilson, Type III functional response in Daphnia. Ecology **89**, 1723–1732 (2008)
103. M. Scheffer, R.J. De Boer, Implications of spatial heterogeneity for the paradox of enrichment. Ecology **76**, 2270–2277 (1995)
104. M. Scheffer, J.M. Baveco, D.L. DeAngelis, K.A. Rose, E.H. Van Nes, Super-individuals a simple solution for modelling large populations on an individual basis. Ecol. Model. **80**, 161–170 (1995)

105. F. Schmitt, L. Seuront, J.-S. Hwang, S. Souissi, L.C. Tseng, Scaling of swimming sequences in copepod behavior: data analysis and simulation. Physica A **364**, 287–296 (2006)
106. F.G. Schmitt, L. Seuront, Multifractal random walk in copepod behavior. Physica A **301**, 375–396 (2001)
107. T. Sekino, N. Yamamura, Diel vertical migration of zooplankton: optimum migrating schedule based on energy accumulation. Evol. Ecol. **13**, 267–282 (1999)
108. L. Seuront, J.-S. Hwang, L.-C. Tseng, F. Schmitt, S. Souissi, C.-K. Wong, Individual variability in the swimming behavior of the sub-tropical copepod Oncaea venusta (Copepoda: Poecilostomatoida). Mar. Ecol. Prog. Ser. **283**, 199–217 (2004)
109. P.E. Smouse, S. Focardi, P.R. Moorcroft, J.G. Kie, J.D. Forester, J.M. Morales, Stochastic modelling of animal movement. Phil. Trans. R. Soc. B **365**, 2201–2211 (2010)
110. M.E. Solomon, The natural control of animal populations. J. Anim. Ecol. **18**, 1–35 (1949)
111. W.J. Sutherland, Aggregation and the "ideal free" distribution. J. Anim. Ecol. **52**, 821–828 (1983)
112. K.S. Tande, U. Bamstedt, Grazing rates of the copepods Calanus glacialis and C. finmarchicus in arctic waters of the Barents Sea. Mar. Biol. **87**, 251–258 (1985)
113. P. Tiselius, P.R. Jonsson, Foraging behaviour of six calanoid copepods: observations and hydrodynamic analysis. Mar. Ecol. Prog. Ser. **66**, 23–33 (1990)
114. P. Tiselius, P.R. Jonsson, P.G. Verity, A model evaluation of the impact of food patchiness on foraging strategy and predation risk in zooplankton. Bull. Mar. Sci. **53**, 247–264 (1993)
115. J.E. Truscott, J. Brindley, Ocean plankton populations as excitable media. Bull. Math. Biol. **56**, 981–998 (1994)
116. L.-C. Tseng, R. Kumar, H.-U. Dahms, Q.-C. Chen, J.-S. Hwang, Copepod gut contents, ingestion rates, and feeding impacts in relation to their size structure in the southeastern Taiwan Strait. Zool. Stud. **47**, 402–416 (2008)
117. A. Tsuda, H. Saito, H. Kasai, Annual variation of occurrence and growth in relation with life cycles of Neocalanus flemingeri and N. plumchrus (Calanoida, Copepoda) in the western subarctic Pacific. Mar. Biol. **135**, 533–544 (1999)
118. S. Tuljapurkar, H. Caswell, *Structured Population Models in Marine, Terrestrial, and Freshwater Systems* (Chapman and Hall, London, 1997), p. 656
119. A. Visser, Lagrangian modelling of plankton motion: from deceptively simple random walks to Fokker–Planck and back again. J. Mar. Syst. **70**, 287–299 (2008)
120. G.M. Viswanathan, V. Afanasyev, S.V. Buldyrev, S. Havlin, M.G.E. da Luz, E.P. Raposo, H.E. Stanley, Lévy flights search patterns of biological organisms. Physica A **295**, 85–88 (2001)
121. J. Woods, A. Perilli, W. Barkmann, Stability and predictability of a virtual plankton ecosystem created with an individual-based model. Progr. Oceanogr. **67**, 43–83 (2005)

Part III
Populations, Communities and Ecosystems

Life on the Move: Modeling the Effects of Climate-Driven Range Shifts with Integrodifference Equations

Ying Zhou and Mark Kot

Abstract Climate change is causing many species to shift their ranges. We analyze an integrodifference equation that combines growth, dispersal, and a shifting habitat in order to assess the impact of climate change on persistence. We apply this model to butterflies and show that over-dispersal and under-dispersal can both lead to extinction. We focus on the critical range-shift speed (for extinction), survey numerical methods for determining this speed, and introduce new analytic approximations for the critical shift speed. Finally, we apply our numerical methods and analytic approximations to a variety of redistribution kernels and show that critical-speed curves shed light on the complicated effects of dispersal on persistence in a changing climatic environment.

1 Introduction

The geographic ranges of wildlife species are constantly responding to changes in climate. Paleoecological studies have documented extensive species range changes during the glacial and interglacial alternations in the Quaternary Period [12, 32]. More recently (1956–2005), the Earth has been warming up at the rate of 0.13 °C/decade [34]. Species were expected to shift their ranges during this period, and this has now been documented [13, 29, 39, 49, 70, 71].

One major difference between the current warming event and earlier climatic changes in the Quaternary is that modern anthropogenic activities have created obstacles that make it difficult for the Earth's biota to survive climate change. Species need to shift their ranges to avoid excessive habitat loss during climate change, but, unlike organisms hundreds of thousands of years ago, species now face

Y. Zhou (✉) · M. Kot
Department of Applied Mathematics, University of Washington, Box 352420, Seattle, WA, USA
e-mail: yzhou@amath.washington.edu; mark_kot@comcast.net

M.A. Lewis et al. (eds.), *Dispersal, Individual Movement and Spatial Ecology*,
Lecture Notes in Mathematics 2071, DOI 10.1007/978-3-642-35497-7_9,
© Springer-Verlag Berlin Heidelberg 2013

severe habitat fragmentation that makes shifting their ranges difficult [30, 64, 78]. Invasive exotics introduced by anthropogenic activities further stress indigenous species, since a changing climate creates new opportunities for invasive species to compete with native species. In general, ecologists must now assess the spatial effects of multiple factors in planning conservation strategies. They must, in particular, determine how multiple factors affect species ranges.

Many quantitative tools help us estimate future range shifts. For example, correlation models such as climate envelope models have been used to predict ranges for various climate change scenarios [4, 25, 36]. These models typically correlate species ranges with climatic variables using statistical tools, and then estimate potential ranges by combining correlational information with projections of future climate. These models cannot, however, easily integrate population dynamics, such as growth, dispersal, and interspecific interactions.

Other mathematical models have been used to bridge this gap. Travis [87] and Best et al. [7] used stochastic, spatially explicit, patch occupancy models to study the effects of range-shift speed on persistence. Potapov and Lewis [72] and Berestycki et al. [5], in turn, used deterministic reaction–diffusion models to study this topic. Reaction–diffusion equations are particularly suitable for organisms with continuous and simultaneous growth and dispersal.

For many organisms, growth and dispersal are discrete episodes in the life cycle. Consider, for example, Edith's checkerspot butterfly (*Euphydryas editha*). In the San Francisco Bay area of California, eggs of this species hatch in April and the larvae feed and develop for 10–14 days. The larvae then enter diapause [27]. Post-diapause larvae feed and grow from December to February. They then pupate and emerge as adults [27]. The adults of *E. editha*, which live some 10 days, emerge, fly, mate, search for oviposition sites, and complete their life cycle in March and April [61]. For this univoltine species, adults are the primary dispersers; their dispersal occurs during a very narrow window of time.

The mortality rate of *E. editha*'s larvae depends on synchrony between their life cycle and that of their host plants: larvae must reach their fourth instar and enter diapause before their host plants senesce [67]. This synchrony is strongly affected by climatic variables such as temperature and precipitation [67]. As a result, the population dynamics of *E. editha* are sensitive to climate change. Parmesan [66] examined populations of *E. editha* throughout its range and found that populations along its southern range boundary suffered extinction rates four times higher than those along its northern range boundary. Population extinction rates at lower elevations, meanwhile, were nearly three times as high as those at higher elevations. These drastic differences in extinction rates clarify why the mean location of *E. editha* populations shifted northward and upward [67].

Many other examples of organisms with discrete growth and dispersal occur in the range-shift literature. For these species, population dynamics are more easily described using discrete-time models. We therefore attack the problem of climate-induced range shifts using integrodifference equations (IDEs). IDEs are discrete-time, continuous-space models that combine growth and dispersal [28, 41, 42, 47, 51,

54, 55, 62, 63, 88]. In this chapter, we will describe and analyze an IDE model that includes climate-driven spatial shifts.

In Sect. 2, we describe our basic model, apply it to butterflies, and show how over-dispersal and under-dispersal can both lead to extinction. In Sect. 3, we introduce the critical range-shift speed and reduce the problem of finding this speed to an eigenvalue problem. In Sect. 4, we survey numerical approaches for solving this problem. In Sects. 5 and 7, we use Legendre series and Taylor series to obtain analytic approximations of the critical shift speed. We apply our numerical and analytic approximations to a toy problem, in Sect. 6, and to more realistic kernels, in Sect. 8. Finally, we summarize and discuss our results in Sect. 9.

2   Is Further Better?

Several recent studies suggest that traits such as dispersal ability affect whether a species can successfully shift its range [11, 20, 65, 75, 77, 86]. For example, Pöyry et al. [73] related observed range shifts of butterflies in Finland to 11 butterfly life-history traits, such as mobility, habitat, and host-plant form. They found that habitat availability and dispersal capacity were the two traits most likely to determine whether a butterfly could keep up with climate change by shifting its range.

There is little argument that a sedentary species with little habitat has poor prospects in a world of climate change. But how about a vagile species whose habitat is shifting and fragmenting as we speak? What are its prospects? And, does higher dispersal ability always lead to greater success? We attempt to answer these questions by applying a recently developed IDE model [93] to butterflies.

Without loss of generality, let us assume that we have a univoltine butterfly that thrives on a suitable, spatially continuous patch of habitat in the Northern Hemisphere. Since ranges are expected to shift polewards, we will assume that this patch is a strip or zone bounded by lines of latitude. For mathematical simplicity, we reduce this strip to the one-dimensional interval $[-L/2, L/2]$. The length of this interval, L, in kilometers, represents the patch size. We assume that climate change shifts the two zonal boundaries northwards, at the speed of c km/year. Thus, t years after the initial time point, the suitable patch is located at $[-L/2 + ct, L/2 + ct]$.

Because the butterfly is univoltine and has well-defined life stages, we can keep track of the population dynamics by censusing the population once a year. If we label the density of freshly oviposited eggs in year t at location x as $n_t(x)$, then the density of eggs in the next year can be written as

$$n_{t+1}(x) = \int_{-\frac{L}{2}+ct}^{\frac{L}{2}+ct} k(x, y) \, f\left[n_t(y)\right] \, dy. \tag{1}$$

The formulation of (1) can be understood by going through the butterfly's life cycle. Most of the life cycle, from egg hatching to pupal eclosion, is a relatively sedentary stage. The function f describes growth during this stage. It maps the density of eggs (on plants) to a new density of eggs (in emerging adults). To construct the growth function f we might, for example, consider density dependence, larval mortality, clutch size, etc. In this paper, we will use the Beverton–Holt [8] recruitment curve,

$$f(n_t) = \frac{R_0 n_t}{1 + [(R_0 - 1)/K] n_t},$$ (2)

as our growth function. Here, $R_0 = f'(0)$ is the net reproductive rate and K is the carrying capacity.

The adult butterflies are the dispersal stage. Eggs produced at y are carried to location x with some probability. For a fixed source of eggs y, we think of the redistribution kernel, $k(x, y)$, as a probability density function for the destination x of propagules. The kernel may depend on only the difference between x and y. In this case, we have a difference kernel, $k(x, y) = k(x - y)$, and we may think of $k(x - y)$ as the probability density function for the displacement (rather than the destination). Kernels can be estimated from mark-release-recapture studies.

Unfortunately, ecologists often lack detailed dispersal data for butterflies. As a result, comparative studies of butterflies often assume mobilities based on expert opinion [17, 40, 73, 84]. Recently, however, Stevens et al. [84] performed a meta-analysis of butterfly studies and summarized dispersal data. They showed that distributions of dispersal distances could be fit with negative exponential curves. These curves engender a 2D probability density function for the deposition of propagules [15, 16]. By taking the marginal distribution of this 2D kernel [51], we obtained [94] the 1D redistribution kernel

$$k(x - y) = \frac{\alpha}{\pi} K_0 (\alpha |x - y|),$$ (3)

for butterflies, where $K_0(x)$ is the modified Bessel function of the second kind of order zero. The parameter α, the reciprocal of the mean dispersal distance, ranges from $0.76 \, \text{km}^{-1}$ for vagile species to $24.25 \, \text{km}^{-1}$ for sedentary species [84].

In model (1), eggs hatch and grow and adults emerge if they are in the climate-shifted patch for year t. Adults then oviposit their eggs both inside and outside the patch. Equation (1) tallies the movement of propagules from sources y within the patch to obtain the density of eggs, $n_{t+1}(x)$, at the start of the next generation.

To explore the effects of dispersal on survival, we numerically iterated equation (1) with growth function (2) and redistribution kernel (3). We set the net reproductive rate, the patch size, and the shift speed to the values $R_0 = 1.9$, $L = 0.5 \, \text{km}$, and $c = 0.1 \, \text{km/year}$. We then chose three values of α from across the spectrum of observed values [84]. Figure 1a–c show the dynamics of the three populations. Populations die out for high ($\alpha = 12 \, \text{km}^{-1}$) and low ($\alpha = 2.5 \, \text{km}^{-1}$) values of α. They survive for intermediate α ($\alpha = 6 \, \text{km}^{-1}$).

The causes of the extinctions for high and low α differed. The highly sedentary population ($\alpha = 12 \, \text{km}^{-1}$) in Fig. 1a was limited by its dispersal ability. It simply

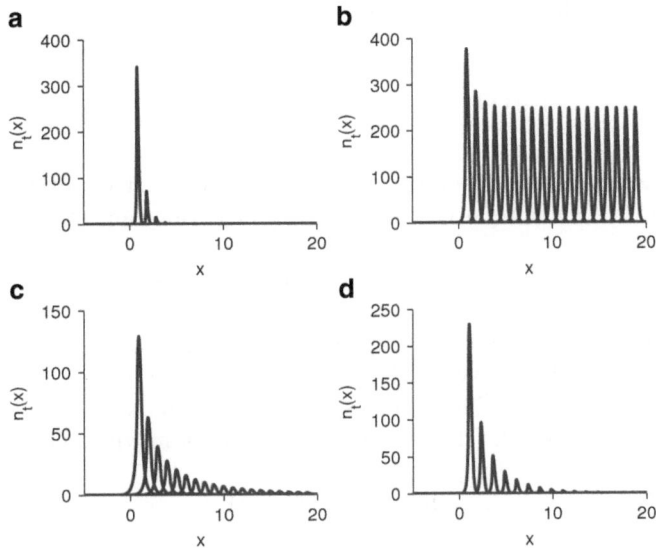

Fig. 1 Simulations of IDE (1) with growth function (2) and kernel (3) show that a species' ability to survive climate change depends on both its dispersal ability and the speed of climate change. For shift speed $c = 0.1$ km/year, (**a**) a sedentary population ($\alpha = 12$ km^{-1}) cannot keep up with its shifting habitat and goes extinct; (**b**) an intermediate population ($\alpha = 6$ km^{-1}) survives; (**c**) a vagile population ($\alpha = 2.5$ km^{-1}) over-disperses and goes extinct. For $c = 0.125$ km/year, (**d**) the intermediate population ($\alpha = 6$ km^{-1}) also goes extinct. For all four subfigures, $L = 0.5$ km, $R_0 = 1.9$, and $K = 1,000$. The initial distribution was $n_0(x) = K \exp(-x^2/2)$. The distribution is displayed every ten generations and was computed using an FFT-assisted implementation of the extended trapezoidal rule with 2^{16} nodes

could not keep up with its shifting habitat. In contrast, the vagile population ($\alpha = 2.5$ km^{-1}) in Fig. 1c over-dispersed and was patch-size (or growth-rate) limited [see also 93, Fig. 5]. Thus, climate change and habitat fragmentation can both be important. Conservation efforts need to integrate both factors.

3 The Critical Range-Shift Speed

Instead of comparing species traits, let us now take a different perspective and look at the severity of climate change. Even if the population in Fig. 1b is doing fine, will it still prosper if the shift speed c is increased? No! Figure 1d demonstrates that our population collapses if we raise the shift speed to $c = 0.125$ km/year.

The shift speed c clearly has an important effect on the viability of our population. Numerical simulations suggest that for each growth rate, patch size, and redistribution kernel, there may be a critical shift speed c^* beyond which the population goes extinct. We will now employ some simple mathematical analyses to determine this critical shift speed.

Consider (1) with a difference kernel,

$$n_{t+1}(x) = \int_{-\frac{L}{2}+ct}^{\frac{L}{2}+ct} k(x-y) \, f\left[n_t(y)\right] dy. \tag{4}$$

For convenience, we will now assume, throughout the remainder of this paper, that our growth curve is nonnegative, monotonically increasing, and that it satisfies

$$f(n) \le f'(0)n, \tag{5}$$

for $n \ge 0$. In particular, we explicitly exclude Allee effects [1]. These conditions are certainly satisfied by Beverton–Holt curve (2) for $R_0 > 1$.

Since the patch moves with constant speed c, we will look for steady states in the moving frame of the patch. This means we will look for moving pulses of the form

$$n_t(x) = n^*(x - ct). \tag{6}$$

Using this ansatz,

$$n_{t+1}(x) = n^*(x - ct - c). \tag{7}$$

Substituting (6) and (7) into (4), we find that the moving pulse satisfies

$$n^*(x - ct - c) = \int_{-\frac{L}{2}+ct}^{\frac{L}{2}+ct} k(x-y) \, f\left[n^*(y - ct)\right] dy. \tag{8}$$

Rewriting (8) in terms of the shifted spatial variables $\bar{x} = x - ct$ and $\bar{y} = y - ct$, we obtain

$$n^*(\bar{x} - c) = \int_{-\frac{L}{2}}^{\frac{L}{2}} k(\bar{x} - \bar{y}) \, f\left[n^*(\bar{y})\right] d\bar{y}. \tag{9}$$

Shifting $\bar{x}$ by c, we find that moving pulse $n^*(\bar{x})$ is a solution of the equation

$$n^*(\bar{x}) = \int_{-\frac{L}{2}}^{\frac{L}{2}} k(\bar{x} + c - \bar{y}) \, f\left[n^*(\bar{y})\right] d\bar{y}. \tag{10}$$

For this derivation to work, our kernel must be a difference kernel.

It is hard, in general, to find closed-form solutions $n^*(\bar{x})$ of (10). We can, however, identify special solutions in special cases. If the growth curve f has the trivial solution as a fixed-point, then

$$n^*(\bar{x}) \equiv 0 \tag{11}$$

is a solution of (10).

For the simple growth functions that we are using, we expect persistence to be equivalent to instability of solution (11) (no bistability). To study the stability of a moving pulse, we add a small, localized perturbation $\xi_t(x)$ to the pulse,

$$n_t(x) = n^*(\bar{x}) + \xi_t(x). \tag{12}$$

We then substitute $n_t(x)$ into integrodifference equation (4), linearize about $n^*(\bar{x})$,

$$\xi_{t+1}(x) = \int_{-\frac{L}{2}+ct}^{\frac{L}{2}+ct} k(x-y) f'\left[n^*(\bar{y})\right] \xi_t(y)\, dy, \tag{13}$$

and study the growth of the perturbation. Equation (13) is difficult to analyze in general, but luckily, for the trivial solution, $f'[n^*(\bar{y})] = f'(0)$ is a constant. As a result, (13) now reduces to

$$\xi_{t+1}(x) = R_0 \int_{-\frac{L}{2}+ct}^{\frac{L}{2}+ct} k(x-y)\, \xi_t(y)\, dy, \tag{14}$$

where $R_0 = f'(0)$ is the net reproductive rate.

Since we are interested in perturbations that persist in the moving frame, we now focus on perturbations that can be written as a product of a growth term λ^t and a traveling term $u(x-ct)$,

$$\xi_t(x) = \lambda^t\, u(\bar{x}) \equiv \lambda^t\, u(x-ct). \tag{15}$$

It now follows that

$$\lambda\, u(\bar{x}) = R_0 \int_{-\frac{L}{2}}^{\frac{L}{2}} k(\bar{x}+c-\bar{y})\, u(\bar{y})\, d\bar{y}. \tag{16}$$

Finally, for notational convenience, but at great risk of confusing the reader, we drop the bars on $\bar{x}$ and $\bar{y}$,

$$\lambda\, u(x) = R_0 \int_{-\frac{L}{2}}^{\frac{L}{2}} k(x+c-y)\, u(y)\, dy. \tag{17}$$

This is the key equation in this paper; we will use it to determine the critical speed c^*. It is a homogeneous Fredholm integral equation of the second kind. Please keep in mind, however, that the x and y in this equation are actually in the moving frame of the habitat. That is, they are really the barred variables.

Equation (17) can also be rewritten as the operator equation,

$$\lambda u(x) = K[u(x)], \tag{18}$$

where K is the linear operator

$$K: \qquad u(x) \quad \rightarrow \quad R_0 \int_{-\frac{L}{2}}^{\frac{L}{2}} k(x + c - y) \, u(y) \, dy. \qquad (19)$$

The parameter λ is an eigenvalue of the operator, while $u(x) \neq 0$ is the corresponding eigenfunction. In general, this eigenvalue problem is nasty, but if the operator K is compact (or completely continuous), the problem simplifies. The eigenvalues of a compact linear operator form a discrete set, the point spectrum $\Lambda = \{\lambda_0, \lambda_1, \lambda_2, \ldots\}$. This set may be finite, countably infinite, or empty [33, 38]. Each eigenvalue has finite multiplicity and eigenvalues can only accumulate at zero. Compact operators are, in many ways, similar to matrices.

What do we need for K to be compact? It helps if the domain of our problem is closed and bounded. So, for mathematical convenience, let us now impose the restriction that $x \in [-L/2, L/2]$ and $y \in [-L/2, L/2]$ for finite L. In addition, it also helps if our kernel $k(x - y)$ is a continuous function [44] or, more generally, if $k(x - y)$ is a continuous function for $x \neq y$ and if there are numbers, m and $a < 1$, such that $|k(x - y)| \leq m|x - y|^{-a}$ [33]. All of the kernels in this chapter satisfy one or the other of these conditions.

In general, the eigenvalues λ of problem (17) are complex. If, however, the kernel is strictly positive, we can take advantage of Jentzsch's (1912) [37] theorem (see also [45] and [31]). This theorem is analogous to the Perron–Frobenius theorem for positive matrices. For our integral operator, it guarantees the existence of a simple, positive eigenvalue of largest modulus that dominates all other eigenvalues. The eigenfunction for this eigenvalue is positive. If the conditions of Jentzsch's theorem are met, the stability of trivial solution (11) changes as the dominant eigenvalue passes through $\lambda = 1$. For our problem, this occurs at the critical shift speed c^*.

The restriction that the kernel is strictly positive is important. If the kernel is only nonnegative, eigenvalues need not exist. If they do exist, the spectral radius of the operator K,

$$r(K) = \max_{\lambda_i \in \Lambda} |\lambda_i|, \qquad (20)$$

is a (positive) eigenvalue with a nonnegative eigenfunction [38]. In this case, stability of the trivial solution is still lost through $\lambda = 1$.

All the redistribution kernels in this chapter are nonnegative. Continuous redistribution kernels with infinite support are positive and satisfy Jentzsch's theorem. Kernels with compact support need not satisfy this theorem. If, however, the radius of support is sufficiently large (relative to the patch size L and speed c), Jentzsch's theorem does apply. We will soon find ourselves approximating kernels of infinite support with kernels of compact support. In these cases, we will try to choose parameters that guarantee that Jentzsch's theorem is still satisfied.

Previously, we [93] showed that eigenvalue problem (17) simplifies to a finite-dimensional problem in linear algebra if its kernel is separable. A separable kernel can be written as a finite sum, with each term in the sum the product of a function of x alone and a function of y alone. Taking advantage of this fact, we determined the critical shift speed c^* for a simple, separable toy problem.

Separable kernels are, however, rare. We now consider more general (numerical and analytical) methods that allow us to calculate the dominant eigenvalue and the critical shift speed c^*.

4 Numerical Approaches

One simple numerical approach for computing the dominant eigenvalue of problem (17) is the power method. As with matrix equations, the basic idea is to take an initial guess for the eigenvector and to repetitively rescale and iterate. More precisely, at each step of the process, we take our current estimate of the eigenfunction, $u_t(x)$, rescale this function using its sup norm over $D = [-L/2, L/2]$,

$$\tilde{u}_t(x) = \frac{u_t(x)}{\sup\limits_{x \in D} u_t(x)}, \tag{21}$$

and use the recurrence relation

$$u_{t+1}(x) = K[\tilde{u}_t(x)], \tag{22}$$

to obtain a new, improved estimate of our eigenfunction. Operator K, (19), accounts for both growth and dispersal. The dominant eigenvalue of the integral operator is now approximated by

$$\sup\limits_{x \in D} u_t(x) \tag{23}$$

for large t. The power method is easy to implement, but it can be computationally inefficient.

More efficient approaches are based on Nyström's method [19,74]. For separable kernels, problem (17) simplifies to finding eigenvalues for a finite-dimensional matrix. It makes sense, therefore, to approximate our integral operator with a matrix. To do this, we first discretize our integral using a quadrature rule.

Let us consider, for example, the repeated trapezoidal rule. We divide the domain of integration, $[-L/2, L/2]$, into N equal subintervals of length $\Delta y = L/N$. Replacing the variable y in (17) with grid points

$$y_j = -\frac{L}{2} + j \cdot \Delta y, \qquad j = 0, 1, \ldots, N, \tag{24}$$

we approximate the integral in (17) using the trapezoidal rule,

$$\int_{-\frac{L}{2}}^{\frac{L}{2}} k(x + c - y)\, u(y)\, dy \;\approx\; \frac{\Delta y}{2} \sum_{j=0}^{N-1} [k(x + c - y_j)\, u(y_j) \tag{25}$$

$$+ k(x + c - y_{j+1})\, u(y_{j+1})].$$

If we now evaluate the function $u(x)$ at the grid points $x_i = y_i$, $i = 0, 1, 2, \ldots, N$, eigenvalue problem (17) reduces to

$$\lambda u(x_i) = R_0 \frac{\Delta y}{2} \sum_{j=0}^{N-1} [k(x_i + c - y_j) u(y_j) + k(x_i + c - y_{j+1}) u(y_{j+1})]. \quad (26)$$

Finally, if we denote $u_i = u(x_i)$ and

$$A_{i0} = \frac{\Delta x}{2} k(x_i + c - y_0), \quad (27)$$

$$A_{ij} = \Delta x \, k(x_i + c - y_j), \qquad 1 \le j \le N - 1,$$

$$A_{iN} = \frac{\Delta x}{2} k(x_i + c - y_N),$$

we obtain the finite-dimensional linear system

$$\lambda u_i = R_0 \sum_{j=0}^{N} A_{ij} u_j, \quad (28)$$

for $i = 0, \ldots, N$.

Thus, by employing Nyström's method, we transform the analysis of the dominant eigenvalue of an integral operator into the analysis of the dominant eigenvalue λ of linear system (28).

We can now analyze linear system (28) in one of two ways. The first approach is to determine the eigenvalues of system (28) directly. The eigenvalues may be obtained using commands such as *eig*, *eigen*, or *spec* in computing environments such as MATLAB, R, or Scilab or by using well-known routines from numerical libraries such as *Numerical Recipes* [74], LAPACK [2], or the GNU Scientific Library [23]. These commands and routines commonly balance a matrix, reduce the balanced matrix to Hessenberg form, and find the eigenvalues of the Hessenberg matrix using a QR algorithm. Please see [74] for further details. Having found the eigenvalues, we now choose the dominant eigenvalue. Since this eigenvalue depends continuously on the parameters of the model, we can find the critical value for c, corresponding to $\lambda = 1$, using a standard root-finding algorithm, such as the method of bisection or Brent's method [9, 74].

As an alternative, set λ, in linear system (28), equal to one. Then, use an efficient algorithm, such as LU decomposition [74], to evaluate the determinant of the system. Finally, use a numerical root finder to locate the value of c that makes the determinant zero. This last approach has the advantage of being simple to implement from scratch, but has the disadvantage that you are not guaranteed that $\lambda = 1$ is the dominant eigenvalue.

5 Analytic Approximations

In addition to solving for c^* numerically, we want analytic estimates of the critical speed. The easiest way to obtain these estimates is to assume that the eigenfunction and the kernel in (17) can be approximated using single or double series of suitably chosen basis functions. These basis functions should be complete and linearly independent. Ideally, they should also be orthogonal. Obvious candidates include trigonometric functions (sines and cosines) and orthogonal polynomials such as Chebyshev, Hermite, Jacobi, Laguerre, or Legendre polynomials.

For convenience, we now expand the kernel $k(x + c - y)$ in the double series

$$k(x + c - y) = \sum_{i=0}^{\infty} \sum_{j=0}^{\infty} A_{ij} \, X_i(x) \, X_j(y) \,, \tag{29}$$

where $X_i(x)$ and $X_j(y)$ are Legendre polynomials relative to the interval $[-L/2, \, L/2]$. (Please see the appendix for a brief introduction to Legendre polynomials). The coefficients A_{ij}, which depend on c, are, by (96),

$$A_{ij} = \frac{(2i+1)(2j+1)}{L^2} \int_{-L/2}^{L/2} \int_{-L/2}^{L/2} k(x + c - y) X_i(x) X_j(y) \, dy \, dx \,. \tag{30}$$

If we insert expansion (29) into eigenvalue equation (17), we see that

$$\lambda \, u(x) = R_0 \sum_{i=0}^{\infty} \left(\sum_{j=0}^{\infty} A_{ij} \int_{-L/2}^{L/2} X_j(y) u(y) \, dy \right) X_i(x) \,. \tag{31}$$

We will treat this equation as an expansion of the eigenfunctions, $u(x)$, in Legendre polynomials relative to the interval $[-L/2, \, L/2]$,

$$u(x) = \sum_{i=0}^{\infty} a_i \, X_i(x) \,. \tag{32}$$

The coefficients a_i clearly satisfy

$$a_i = \frac{R_0}{\lambda} \sum_{j=0}^{\infty} A_{ij} \int_{-L/2}^{L/2} X_j(y) u(y) \, dy \,, \quad i = 0, 1, 2, \dots . \tag{33}$$

Expansion (32) presents the eigenfunctions as a linear combination of the orthogonal polynomials $X_i(x)$. If we use this expansion to eliminate $u(y)$ in coefficient equation (33), we see that

$$\lambda\, a_i \;=\; R_0 \sum_{j=0}^{\infty} A_{ij} \int_{-L/2}^{L/2} X_j(y) \sum_{k=0}^{\infty} a_k\, X_k(y)\, dy \tag{34}$$

$$=\; R_0 \sum_{k=0}^{\infty} \left(\sum_{j=0}^{\infty} A_{ij} \int_{-L/2}^{L/2} X_j(y)\, X_k(y)\, dy \right) a_k \,.$$

Since our Legendre polynomials are orthogonal and satisfy

$$\int_{-L/2}^{L/2} [X_i(x)]^2\, dx \;=\; \frac{L}{2i+1}\,, \tag{35}$$

it follows that

$$\lambda\, a_i \;=\; R_0\, L \sum_{j=0}^{\infty} \frac{A_{ij}}{2j+1}\, a_j \tag{36}$$

for $i = 0, 1, 2, \ldots$ and for $j = 0, 1, 2, \ldots$.

System (36) is an infinite-dimensional system of linear algebraic equations for the eigenvalues λ of the trivial solution. We can approximate the eigenvalues of largest modulus by truncating this system so that $i = 0, 1, \ldots, N$ and $j = 0, 1, \ldots, N$ for finite N. In many cases, N need not be large.

Indeed, in some cases, $N = 0$ will suffice. For $N = 0$, we treat eigenfunctions, by (32), as constants,

$$u(x) \;\approx\; a_0\, X_0(x) \;=\; a_0\,. \tag{37}$$

System (36), in turn, reduces to

$$\lambda\, a_0 \;\approx\; R_0\, L\, A_{00}\, a_0\,. \tag{38}$$

After dividing both sides of this last equation by a_0, we obtain

$$\lambda \;\approx\; R_0\, L\, A_{00}\,, \tag{39}$$

where, by coefficient equation (30),

$$A_{00} \;=\; \frac{1}{L^2} \int_{-L/2}^{L/2} \int_{-L/2}^{L/2} k(x+c-y)\, dy\, dx\,. \tag{40}$$

At the critical speed $c = c^*$, $\lambda = 1$, and we conclude that

$$1 \;=\; \frac{R_0}{L} \int_{-L/2}^{L/2} \int_{-L/2}^{L/2} k(x+c^*-y)\, dy\, dx\,. \tag{41}$$

We can often use this last equation to obtain good estimates of the critical speed c^*.

We will refer to (39) as our $N = 0$ eigenvalue approximation. Equation (41) is our $N = 0$ critical-speed equation. The right hand side of these equations is the product of the net reproductive rate and the average dispersal success [56, 88, 89] of our shifted kernel.

A more precise estimate of the critical speed can be obtained by letting $N = 1$. For $N = 1$, we treat eigenfunctions as linear functions,

$$u(x) \approx a_0 X_0(x) + a_1 X_1(x) = a_0 + \frac{2a_1}{L} x .$$ (42)

System (36) now reduces to $N = 1$ approximation

$$\lambda \begin{bmatrix} a_0 \\ a_1 \end{bmatrix} = R_0 L \begin{bmatrix} A_{00} & \dfrac{A_{01}}{3} \\ A_{10} & \dfrac{A_{11}}{3} \end{bmatrix} \begin{bmatrix} a_0 \\ a_1 \end{bmatrix}.$$ (43)

At the critical speed, $c = c^*$, $\lambda = 1$, and we have that

$$\begin{bmatrix} R_0 L A_{00} - 1 & \dfrac{1}{3} R_0 L A_{01} \\ R_0 L A_{10} & \dfrac{1}{3} R_0 L A_{11} - 1 \end{bmatrix} \begin{bmatrix} a_0 \\ a_1 \end{bmatrix} = \begin{bmatrix} 0 \\ 0 \end{bmatrix}.$$ (44)

We want nontrivial eigenvectors, and so we require that this system be singular,

$$\begin{vmatrix} R_0 L A_{00} - 1 & \dfrac{1}{3} R_0 L A_{01} \\ R_0 L A_{10} & \dfrac{1}{3} R_0 L A_{11} - 1 \end{vmatrix} = 0.$$ (45)

This is our $N = 1$ critical-speed equation. It can be used to obtain improved estimates of the critical speed c^*.

We can proceed, in a similar way, for higher N. Often, however, low-order estimates of the critical speed do surprisingly well.

6   A Simple Example

We illustrate the above approximation scheme with a simple example. This is a toy problem that we have chosen for its analytic tractability. We will consider more realistic kernels in a later section.

Consider the symmetric, quadratic kernel

$$k(x) = \begin{cases} \dfrac{3}{4b}\left(1 - \dfrac{x^2}{b^2}\right), & |x| \le b, \\[2mm] 0, & |x| > b, \end{cases} \qquad (46)$$

for $b > 0$. The coefficient at the front of this kernel has been chosen to guarantee that the kernel integrates to one.

If we add the restriction that

$$-b < x + c - y < b \qquad (47)$$

for all x and y in the closed interval $[-L/2, L/2]$, our kernel is positive over the patch and the conditions for Jentzsch's theorem are satisfied. Eigenvalue problem (17) now reduces to

$$\lambda\, u(x) = R_0 \int_{-L/2}^{L/2} \frac{3}{4b}\left[1 - \frac{(x + c - y)^2}{b^2}\right] u(y)\, dy\,. \qquad (48)$$

The kernel of this eigenvalue problem is, in fact, separable. It is easy to show that all eigenfunctions $u(x)$ of this problem are quadratic in x and that system

$$\lambda \begin{bmatrix} a_0 \\ a_1 \\ a_2 \end{bmatrix} = R_0\, L \begin{bmatrix} A_{00} & \dfrac{A_{01}}{3} & \dfrac{A_{02}}{5} \\[2mm] A_{10} & \dfrac{A_{11}}{3} & \dfrac{A_{12}}{5} \\[2mm] A_{20} & \dfrac{A_{21}}{3} & \dfrac{A_{22}}{5} \end{bmatrix} \begin{bmatrix} a_0 \\ a_1 \\ a_2 \end{bmatrix}, \qquad (49)$$

with coefficients

$$A_{00} = \frac{3}{4b} - \frac{1}{8b^3}(L^2 + 6c^2), \qquad (50)$$

$$A_{10} = -\frac{3cL}{4b^3}, \quad A_{01} = \frac{3cL}{4b^3},$$

$$A_{20} = -\frac{L^2}{8b^3}, \quad A_{11} = \frac{3L^2}{8b^3}, \quad A_{02} = -\frac{L^2}{8b^3},$$

$$A_{21} = A_{12} = A_{22} = 0$$

gives exact eigenvalues.

Nevertheless, (39) and (43) give us good approximations for the dominant eigenvalue. For $N = 0$ approximation (39),

$$\lambda \approx R_0\, L\, A_{00} = R_0\, L \left[\frac{3}{4b} - \frac{1}{8b^3}(L^2 + 6c^2)\right]. \qquad (51)$$

Setting $\lambda = 1$ and solving for c gives us

$$c^* \approx \pm \sqrt{b^2 \left(1 - \frac{4b}{3R_0L}\right) - \frac{L^2}{6}}. \tag{52}$$

For two-row, two-column ($N = 1$) approximation (43),

$$\lambda \begin{bmatrix} a_0 \\ a_1 \end{bmatrix} = R_0 L \begin{bmatrix} \dfrac{3}{4b} - \dfrac{1}{8b^3}(L^2 + 6c^2) & \dfrac{cL}{4b^3} \\ -\dfrac{3cL}{4b^3} & \dfrac{L^2}{8b^3} \end{bmatrix} \begin{bmatrix} a_0 \\ a_1 \end{bmatrix}, \tag{53}$$

the characteristic equation is

$$\lambda^2 + \frac{3R_0L}{4b^3}(c^2 - b^2)\lambda + \frac{R_0^2 L^4}{64b^6}\left[6(c^2 + b^2) - L^2\right] = 0. \tag{54}$$

Setting $\lambda = 1$ and solving for c gives us

$$c^* \approx \pm \sqrt{\frac{(R_0 L^3 - 8 b^3)(R_0 L^3 - 6 b^2 R_0 L + 8 b^3)}{6 R_0 L (R_0 L^3 + 8 b^3)}}. \tag{55}$$

Finally, for $N = 2$, the characteristic equation is

$$\lambda^3 + \frac{3(c^2 - b^2)R_0 L}{4 b^3}\lambda^2 + \frac{3R_0^2 L^4 \left[5(c^2 + b^2) - L^2\right]}{160 b^6}\lambda + \frac{R_0^3 L^9}{2560 b^9} = 0. \tag{56}$$

Setting $\lambda = 1$ and solving for c gives us

$$c^* = \pm \sqrt{\frac{(8 b^3 - R_0 L^3)(R_0^2 L^6 - 40 b^3 R_0 L^3 + 240 b^5 R_0 L - 320 b^6)}{240 b^3 R_0 L(R_0 L^3 + 8 b^3)}}. \tag{57}$$

In Fig. 2, $N = 0$ approximation (52), $N = 1$ approximation (55), and the exact ($N = 2$) value of the critical speed, (57), are plotted against the net reproductive rate R_0. Since kernel (46) is symmetric, the curves in Fig. 4 are symmetric with respect to the R_0 axis. Shifting the patch to the right or left has the same effect on population persistence. For $N = 0$, (52) gives a good approximation to the true critical-speed curve for R_0 small. $N = 1$ approximation (55) is, in turn, visually indistinguishable from the exact critical-speed curve for both low and high values of R_0.

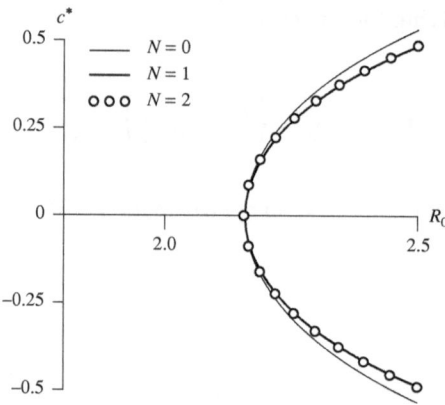

Fig. 2 The critical speed c^* plotted as a function of the net reproductive rate R_0 for quadratic kernel (46). The three curves depict $N = 0$ approximation (52), $N = 1$ approximation (55), and exact solution (57). Since kernel (46) is symmetric, these curves are symmetric about the R_0 axis. The $N = 1$ approximation is visually indistinguishable from the exact solution. Here, $b = 1.5$ km, and $L = 1$ km; we restricted $|c| < 0.5$ km/year in order to satisfy (47)

7 A Simplifying Approximation

In the above example, our $N = 0$ and $N = 1$ expansions both generated good estimates of the critical range-shift speed. Our kernel, moreover, were sufficiently simple that we could easily solve for c^*. For many kernels, unfortunately, it is much harder to solve for c^*.

For sufficiently smooth kernels, we will therefore expand the kernel $k(x + c - y)$ in a Taylor series in $c^* - y$. If the kernel is locally quadratic, we can then keep the first three terms in our Taylor series and approximate c^* using the quadratic formula. We illustrate the procedure for $N = 0$. This procedure gives us a simple formula for c^* that should be accurate for small critical speeds.

For $N = 0$,

$$\lambda \approx \frac{R_0}{L} \int_{-L/2}^{L/2} \int_{-L/2}^{L/2} k(x + c - y) \, dy \, dx \,. \tag{58}$$

Rewriting $k(x + c - y)$ as a Taylor series about x yields

$$\lambda \approx \frac{R_0}{L} \sum_{n=0}^{\infty} \frac{1}{n!} \left[\int_{-L/2}^{L/2} \int_{-L/2}^{L/2} k^{(n)}(x)(c - y)^n \, dy \, dx \right]. \tag{59}$$

If we keep the first three terms in the Taylor series and set $\lambda = 1$, we obtain

$$1 \approx R_0 \int_{-L/2}^{L/2} k(x)\, dx + \frac{R_0}{L} \int_{-L/2}^{L/2} k'(x)\, dx \left[\int_{-L/2}^{L/2} (c^* - y)\, dy \right] \qquad (60)$$

$$+ \frac{R_0}{L} \int_{-L/2}^{L/2} k''(x)\, dx \left[\frac{1}{2} \int_{-L/2}^{L/2} (c^* - y)^2\, dy \right].$$

After calculating the integrals in y, we obtain the quadratic equation (in c^*)

$$\alpha c^{*2} + 2\beta c^* + \gamma + \frac{L^2 \alpha}{12} - \frac{1}{R_0} \approx 0, \qquad (61)$$

where

$$\alpha = \frac{1}{2} \left[k'\left(\frac{L}{2}\right) - k'\left(-\frac{L}{2}\right) \right], \qquad \beta = \frac{1}{2} \left[k\left(\frac{L}{2}\right) - k\left(-\frac{L}{2}\right) \right], \qquad (62)$$

and

$$\gamma = \int_{-\frac{L}{2}}^{\frac{L}{2}} k(x)\, dx . \qquad (63)$$

We can now use the quadratic formula to solve for the critical speed

$$c^* \approx -\frac{\beta}{\alpha} \pm \frac{1}{\alpha} \sqrt{\beta^2 - \alpha(\gamma + L^2\alpha/12 - 1/R_0)} . \qquad (64)$$

The coefficient β of the linear term in quadratic equation (61) vanishes when the kernel $k(x)$ is symmetric. Indeed, for symmetric kernels,

$$k\left(\frac{L}{2}\right) = k\left(-\frac{L}{2}\right), \qquad \text{and} \qquad k'\left(\frac{L}{2}\right) = -k'\left(-\frac{L}{2}\right). \qquad (65)$$

Approximation (61) then reduces to

$$\alpha\, c^{*2} + \gamma + \frac{L^2 \alpha}{12} - \frac{1}{R_0} \approx 0 . \qquad (66)$$

Solving for c^*, we obtain the remarkably simple formula

$$c^* \approx \pm \sqrt{\frac{1}{R_0 \alpha} - \frac{\gamma}{\alpha} - \frac{L^2}{12}} . \qquad (67)$$

Thus, for symmetric kernels, our quadratic approximation preserves the symmetry of c^* with respect to R_0.

8 Realistic Kernels

Let us now apply our numerical procedures and/or our analytical approximations to some realistic redistribution kernels. We will look at three well-known workhorses: the Gaussian, Laplace, and Cauchy distributions. In addition, we will look at modified Bessel kernel (3).

8.1 Gaussian Distribution

Let us first consider the Gaussian kernel

$$k(x) = \frac{1}{\sqrt{2\pi\sigma^2}} e^{-x^2/(2\sigma^2)} , \tag{68}$$

with standard deviation $\sigma > 0$. The Gaussian is the archetypal mesokurtic distribution; it has a special role in the theory of dispersal because of its strong connection to both the diffusion equation and the central limit theorem.

For this kernel, eigenvalue problem (17) reduces to

$$\lambda u(x) = R_0 \int_{-L/2}^{L/2} \frac{1}{\sqrt{2\pi\sigma^2}} e^{-(x+c-y)^2/(2\sigma^2)} u(y) \, dy . \tag{69}$$

The Gaussian kernel is not separable; we must, therefore, rely on numerical or analytical approximations to calculate the critical range-shift speed for this kernel [93].

Since kernel (68) is symmetric, let us first consider Taylor-series approximation (67). For this symmetric kernel,

$$\alpha = k'\left(\frac{L}{2}\right) = -\frac{L}{2\sigma^3\sqrt{2\pi}} e^{-L^2/(8\sigma^2)} \tag{70}$$

and

$$\gamma = \int_{-L/2}^{L/2} k(x) \, dx = \text{erf}\left(\frac{\sqrt{2}L}{4\sigma}\right) , \tag{71}$$

where the error function, erf(x), is given by the integral

$$\text{erf}(x) = \frac{2}{\sqrt{\pi}} \int_0^x e^{-z^2} \, dz . \tag{72}$$

Formula (67) thus yields

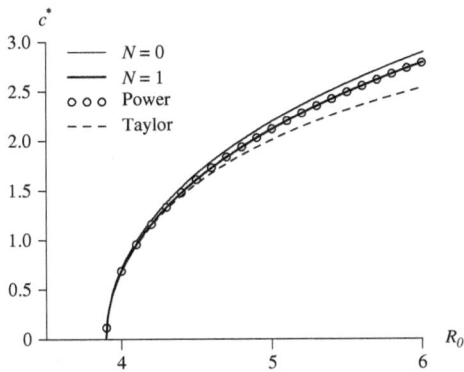

Fig. 3 The critical speed c^* plotted as a function of the net reproductive rate R_0 for Gaussian kernel (68). The four curves depict roots of $N = 0$ and $N = 1$ critical-speed equations (41) and (45), the numerical output of the power method, and Taylor-series approximation (73). Here, $\sigma = 3.0$ km, and $L = 2.0$ km. The $N = 0$ roots overestimate, but the $N = 1$ roots agree with, the power-method curve. Taylor-series approximation (73) provides good estimates of c^* for small critical speeds, but underestimates c^* for large critical speeds

$$c^* \approx \pm \sqrt{\frac{2\sigma^3 \sqrt{2\pi} \left[R_0 \, \mathrm{erf}\left(\sqrt{2} \, L/(4\sigma) \right) - 1 \right]}{R_0 \, L \, e^{-L^2/(8\sigma^2)}} - \frac{L^2}{12}}. \tag{73}$$

For the Gaussian distribution, we can no longer extract c^* from critical-speed equations (41) and (45) analytically. We did, however, extract critical speeds from these equations numerically, for comparison, by evaluating the coefficients A_{00}, A_{10}, A_{01}, and A_{11} as Riemann sums and by then solving for c^* using a simple root-finding method, the method of bisection. The Riemann sums were calculated using a 100×100 grid; the method of bisection was run with a tolerance of 1×10^{-6}.

Finally, we determined c^* from eigenvalue problem (17) numerically, using both the power method and Nyström's method (see Sect. 4). For the power method, we iterated 250 times, for each value of R_0, using an FFT-assisted implementation of the extended trapezoidal rule with 2^{10} nodes. For Nyström's method, we used MATLAB's *eigs* command, with 100 grid points, and a root finder, the method of bisection, with tolerance 10^{-8}. The two numerical approaches gave identical answers; we illustrate our results using output from the power method.

Figure 3 shows plots of the critical speed c^*, as a function of R_0, for a Gaussian kernel with standard deviation $\sigma = 3.0$ km and patch width $L = 2.0$ km. The curves were obtained using, from top to bottom, $N = 0$ and $N = 1$ critical-speed equations (41) and (45), the power method, and Taylor-series approximation (73). Roots of the $N = 0$ equation slightly overestimate the true critical speed, but the roots of the $N = 1$ equation are visually indistinguishable from the numerical output of the power method. Taylor-series approximation (73) provides good estimates of c^* for small critical speeds, but underestimates c^* for large critical speeds.

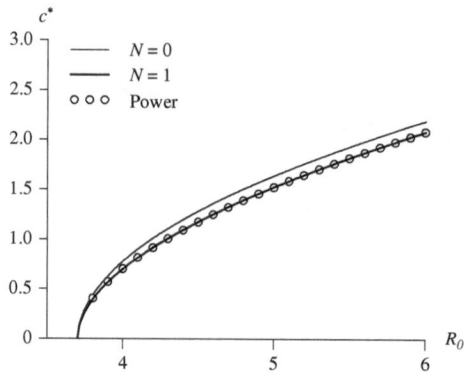

Fig. 4 The critical speed c^* plotted as a function of the net reproductive rate R_0 for Laplace kernel (74). The three curves depict roots of $N = 0$ and $N = 1$ critical-speed equations (41) and (45) and the numerical output of the power method. Here, $b = 3.0\,\text{km}$, and $L = 2.0\,\text{km}$. The $N = 0$ roots overestimate, but the $N = 1$ roots agree with, the power-method curve. Since the Laplace distribution is not differentiable at the origin, we do not plot a Taylor-series approximation for this example

8.2 Laplace Distribution

The Laplace distribution,

$$k(x) = \frac{1}{2b}e^{-|x|/b} , \tag{74}$$

is a symmetric, leptokurtic dispersal kernel that is frequently encountered in empirical studies [e.g., 60,83,85]. It can also arise, in models, from a combination of diffusion and settling or advection and settling [57,63]. Kotz et al. [43] have written the definitive treatise analyzing the Laplace distribution from a statistical viewpoint.

For the Laplace distribution, eigenvalue problem (17) reduces to

$$\lambda u(x) = R_0 \int_{-L/2}^{L/2} \frac{1}{2b}\, e^{-|x+c-y|/b}\, u(y)\, dy . \tag{75}$$

Since the Laplace distribution is not separable and is not differentiable at the origin, we must rely on numerical methods to calculate the critical range-shift speed.

Figure 4 shows plots of the critical speed c^*, as a function of R_0, for a Laplace dispersal kernel with $b = 3.0\,\text{km}$ and patch width $L = 2.0\,\text{km}$. These curves were obtained using (from top to bottom) $N = 0$ and $N = 1$ critical-speed equations (41) and (45) and the power method. Since the Laplace distribution is not differentiable at the origin, we do not plot Taylor-series approximation (67) for this distribution. The plotted curves were produced in the same manner as for the Gaussian kernel. Roots of the $N = 0$ equation slightly overestimate the true critical speed, but roots of the $N = 1$ equation are, once again, visually indistinguishable from the numerical output of the power method.

8.3   Cauchy Distribution

The Cauchy distribution,

$$k(x) = \frac{1}{\pi b \left(1 + \frac{x^2}{b^2}\right)}, \tag{76}$$

is a symmetric, fat-tailed distribution with no mean, variance, or higher moments. The Cauchy distribution, and related power law models, have proven important in studies of spore dispersal gradients [22], long-distance dispersal [82], and the spread of plant diseases [10, 58, 81].

For the Cauchy distribution, eigenvalue problem (17) reduces to

$$\lambda u(x) = R_0 \int_{-L/2}^{L/2} \frac{u(y)}{\pi b \left[1 + \frac{(x+c-y)^2}{b^2}\right]} \, dy \,. \tag{77}$$

Since the Cauchy distribution is not separable, we must rely on numerical or analytical approximations to calculate the critical range-shift speed.

In this instance,

$$\alpha = k'\left(\frac{L}{2}\right) = -\frac{bL}{\pi \left(b^2 + \frac{L^2}{4}\right)^2} \,. \tag{78}$$

and

$$\gamma = \int_{-L/2}^{L/2} k(x) \, dx = \frac{2}{\pi} \arctan \frac{L}{2b} \,. \tag{79}$$

Taylor-series approximation (67) now produces

$$c^* \approx \pm \sqrt{\frac{[2R_0 \arctan(L/2b) - \pi] (4b^2 + L^2)^2}{16 R_0 \, bL} - \frac{L^2}{12}}. \tag{80}$$

Figure 5 shows plots of the critical speed c^*, as a function of R_0, for a Cauchy dispersal kernel with $b = 2.0$ km and patch width $L = 2.0$ km. These curves were obtained using (from top to bottom) $N = 0$ and $N = 1$ critical-speed equations (41) and (45), the power method, and Taylor-series approximation (80). These curves were produced in the same manner as for the Gaussian kernel. Roots of the $N = 0$ equation slightly overestimate the true critical speed, but roots of the $N = 1$ equation are, once again, visually indistinguishable from the numerical output of the power method. The Taylor-series approximation, (80), once again provides good estimates of c^* for small critical speeds, but underestimates c^* for large critical speeds.

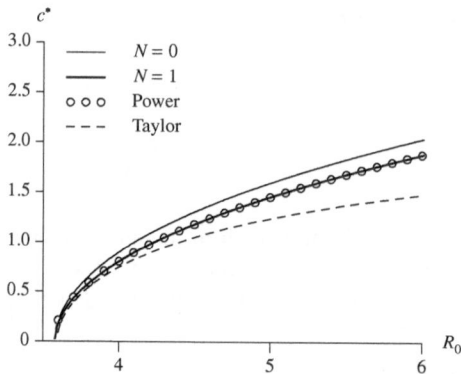

Fig. 5 The critical speed c^* plotted as a function of the net reproductive rate R_0 for Cauchy kernel (76). The four curves depict roots of $N = 0$ and $N = 1$ critical-speed equations (41) and (45), the numerical output of the power method, and Taylor-series approximation (80). Here, $b = 2.0$ km, and $L = 2.0$ km. The $N = 0$ roots overestimate, but the $N = 1$ roots agree with, the power-method curve. Taylor-series approximation (80) provides good estimates of c^* for small critical speeds, but underestimates c^* for large critical speeds

8.4 Modified Bessel Distribution

The modified Bessel distribution,

$$k(x) = \frac{\alpha}{\pi} K_0(\alpha|x|), \tag{81}$$

arises as the marginal distribution of a 2D distribution whose 1D distribution of dispersal distances is the exponential distribution [94]. More generally, this distribution is the product distribution for two normally distributed variates [18, 21].

For this kernel, eigenvalue problem (17) reduces to

$$\lambda u(x) = R_0 \int_{-L/2}^{L/2} \frac{\alpha}{\pi} K_0\left(\alpha|x + c - y|\right) u(y)\, dy. \tag{82}$$

Since the modified Bessel distribution is not separable, we must again rely on numerical or analytical approximations to calculate the critical range-shift speed.

Figure 6 shows the critical speed c^*, as a function of R_0, for modified Bessel function (81) for $\alpha = 2.5, 6, 12$ km^{-1} and patch width $L = 0.5$ km. These curves were obtained using Nyström's method with MATLAB's *eigs* command, with 100 grid points, and a root finder, the method of bisection, with tolerance 10^{-8}. The curves cross in several places and are consistent with the dynamics in Fig. 1.

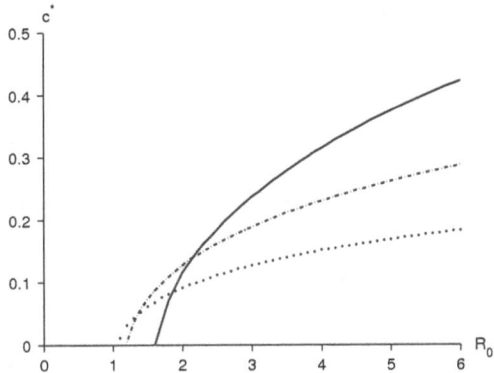

Fig. 6 The critical speed c^* plotted as a function of the net reproductive rate R_0 for modified Bessel kernel (81) and patch size $L = 0.5$ km. The *solid curve* represents a vagile population ($\alpha = 2.5$ km^{-1}), the *dot-dashed curve* represents an intermediate population ($\alpha = 6$ km^{-1}), and the *dotted curve* is a sedentary population ($\alpha = 12$ km^{-1}). The *dotted curve* has both a smaller R_0 intercept and a lower c^* asymptote, implying that the sedentary population does comparatively better for small shift speeds but comparatively worse for high shift speeds. These *curves* were obtained using Nyström's method with MATLAB's *eigs* command, with 100 grid points, and a root finder, the method of bisection, with tolerance 10^{-8}

9 Discussion

Climate change is altering the distributions of wildlife species at a fast rate. To estimate this rate, Loarie et al. [53] introduced and estimated an index of velocity of temperature change; this index has a global (geometric) mean of 0.42 km/year. Parmesan and Yohe [69], in turn, analyzed data for 1,700 species and estimated the average speed of significant poleward range shifts to be 6.1 km/decade. Can species keep up with this rate of climate change? In this chapter, we used an integrodifference equation to model the dynamics of a population that resides in a patch that shifts with speed c. In describing our model, we focused on butterflies, but our model can be applied to many animals and plants. We found that our model has a critical shift speed c^* beyond which the population goes extinct. The critical shift speed c^* depends sensitively on the dispersal ability of the species.

In Sect. 3, we reduced the problem of finding the critical shift speed c^* to an eigenvalue problem. We then showed that c^* can be estimated using numerical methods (Sect. 4) or analytical approximations (Sects. 5 and 7). Simple analytic approximations frequently yield results that are surprisingly close to numerical output. The biggest drawback of our analytical approach is that we must approximate our redistribution kernels with positive functions if we wish to satisfy Jentzsch's theorem. This positivity is easily broken.

Our analyses of the critical speed c^* show that a species' dispersal ability has a profound but complicated impact on its success. When we plotted the critical speed c^* with respect to the net reproductive rate R_0 for butterflies, in Sect. 8.4, we found that the curves (for different dispersal parameters) cross at many points (Fig. 6). Sedentary populations do better (can survive for lower R_0) for low shift speeds, intermediate populations do better for intermediate shift speeds, and vagile populations do better at high shift speeds, but the loci of these transitions depend on where the curves intersect.

These observations suggest different conservation strategies for different dispersal classes. Sedentary species experiencing rapid habitat shifts may benefit from assisted dispersal. For vagile species experiencing slow habitat shifts but severe habitat fragmentation, restoration of degraded habitat may be of greater benefit.

The shape of the redistribution kernel also matters. The kernels in Figs. 3–5 have similar median absolute deviations (MAD $= \sigma/1.4826 = 2.02$, MAD $= b \ln 2 = 2.08$, and MAD $= b = 2$). In spite of this, if we superimpose the critical speed curves for the three kernels (data not shown), they intersect at numerous points. Cauchy distribution (76) and Laplace distribution (74) have smaller R_0 intercepts, presumably because more propagules pile up near the origin for these two distributions.

We hope to extend our analyses in new directions. In Sect. 3, we performed a linear stability analysis for the trivial solution $n^*(\bar{x}) = 0$. This was appropriate because we were only interested in population persistence. For more complicated growth functions, studying the stability of nontrivial traveling pulses may also reveal interesting dynamics. Needless to say, this is a harder problem. One must first determine the nontrivial pulse.

In addition, we have only analyzed a single-species model. Recent studies suggest that species-specific responses to climate change may sunder ecological communities [6,79]. Studying multi-species models should thus prove interesting.

Finally, in our model, eggs hatch and grow and adults emerge only if they are in the climate-shifted patch for year t. Thus climate change only affects reproductive processes. We have built our model in this way because of the demonstrated effect of climate change on phenology and reproductive biology [14,24,50,59,68,90]. The observed dynamics could be quite different if climate change has a direct effect on dispersal. These are all interesting and open problems that merit future research.

Appendix: Legendre Polynomials

The Legendre polynomials are the orthogonal polynomials formed by applying the Gram–Schmidt orthogonalization process to the functions $1, x, x^2, \ldots$ on the interval $[-1, 1]$ with the usual inner product. These polynomials are commonly used to approximate probability density functions (along with their derivatives, integrals, and convolutions) [3,26,52,80] and to solve integral equations [46,48,91,92].

The Legendre polynomials are given by

$$P_i(x) = \frac{1}{2^i i!} \frac{d^i}{dx^i} [(x^2 - 1)^i] \tag{83}$$

for $i = 0, 1, 2, \ldots$ The first few Legendre polynomials are

$$P_0(x) = 1, \quad P_1(x) = x, \quad P_2 = \frac{1}{2}(3x^2 - 1), \quad P_3(x) = \frac{1}{2}(5x^3 - 3x). \quad (84)$$

Additional Legendre polynomials can be generated using the recurrence relation

$$P_{i+1}(x) = \frac{2i+1}{i+1} x P_i - \frac{i}{i+1} P_{i-1}(x). \quad (85)$$

The Legendre polynomials are even functions for n even and odd functions for i odd. Each of the $P_i(x)$ has i distinct and real roots on the interval $(-1, 1)$. For our purposes, the most important property of the Legendre polynomials is that they form an orthogonal system on $[-1, 1]$ with

$$\int_{-1}^{1} P_i(x) P_j(x) \, dx = \frac{2}{2i+1} \delta_{ij}. \quad (86)$$

Here, δ_{ij} is the Kronecker delta, which equals 1 if $i = j$ and 0 if $i \neq j$.

Because the Legendre polynomials form a complete, orthogonal system over the interval $[-1, 1]$, we may expand a function $f(x)$ on this interval in a Legendre (or Fourier–Legendre) series of the form

$$f(x) = \sum_{i=0}^{\infty} a_i P_i(x). \quad (87)$$

Please see [35,76] for convergence conditions. It is easy to show, using orthogonality condition (86), that the coefficients a_i satisfy

$$a_i = \frac{2i+1}{2} \int_{-1}^{1} f(x) P_i(x) \, dx. \quad (88)$$

In a similar way, we can expand a bivariate function, $f(x, y)$, in a double series of the form

$$f(x, y) = \sum_{i=0}^{\infty} \sum_{j=0}^{\infty} A_{ij} P_i(x) P_j(y). \quad (89)$$

The coefficients A_{ij} in this series satisfy

$$A_{ij} = \frac{(2i+1)(2j+1)}{4} \int_{-1}^{1} \int_{-1}^{1} f(x, y) P_i(x) P_j(y) \, dy \, dx. \quad (90)$$

The problem that we are interested in (17), involves an integral over the interval $[-L/2, L/2]$. Rather than rescaling our problem, we find it convenient to follow [76] by introducing Legendre polynomials relative to the interval $[-L/2, L/2]$,

$$X_i(x) = P_i\left(\frac{2x}{L}\right). \tag{91}$$

In light of this transformation, our orthogonality condition, (86), now takes the form

$$\int_{-L/2}^{L/2} X_i(x)\, X_j(x)\, dx = \frac{L}{2i+1}\,\delta_{ij}. \tag{92}$$

When we write a function, $f(x)$, in a series of Legendre polynomials relative to the interval $[-L/2,\ L/2]$,

$$f(x) = \sum_{n=0}^{\infty} a_i\, X_i(x), \tag{93}$$

the coefficients a_i now satisfy

$$a_i = \frac{2i+1}{L} \int_{-L/2}^{L/2} f(x)\, X_i(x)\, dx. \tag{94}$$

Likewise, when we expand a bivariate function, $f(x,\ y)$, in a double series of Legendre polynomials relative to the interval $[-L/2,\ L/2]$,

$$f(x,\ y) = \sum_{i=0}^{\infty}\sum_{j=0}^{\infty} A_{ij}\, X_i(x)\, X_j(y), \tag{95}$$

the coefficients A_{ij} now satisfy

$$A_{ij} = \frac{(2i+1)(2j+1)}{L^2} \int_{-L/2}^{L/2}\int_{-L/2}^{L/2} f(x,\ y)\, X_i(x)\, X_j(y)\, dy\, dx. \tag{96}$$

References

1. W.C. Allee, *The Social Life of Animals* (Norton, New York, 1938)
2. E. Anderson, Z. Bai, C. Bischof, S. Blackford, J. Demmel, J. Dongarra, J.D. Croz, A. Greenbaum, S. Hammarling, A. McKenney, D. Sorensen, *LAPACK User's Guide* (Society for Industrial and Applied Mathematics, Philadelphia, 1999)
3. R.D. Badinelli, Approximating probability density functions and their convolutions using orthogonal polynomials. Eur. J. Oper. Res. **95**, 211–230 (1996)
4. M. Bakkenes, J. Alkemade, F. Ihle, R. Leemans, J. Latour, Assessing effects of forecasted climate change on the diversity and distribution of European higher plants for 2050. Glob. Change Biol. **8**, 390–407 (2002)
5. H. Berestycki, O. Diekmann, C.J. Nagelkerke, P.A. Zegeling, Can a species keep pace with a shifting climate? Bull. Math. Biol. **71**, 399–429 (2009)

6. M.P. Berg, E.T. Kiers, G. Driessen, M. Van Der Heijden, B.W. Kooi, F. Kuenen, M. Liefting, H.A. Verhoef, J. Ellers, Adapt or disperse: understanding species persistence in a changing world. Glob. Change Biol. **16**, 587–598 (2010)

7. A.S. Best, K. Johst, T. Münkemüller, J.M. Travis, Which species will succesfully track climate change? The influence of intraspecific competition and density dependent dispersal on range shifting dynamics. Oikos **116**, 1531–1539 (2007)

8. R.J.H. Beverton, S.J. Holt, *On the Dynamics of Exploited Fish Populations* (Her Majesty's Stationery Office, London, 1957)

9. R.P. Brent, *Algorithms for Minimization Without Derivatives* (Prentice-Hall, Englewood Cliffs, 1973)

10. J.K.M. Brown, M.S. Hovmoller, Aerial dispersal of pathogens on the global and continental scales and its impact on plant disease. Science **297**, 537–541 (2002)

11. L.B. Buckley, Linking traits to energetics and population dynamics to predict lizard ranges in changing environments. Am. Nat. **171**, E1–E19 (2008)

12. M.B. Bush, H. Hooghiemstra, Tropical biotic responses to climate change, in *Climate Change and Biodiversity*, ed. by T.E. Lovejoy, L. Hannah (Yale University Press, New Haven, 2005), pp. 125–137

13. I.C. Chen, J.J. Shiu, B. Suzan, J.D. Holloway, V.K. Chey, H.S. Barlow, J.K. Hill, C.D. Thomas, Elevation increases in moth assemblages over 42 years on a tropical mountain. Proc. Natl. Acad. Sci. USA **106**, 1479–1483 (2009)

14. I. Chuine, Why does phenology drive species distribution? Phil. Trans. R. Soc. B **365**, 3149–3160 (2010)

15. R. Cousens, C. Dytham, R. Law, *Dispersal in Plants: A Population Perspective* (Oxford University Press, Oxford, 2008)

16. R.D. Cousens, When will plant morphology affect the shape of a seed dispersal "kernel"? J. Theor. Biol. **211**, 229–238 (2001)

17. M.J.R. Cowley, C.D. Thomas, D.B. Roy, R.J. Wilson, J.L. Leon-Cortes, D. Gutierrez, C.R. Bulman, R.M. Quinn, D. Moss, K.J. Gaston, Density-distribution relationships in British butterflies. I. The effect of mobility and spatial scale. J. Anim. Ecol. **70**, 410–425 (2001)

18. C.C. Craig, On the frequency function of xy. Ann. Math. Stat. **7**, 1–15 (1936)

19. L.M. Delves, J. Walsh, *Numerical Solution of Integral Equations* (Clarendon Press, Oxford, 1974)

20. R. Engler, A. Guisan, MIGCLIM: predicting plant distribution and dispersal in a changing climate. Divers. Distrib. **15**, 590–601 (2009)

21. B. Epstein, Some applications of the Mellin transform in statistics. Ann. Math. Stat. **19**, 370–379 (1948)

22. B.D.L. Fitt, P.H. Gregory, A.D. Todd, H.A. McCartney, O.C. MacDonald, Spore dispersal and plant disease gradients: a comparison between two empirical models. J. Phytopathol. **118**, 227–242 (1987)

23. M. Galassi, J. Davies, J. Theiler, B. Gough, G. Jungman, P. Aiken, M. Booth, F. Rossi, *GNU Scientific Library: Reference Manual* (Network Theory Ltd., Bristol, 2009)

24. K.J. Gaston, *The Structure and Dynamics of Geographic Ranges* (Oxford University Press, Oxford, 2003)

25. A. Guisan, W. Thuiller, Predicting species distribution: offering more than simple habitat models. Ecol. Lett. **8**, 993–1009 (2005)

26. P. Hall, Comparison of two orthogonal series methods of estimating a density and its derivatives on an interval. J. Multivar. Anal. **12**, 432–449 (1982)

27. S. Harrison, D. Murphy, P. Ehrlich, Distribution of the bay checkerspot butterfly, *Euphydryas editha bayensis*: Evidence for a metapopulation model. Am. Nat. **132**, 360–382 (1988)

28. A. Hastings, K. Higgins, Persistence of transients in spatially structured ecological models. Science **263**, 1133–1136 (1994)

29. R. Hickling, D.B. Roy, J.K. Hill, R. Fox, C.D. Thomas, The distributions of a wide range of taxonomic groups are expanding polewards. Glob. Change Biol. **12**, 450–455 (2006)

30. O. Honnay, K. Verheyen, J. Butaye, H. Jacquemyn, B. Bossuyt, M. Hermy, Possible effects of habitat fragmentation and climate change on the range of forest plant species. Ecol. Lett. **5**, 525–530 (2002)
31. T. Horiguchi, Y. Fukui, A variation of the Jentzsch theorem for a symmetric integral kernel and its application. Interdiscip. Inf. Sci. **2**, 139–144 (1996)
32. B. Huntley, North temperate responses, in *Climate Change and Biodiversity*, ed. by T. Lovejoy, L. Hannah (Yale University Press, New Haven, 2005), pp. 109–124
33. V. Hutson, J.S. Pym, *Applications of Functional Analysis and Operator Theory* (Academic, London, 1980)
34. IPCC, Climate Change 2007: Synthesis Report. Contribution of Working Groups I, II and III to the Fourth Assessment Report of the Intergovernmental Panel on Climate Change. Core Writing Team and R.K. Pachauri, A. Reisinger, IPCC, Geneva (2007)
35. D. Jackson, *The Theory of Approximation* (American Mathematical Society, New York, 1930)
36. C. Jeffree, E. Jeffree, Redistribution of the potential geographical ranges of mistletoe and Colorado beetle in Europe in response to the temperature component of climate change. Funct. Ecol. **10**, 562–577 (1996)
37. R. Jentzsch, Über integralgleichungen mit positivem kern. J. die Reine Angew. Math. **141**, 235–244 (1912)
38. S. Karlin, The existence of eigenvalues for integral operators. Trans. Am. Math. Soc. **113**, 1–17 (1964)
39. A.E. Kelly, M.L. Goulden, Rapid shifts in plant distribution with recent climate change. Proc. Natl. Acad. Sci. USA **105**, 11823–11826 (2008)
40. A. Komonen, A. Grapputo, V. Kaitala, J.S. Kotiaho, J. Päivinen, The role of niche breadth, resource availability and range position on the life history of butterflies. Oikos **105**, 41–54 (2004)
41. M. Kot, W.M. Schaffer, Discrete-time growth-dispersal models. Math. Biosci. **80**, 109–136 (1986)
42. M. Kot, M.A. Lewis, P. Van Den Driessche, Dispersal data and the spread of invading organisms. Ecology **77**, 2027–2042 (1996)
43. S. Kotz, T.J. Kozubowski, K. Podgorski, *The Laplace Distribution and Generalizations: A Revisit with Applications to Communications, Economics, Engineering, and Finance* (Birkhauser, Boston, 2001)
44. E. Kreyszig, *Introductory Functional Analysis with Applications* (Wiley, New York, 1978)
45. S. Krzemiński, Comment on 'A simple proof of the Perron–Frobenius theorem for positive symmetric matrices'. J. Phys. A. Math. Gen. **10**, 1437–1438 (1977)
46. H. Kschwendt, Numerical solution of integral equations using Legendre polynomials. J. Math. Phys. **10**, 1964–1968 (1969)
47. J. Latore, P. Gould, A.M. Mortimer, Spatial dynamics and critical patch size of annual plant populations. J. Theor. Biol. **190**, 277–285 (1998)
48. T.T. Lee, Y.F. Chang, Solutions of convolution integral and integral equations via double general orthogonal polynomials. Int. J. Syst. Sci. **19**, 415–430 (1988)
49. J. Lenoir, J.C. Gégout, P.A. Marquet, P. de Ruffray, H. Brisse, A significant upward shift in plant species optimum elevation during the 20th Century. Science **320**, 1768–1771 (2008)
50. T.M. Letcher, *Climate Change: Observed Impacts on Planet Earth* (Elsevier, Amsterdam, 2009)
51. M.A. Lewis, M.G. Neubert, H. Caswell, J.S. Clark, K. Shea, A guide to calculating discrete-time invasion rates from data, in *Conceptual Ecology and Invasions Biology: Reciprocal Approaches to Nature*, ed. by M.W. Cadotte, S.M. McMahon, T. Fukami (Springer, Dordrecht, 2006), pp. 169–192
52. X.B. Li, F.Q. Gong, A method for fitting probability distributions to engineering properties of rock masses using Legendre orthogonal polynomials. Struct. Saf. **31**, 335–343 (2009)
53. S. Loarie, P. Duffy, H. Hamilton, G.P. Asner, C.B. Field, D.D. Ackerly, The velocity of climate change. Nature **462**(24), 1052–1055 (2009)

54. D.R. Lockwood, A. Hastings, L.W. Botsford, The effects of dispersal patterns on marine reserves: does the tail wag the dog? Theor. Popul. Biol. **61**, 297–309 (2002)
55. F. Lutscher, Density-dependent dispersal in integrodifference equations. J. Math. Biol. **56**, 499–524 (2008)
56. F. Lutscher, M.A. Lewis, Spatially-explicit matrix models. J. Math. Biol. **48**, 293–324 (2004)
57. F. Lutscher, E. Pachepsky, M.A. Lewis, The effect of dispersal patterns on stream populations. SIAM J. Appl. Math. **65**, 1305–1327 (2005)
58. L.V. Madden, G. Hughes, F. van den Bosch, *The Study of Plant Disease Epidemics* (American Phytopathological Society, St. Paul, 2007)
59. J. McCarty, Ecological consequences of recent climate change. Conserv. Biol. **15**, 320–331 (2001)
60. T.E.X. Miller, B. Tenhumberg, Contributions of demography and dispersal parameters to the spatial spread of a stage-structured insect invasion. Ecol. Appl. **20**, 620–633 (2010)
61. D. Murphy, N. Wahlberg, I. Hanski, P.R. Ehrlich, Introducing checkerspots: Taxonomy and ecology, in *On the Wings of Checkerspots: A Model System for Population Biology*, ed. by P.R. Ehrlich, I. Hanski (Oxford University Press, Oxford, 2004), pp. 17–33
62. M.G. Neubert, H. Caswell, Demography and dispersal: calculation and sensitivity analysis of invasion speed for structured populations. Ecology **81**, 1613–1628 (2000)
63. M.G. Neubert, M. Kot, M.A. Lewis, Dispersal and pattern formation in a discrete-time predator-prey model. Theor. Popul. Biol. **48**, 7–43 (1995)
64. P. Opdam, D. Wascher, Climate change meets habitat fragmentation: linking landscape and biogeographical scale levels in research and conservation. Biol. Conserv. **117**, 285–297 (2004)
65. W.A. Ozinga, C. Römermann, R.M. Bekker, A. Prinzing, W.L.M. Tamis, J.H.J. Schaminée, S.M. Hennekens, K. Thompson, P. Poschlod, M. Kleyer, J.P. Bakker, J.M. van Groenendael, Dispersal failure contributes to plant losses in NW Europe. Ecol. Lett. **12**, 66–74 (2009)
66. C. Parmesan, Climate and species' range. Nature **382**, 765–766 (1996)
67. C. Parmesan, Detection at multiple levels: *Euphydryas editha* and climate change, in *Climate Change and Biodiversity*, ed. by T.E. Lovejoy, L. Hannah (Yale University Press, New Haven, 2005), pp. 56–60
68. C. Parmesan, Ecological and evolutionary responses to recent climate change. Annu. Rev. Ecol. Evol. Syst. **37**, 637–669 (2006)
69. C. Parmesan, G. Yohe, A globally coherent fingerprint of climate change impacts across natural systems. Nature **421**, 37–42 (2003)
70. C. Parmesan, N. Ryrholm, C. Stefanescu, J. Hill, C. Thomas, H. Descimon, B. Huntley, L. Kaila, J. Kullberg, T. Tammaru, W. Tennent, J. Thomas, M. Warren, Poleward shifts in geographical ranges of butterfly species associated with regional warming. Nature **399**, 579–583 (1999)
71. A.L. Perry, P.J. Low, J.R. Ellis, J.D. Reynolds, Climate change and distribution shifts in marine fishes. Science **308**, 1912–1915 (2005)
72. A.B. Potapov, M.A. Lewis, Climate and competition: the effect of moving range boundaries on habitat invasibility. Bull. Math. Biol. **66**, 975–1008 (2004)
73. J. Pöyry, M. Luoto, R.K. Heikkinen, M. Kuussaari, K. Saarinen, Species traits explain recent range shifts of Finnish butterflies. Glob. Change Biol. **15**, 732–743 (2009)
74. W.H. Press, S.A. Teukolsky, W.T. Vetterling, B.P. Flannery, *Numerical Recipes in C: The Art of Scientific Computing* (Cambridge University Press, Cambridge, 1992)
75. R.B. Primack, S.L. Miao, Dispersal can limit local plant distribution. Conserv. Biol. **6**, 513–519 (1992)
76. G. Sansone, *Orthogonal Functions* (Dover Publications, Mineola, 2004)
77. F.M. Schurr, G.F. Midgley, A.G. Rebelo, G. Reeves, P. Poschlod, S.I. Higgins, Colonization and persistence ability explain the extent to which plant species fill their potential range. Glob. Ecol. Biogeogr. **16**, 449–459 (2007)
78. M.W. Schwartz, Modelling effects of habitat fragmentation on the ability of trees to respond to climatic warming. Biodivers. Conserv. **2**, 51–61 (1992)

79. O. Schweiger, J. Settele, O. Kudrna, S. Klotz, I. Kühn, Climate change can cause spatial mismatch of trophically interacting species. Ecology **89**, 3472–3479 (2008)
80. T.A. Severini, *Elements of Distribution Theory* (Cambridge University Press, New York, 2005)
81. M. Shaw, Modeling stochastic processes in plant pathology. Annu. Rev. Phytopathol. **32**, 523–544 (1994)
82. M.W. Shaw, Simulation of population expansion and spatial pattern when individual dispersal distributions do not decline exponentially with distance. Proc. R. Soc. Lond. B **259**, 243–248 (1995)
83. C.A. Smith, I. Giladi, Y.S. Lee, A reanalysis of competing hypotheses for the spread of the California sea otter. Ecology **90**, 2503–2512 (2009)
84. V. Stevens, C. Turlure, M. Baguette, A meta-analysis of dispersal in butterflies. Biol. Rev. **85**, 625–642 (2010)
85. M.T. Tinker, D.F. Doak, J.A. Estes, Using demography and movement behavior to predict range expansion of the southern sea otter. Ecol. Appl. **18**, 1781–1794 (2008)
86. A. Trakhtenbrot, R. Nathan, G. Perry, D.M. Richardson, The importance of long-distance dispersal in biodiversity conservation. Divers. Distrib. **11**, 173–181 (2005)
87. J.M. Travis, Climate change and habitat destruction: a deadly anthropogenic cocktail. Proc. R. Soc. Lond. B **270**, 467–473 (2003)
88. R.W. Van Kirk, M.A. Lewis, Integrodifference models for persistence in fragmented habitats. Bull. Math. Biol. **59**, 107–137 (1997)
89. R.W. Van Kirk, M.A. Lewis, Edge permeability and population persistence in isolated habitat patches. Nat. Resour. Model. **12**, 37–64 (1999)
90. G. Walther, E. Post, P. Convey, A. Menzel, C. Parmesan, T. Beebee, J. Fromentin, O. Hoegh-Guldberg, F. Bairlein, Ecological responses to recent climate change. Nature **416**, 389–395 (2002)
91. M.L. Wanga, R.Y. Changa, S.Y. Yanga, Double generalized orthogonal polynomial series for the solution of integral equations. Int. J. Syst. Sci. **19**, 459–470 (1988)
92. S. Yalcinbas, M. Aynigul, T. Akkaya, Legendre series solutions of Fredholm integral equations. Math. Comput. Appl. **15**, 371–381 (2010)
93. Y. Zhou, M. Kot, Discrete-time growth-dispersal models with shifting species ranges. Theor. Ecol. **4**, 13–25 (2011)
94. Y. Zhou, M. Kot, The role of the modified Bessel distribution in dispersal models (2013) (in preparation)

Control of Competitive Bioinvasion

Horst Malchow, Alex James, and Richard Brown

Abstract The invasion of alien and displacement of indigenous species is a crucial ecological and economical problem of even increasing significance. Measures to control and perhaps to stop and reverse such invasive processes are urgently needed. Mathematical models are a suitable tool to preview the impact of control measures before utilizing them in nature. Here, a reaction-diffusion model is used to describe the competition and dispersal of invasive and native species. Not only the environment is changing but also growth, harvesting and dispersal of the two competitors vary in space and time. Extreme events such as fires or landslides or any other processes yielding bare re-invadable ground lead to temporary extinction of both species at a randomly chosen time and spatial range. The spatiotemporal dimension of these extreme fragmentation events, the ratio of the dispersal rates of the competing species as well as the selective removal of the invader turn out to be the crucial driving forces of the system dynamics. Finally, the controlling effect of a targeted infection of the invasive species with a specific pathogen is studied in an eco-epidemiological competition-diffusion model.

H. Malchow (✉)
Institute of Environmental Systems Research, University of Osnabrück
Barbarastr. 12, 49076 Osnabrück, Germany
e-mail: horst.malchow@uni-osnabrueck.de

A. James · R. Brown
Department of Mathematics and Statistics, University of Canterbury, Private Bag 4800, Christchurch, New Zealand
e-mail: a.james@math.canterbury.ac.nz; r.brown@math.canterbury.ac.nz

M.A. Lewis et al. (eds.), *Dispersal, Individual Movement and Spatial Ecology*,
Lecture Notes in Mathematics 2071, DOI 10.1007/978-3-642-35497-7_10,
© Springer-Verlag Berlin Heidelberg 2013

1 Introduction

The negative econo-ecological effects of bioinvasions including the spread of infectious diseases [8, 36] have led to a remarkable push of bioinvasion science. Not only an increasing number of laboratory and field studies but also the rapid development of theoretical methods to describe bioinvasions and their control could be noticed during recent years, cf. [17, 31, 39, 50]. Mathematical and computational methods are meanwhile recognized tools to investigate the dynamics of invasions, both supplementary to and initiating field studies as well as control measures. Related summaries and overview publications are for instance [7, 18, 26, 32, 35, 38, 41] as well as [43].

Here, to model the invasion of alien species such as weeds and their competition with indigenous plants, the textbook model of Lotka–Volterra type with diffusion is used. Carrying capacities are not explicitly defined. Growth, selective harvesting of the invading weed as well as spatial spread undergo seasonal cycles. Furthermore, extreme events such as fires or landslides or any other processes yielding bare re-invadable ground lead to temporary extinction of both species at a randomly chosen time and spatial range. In a previous paper [27], it has been shown that, without seasonal cycles of the mentioned parameters, the frequency and spatial dimension of these extreme fragmentation events, the ratio of the dispersal rates of the competing species as well as the efficiency of selective removal of the invader turn out to be the crucial driving forces of the system dynamics. In the first part of the present paper, the robustness of these results against those seasonal cycles is studied.

Furthermore, in the second part, the targeted infection of the invader with a specific pathogen is considered as biological control measure. There are applications of biological methods of bioinvasion control for more than half a century and it has been a changeful history of magnificent successes and risky failures [1, 11, 16, 20, 30, 51]. Ecological and epidemiological models are known since more than 200 years. But it is only about 25 years ago that first attempts to merge these models have been published, cf. [2, 12, 14, 15] as well as [47]. In this paper, the invading model weed will be controlled by a frequency-dependently transmitted fungus infection.

2 A Competition-Diffusion Model with Annual Cycles and Random Extreme Events

For the description of the spatiotemporal invasion of a resident population by a competing alien, the Lotka–Volterra competition-diffusion model is used, i.e.,

$$\frac{\partial N_i(\mathbf{x}, t)}{\partial t} = r_i N_i - N_i \sum_{j=1}^{2} c_{ij} N_j + D_i \Delta N_i \, ; i = 1, 2 \, , \tag{1}$$

where N_1 and N_2 are resident and invader densities at position $\mathbf{x} = \{x_1, x_2\}$ and time t respectively. Carrying capacities will not explicitly be introduced because they can suppress a higher variety of solutions and rather appear as emergent property of the system [13, 22, 23]. The r's stand for the growth rates that can be thought as superposition of biomass generation and loss rates b_1, b_2 and m_1, m_2 respectively as well as density-dependent harvesting h_1, h_2, i.e.,

$$r_i = b_i - m_i - h_i = r_i^* - h_i ; \quad i = 1, 2. \tag{2}$$

The c's are the inter- and intraspecific competition coefficients and the D's the diffusivities. $\Delta = \partial^2/\partial x_1^2 + \partial^2/\partial x_2^2$ is the Laplacian for the considered horizontal processes.

2.1 Existence and Stability Ranges of Spatially Uniform Stationary Solutions

There are the four stationary solutions with their stability ranges:

1.	$(0, 0)$	always unstable,	
2.	$\left(\dfrac{r_1}{c_{11}}, 0\right)$	stable for	$\dfrac{r_2}{r_1} < \dfrac{c_{22}}{c_{12}}; \dfrac{r_2}{r_1} < \dfrac{c_{21}}{c_{11}},$
3.	$\left(0, \dfrac{r_2}{c_{22}}\right)$	stable for	$\dfrac{r_2}{r_1} > \dfrac{c_{22}}{c_{12}}; \dfrac{r_2}{r_1} > \dfrac{c_{21}}{c_{11}},$
4.	$\left(\dfrac{r_1 c_{22} - r_2 c_{12}}{c_{11} c_{22} - c_{12} c_{21}}, \dfrac{r_2 c_{11} - r_1 c_{21}}{c_{11} c_{22} - c_{12} c_{21}}\right)$	stable for	$\dfrac{c_{22}}{c_{12}} > \dfrac{r_2}{r_1} > \dfrac{c_{21}}{c_{11}},$
2./3.	Bistability of extinction states (2,3)	for	$\dfrac{c_{22}}{c_{12}} < \dfrac{r_2}{r_1} < \dfrac{c_{21}}{c_{11}}.$

In the extinction states (2.,3.), the surviving population k is at its emergent carrying capacity r_k/c_{kk}. Later on, the bistability range of both extinction states from the last row is used for modelling strong competition in time and space.

2.2 Annual Cycles of Growth, Harvesting and Diffusion

The growth processes undergo an annual cycle approximated by a cosine

$$r_i^*(t) = r_{i,min}^* + \frac{1}{2}\left(r_{i,max}^* - r_{i,min}^*\right)\left[1 + \cos\left(\frac{2\pi}{a}t\right)\right] ; \quad i = 1, 2. \tag{3}$$

where $r_{i,max}^*$ and $r_{i,min}^*$ are the corresponding summer and winter extrema.

Only the invading species is harvested from spring to autumn, usually through manual removal, i.e., $h_1 = 0$ and $h_2 \geq 0$. Harvesting is not possible instantaneously on the whole managed field of size $L \times L$. The search for the weed starts at one of the field boundaries, say at $\{x_1 \in [0, L], x_2 = 0\}$. Once a weed patch is found, harvesting begins in x_2 direction on a stripe of size $w \times L$ where w is the width of the patch that is removed within time δt. Then, the search continues and after arriving at the other side of the field, $\{x_1 \in [0, L], x_2 = L\}$, it perhaps restarts at the initially chosen boundary. If the search is assumed always in direction of x_2, this procedure is modelled through

$$h_2^*(x_2, t) = h_2 \, g(x_2, t) \, \Theta \left[\epsilon + cos \left(\frac{2\pi}{a} t \right) \right], \tag{4}$$

where $\Theta[.]$ is the Heaviside function and for $g(x_2, t)$ applies

$$\text{if} \quad \int_0^L N_2(x_1, x_{20}, t_0) dx_1 > 0$$

$$\text{then} \quad g(x_2, t) = 1 \text{ for } x_2 \in [x_{20}, x_{20} + w] \text{ and } t \in [t_0, t_0 + \delta t] \tag{5}$$

$$\text{else} \quad g(x_2, t) = 0.$$

The reduced mobility from late autumn to early spring is considered by a corresponding seasonality of the diffusivities

$$D_i^*(t) = D_i \, \Theta \left[\epsilon + cos \left(\frac{2\pi}{a} t \right) \right] ; \, i = 1, 2. \tag{6}$$

2.3 Random Extreme Events and Assisted Long-Distance Transport

As in the previous paper [27], extreme events such as fires or landslides or any other occurrence yielding bare re-invadable ground lead to temporary extinction of both species. It is assumed that these events may randomly take place within certain time intervals and spatial ranges throughout the year.

Furthermore, wind-born or however assisted long-distance transport of seeds is considered in the spring-summer season, cf. (6). At random times raised and transported seeds settle down at a randomly chosen location and form small population patches of random size having in mind typical dispersal distance kernels [6, 33]. If the location falls into a hostile zone the patches have to fight the surrounding enemies. This type of transport has also been called stratified diffusion [42].

The modelling and simulation of the above mentioned random processes is not equation- but rather rule-based similar to formerly developed models of rule-based fish school motion coupled to equation-based resource dynamics [25].

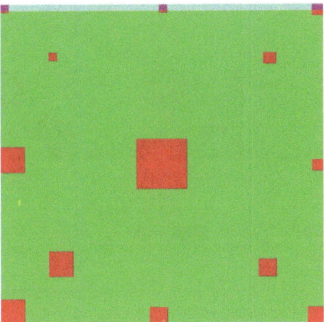

Fig. 1 Initial condition, cf. text

2.4 Numerical Simulations I

As in the previous work [27], the initial conditions have been arbitrarily chosen and are the same for all runs in this section. Because weed patchiness is rather generic e.g. in crops [5,49,54,55], a number of (red) initial invader patches at their emerging carrying capacities of different size has been distributed in a certain artificial way on the habitat of the (green) native species, cf. Fig. 1. At the top, the (teal) initial harvested stripe can be identified. Zero-flux boundary conditions have been applied.

One expected result is that initial invader patches smaller than a certain critical size will be immediately recaptured by the native species. This spatial critical size problem in spatially two- and three-dimensional systems with multiple steady states is known from nucleation theory [9,10,24,34]. Though it is necessary to be stronger or fitter, it is not sufficient to win the competition. One must also have occupied a sufficiently large spatial range.

It is assumed that landslips may randomly take place within intervals of 20 time units and clear areas of up to 50×50 spatial units of a total of 200×200. The time-lag and the size of the landslips are control parameters of the system, the shorter the interval and the greater the spatial dimension the stronger the landslide's impact on the spatiotemporal competition of natives and invaders.

Time is measured in days, space in meters. Hence, denoting the plant's dry weight by dw, the N's are given in $\mathrm{kg\,dw\,m^{-2}}$, the r's and h's in d^{-1}, the c's in $\mathrm{kg\,dw^{-1}\,m^2\,d^{-1}}$ and the D's in $\mathrm{m^2\,d^{-1}}$. ϵ is a dimensionless quantity. The following parameter values have been used:

$$r^*_{1,max} = 1.0\,, \ r^*_{1,min} = 0.3\,, \ r^*_{2,max} = 1.0\,, \ r^*_{2,min} = 0.35\,,$$

$$c_{11} = 1.0\,, \ c_{12} = 1.3\,, \ c_{21} = 1.2\,, \ c_{22} = 1.0\,, \tag{7}$$

$$\epsilon = 0.2\,, \ h_1 = 0.0\,, \ h_2 = 0.35\,, \ D_2 = 22.5\,, \ L = 3000.0\,, \ w = 60.0\,.$$

The competition coefficients of both species have been raised away from the critical value of unity. The invader is assumed to be the stronger competitor, following the enemy release hypothesis [21].

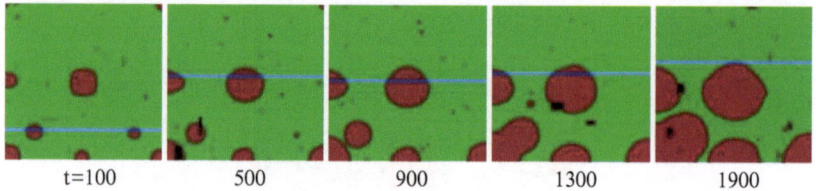

| t=100 | 500 | 900 | 1300 | 1900 |

Fig. 2 Sample simulation 1: Parameters as given in (7), $D_1 = D_2$. Black spots have been cleared due to eradicating extreme event

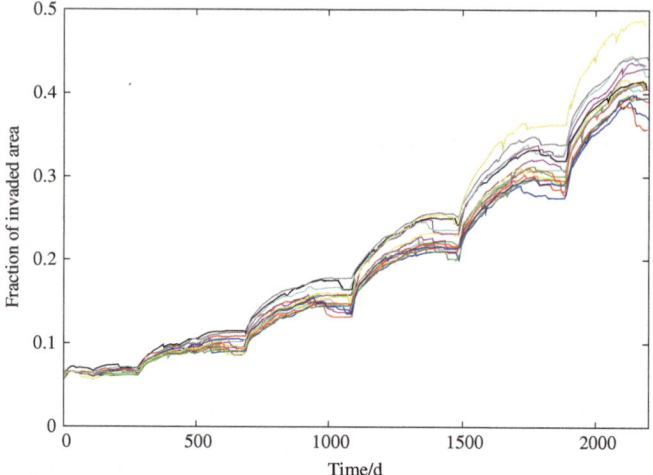

Fig. 3 Results of 15 simulation runs: slow but continuous invasion of the resident's area

At first, it is assumed that both species have the same diffusivities. It is seen that the competitive advantage of the alien leads to a slow but continuous displacement of the resident, cf. Fig. 2. As to be expected, the long-distance transport does not help either of the species because landing in the hostile environment inevitably leads to extinction because the formed patches are not larger than the required critical size.

The results of 15 simulations with different seeds of the random number generator [28] are collected in Fig. 3. Compared to the results for a constant environment, the periodicities in the selected parameters slow down the invasion. However, finally the invading weed wins.

The outcome immediately changes when the resident is twice as fast as the alien. The disadvantage in direct contact competition still exists, however, the higher mobility becomes the essential advantage over the intruder. An illustration is presented in Figs. 4 and 5. In a constant environment the resident needed a four times higher mobility to overcome the invader.

It can be preliminary summarized that the more realistic periodically changing environment stabilizes the resident's living conditions and reduces the invasibility. However, from qualitative point of view, a higher mobility remains a crucial competitive edge.

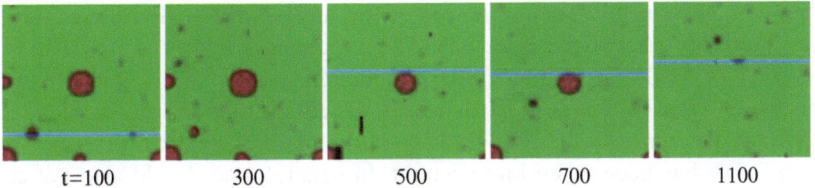

Fig. 4 Sample simulation 2: Parameters as given in (7), $D_1 = 2D_2$. Black spots have been cleared due to eradicating extreme event

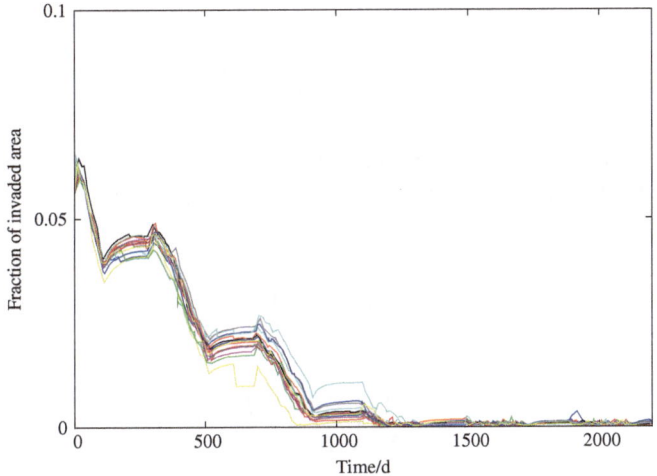

Fig. 5 Results of 15 simulation runs: rapid reversal and extinction of invasion

3 A Competition-Diffusion Model with Infected Invader

A specific infection of the invading population can be used as biocontrol measure to stop and reverse the invasion, cf. [16, 20, 30, 37]. To model this, the invader population is split into susceptibles S and infecteds I,

$$N_2 = S + I.$$

The model of the local dynamics then reads

$$\frac{dN_1}{dt} = r_1 N_1 - c_{11} N_1^2 - c_{12} N_1 (S + I), \tag{8}$$

$$\frac{dS}{dt} = r_S S - c_{22} S(S + I) - c_{21} N_1 S - \lambda \frac{SI}{(S + I)^k}, \tag{9}$$

$$\frac{dI}{dt} = r_I I - c_{22} I(S + I) - c_{21} N_1 I + \lambda \frac{SI}{(S + I)^k} - \mu I, \tag{10}$$

where λ is the transmission coefficient of the disease and μ the disease-induced higher mortality rate of the infecteds. The exponent k allows to describe mass-action type ($k = 0$) and frequency-dependent transmission ($k = 1$) of the disease respectively [2, 29]. For $k = 0$, disease-induced oscillations have been found [45, 48]. A difference in the growth rates of susceptibles r_S and infecteds r_I with $0 \leq r_I < r_S$ has been taken into account. In general, one should also not expect that the competition intensities of susceptibles and infecteds are the same. However, for demonstrating the effect of the invader infection this rough model structure is sufficient.

3.1   Local Dynamics with Infection

For convenience, the model of the local dynamics is not analysed in terms of N_1, S and I but rather in N_1, i and N_2 where i is the prevalence, i.e., the infected fraction of the total invader population N_2 [19],

$$i = \frac{I}{S + I} = \frac{I}{N_2} \text{ with } 0 \leq i \leq 1.$$

Having in mind that

$$\frac{di}{dt} = \frac{1}{N_2}\left(\frac{dI}{dt} - i\frac{dN_2}{dt}\right),\tag{11}$$

it follows

$$\frac{dN_1}{dt} = r_1 N_1 - c_{11}N_1^2 - c_{12}N_1 N_2,\tag{12}$$

$$\frac{di}{dt} = \left(r_I - r_S + \lambda N_2^{1-k} - \mu\right)(1-i)i,\tag{13}$$

$$\frac{dN_2}{dt} = [r_S(1-i) + r_I i]N_2 - c_{21}N_1 N_2 - c_{22}N_2^2 - \mu i N_2.\tag{14}$$

A prominent example of the control of a weed by a fungal disease is the fight against the yellow starthistle in the United States [44, 52, 53]. In phytopathology, the transmission of especially fungal diseases is described with standard incidence [46]. A corresponding model of the invasion of a fungal disease over a vineyard has been investigated by [4]. Further on, only the standard incidence is considered, i.e., $k = 1$. More details of the fungus disease cycle like latent and infectious periods and corresponding compartments [40] are neglected because this is out of scope of the present work. The reduction to the above formulated $S - I$ model (9,10) is sufficient to find the effect of the infection on the invasion.

It is readily seen that the sign of the first factor in (13) determines the dynamics of the system, it reads for $k = 1$

$$r_I - r_S + \lambda - \mu \lesseqgtr 0 . \tag{15}$$

- If it is less than zero, the prevalence approaches zero, i.e., the infecteds go extinct and one obtains a standard Lotka–Volterra system with $r_2 = r_S$.
- If the factor is greater than zero, the prevalence approaches unity, i.e, the susceptibles go extinct and one obtains a standard Lotka–Volterra system with $r_2 = r_I - \mu$.
- Finally, if it is equal to zero, the prevalence will remain at its initial value $i = i_0$, and one finds a standard Lotka–Volterra system with $r_2 = r_S - \lambda i_0$.

Hence, any of the cases results in a standard Lotka–Volterra system and the table of the stability properties of Sect. 2.1 can be simply adopted.

3.2 Spatiotemporal Dynamics with Infection

For modelling the dynamics in time and space, one has to come back to (8)–(10) and to add the diffusion terms, i.e.,

$$\frac{\partial N_1}{\partial t} = r_1 N_1 - c_{11} N_1^2 - c_{12} N_1 (S + I) + D_1 \Delta N_1, \tag{16}$$

$$\frac{\partial S}{\partial t} = r_S S - c_{22} S(S + I) - c_{21} N_1 S - \lambda \frac{SI}{S + I} + D_S \Delta S , \tag{17}$$

$$\frac{\partial I}{\partial t} = (r_I - \mu)I - c_{22} I(S + I) - c_{21} N_1 I + \lambda \frac{SI}{S + I} + D_I \Delta I . \tag{18}$$

The mechanisms of diffusion of infected and healthy plants are quite different because the spread of spores also has to be taken into account. However, the most simple Fickian formulation has been chosen for simplicity. The differences have been considered by different numerical values. The seasonality of system parameters is omitted here because the role of the infection should not be masked.

3.3 Numerical Simulations II

For the numerical simulations of system (16)–(18) a slightly different initial condition is used without qualitatively changing the outcomes. Again, Neumann zero-flux boundary conditions are applied. Initially, the infection is absent. It is

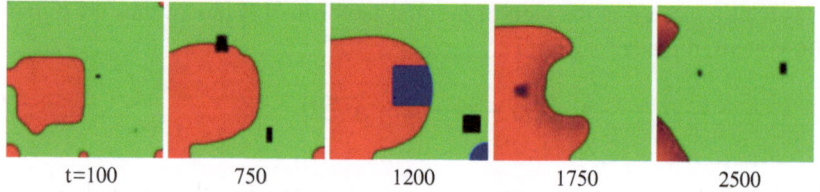

| t=100 | 750 | 1200 | 1750 | 2500 |

Fig. 6 Sample simulation 3: Parameters as given in (19). Black spots have been cleared due to eradicating extreme event. The repelling of the invasion after introducing the infection at $t =$ 1, 200 is obvious

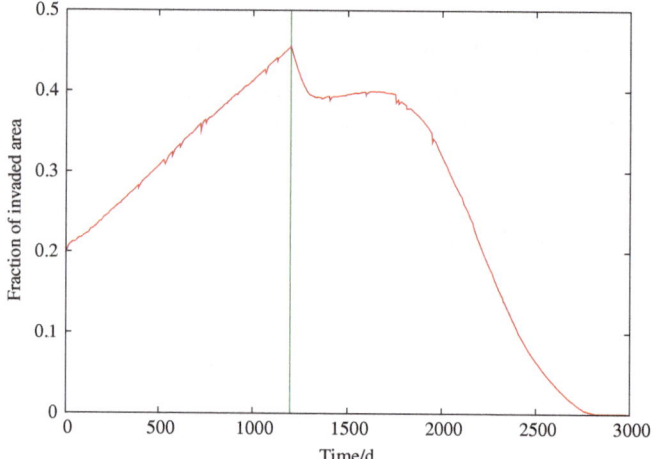

Fig. 7 Presentation of one sample run, also showing the sharp decline of invasion after infecting the invader at $t = 1200$

rather selectively introduced after the invading weed has overtaken a significant portion of the model area, cf. Fig. 6 below.

The following parameter values have been used:

$$r_1 = 1.0, \; r_S = 1.0, \; r_I = 0.8, \; \lambda = 0.405, \; \mu = 0.2,$$
$$c_{11} = 1.0, \; c_{12} = 1.3, \; c_{21} = 1.2, \; c_{22} = 1.0, \tag{19}$$
$$D_1 = 45.0, \; D_S = 22.5, \; D_I = 45.0, \; L = 3000.0.$$

The clear model result is that a targeted infection of the invading weed is a reliable strategy to win the fight against the bioinvasion. It is robust against different ratios of diffusivities as well as periodicities of growth and dispersal such as described in Sect. 2 if the random perturbations are alike (Fig. 7).

4 Concluding Remarks

As in constant environments, in the bistable parameter range, population patches of subcritical size disappear as expected from nucleation theory. The driving force of the competition process remains the temporary erosion in combination with a sufficiently effective harvesting of the invader and different mobilities of the species. However, the latter effect of different mobilities is tempered in the periodically changing environment. But still, even a strongly competitive alien has no chance to invade if the mobility of the indigenous species is sufficiently high.

It has turned out that the most efficient biological control measure is the specific partial infection of the invading population. In laboratory and field studies, however, it has been found that it can be hard to find such a specific agent, cf. [3]. Forthcoming work has to clear the role of non-symmetric competition of susceptible and infected alien species among themselves as well as with the native species. Also the impact of seasonality of system parameters including the infection rate has to be studied.

Acknowledgements H.M. would like to dedicate this paper to his academic teacher and friend Werner Ebeling (Berlin) on occasion of his 75th birthday on 15 September 2011. Furthermore, he is thankful to the Erskine Foundation at the University of Canterbury at Christchurch, New Zealand, for a Visiting Erskine Fellowship.

References

1. H.M. Alexander, R.D. Holt, The interaction between plant competition and disease. Perspect. Plant Ecol. Evol. Syst. **1/2**, 206–220 (1998)
2. R.M. Anderson, R.M. May, The invasion, persistence and spread of infectious diseases within animal and plant communities. Phil. Trans. R. Soc. Lond. B **314**, 533–570 (1986)
3. W. Bruckart, C. Cavin, L. Vajna, I. Schwarczinger, F.J. Ryan, Differential susceptibility of Russian thistle accessions to *Colletotrichum gloeosporioides*. Biol. Contr. **30**, 306–311 (2004)
4. J.B. Burie, A. Calonnec, M. Langlais, Modeling of the invasion of a fungal disease over a vineyard, in *Mathematical Modeling of Biological Systems*, vol. II, ed. by A. Deutsch, R.B. de la Parra, R.J. de Boer, O. Diekmann, P. Jagers, E. Kisdi, M. Kretzschmar, P. Lansky, H. Metz. Epidemiology, Evolution and Ecology, Immunology, Neural Systems and the Brain, and Innovative Mathematical Methods, Modeling and Simulation in Science, Engineering and Technology (Birkhäuser, Boston, 2008), pp. 11–21
5. J. Cardina, G.A. Johnson, D.H. Sparrow, The nature and consequence of weed spatial distribution. Weed Sci. **45**, 364–373 (1997)
6. R. Cousens, C. Dytham, R. Law, Dispersal in plants. A population perspective, in *Oxford Biology* (Oxford University Press, Oxford, 2008)
7. O. Diekmann, J.A.P. Heesterbeck (eds.), Mathematical epidemiology of infectious diseases. Model building, analysis and interpretation, in *Wiley Series in Mathematical and Computational Biology* (Wiley, Chichester, 2000)
8. J.A. Drake, H.A. Mooney (eds.), Biological invasions: a global perspective, in *SCOPE*, vol. 27 (Wiley, Chichester, 1989)
9. W. Ebeling, L. Schimansky-Geier, Stochastic dynamics of a bistable reaction system. Phys. A **98**, 587–600 (1979)

10. W. Ebeling, C. Ivanov, L. Schimansky-Geier, Stochastic theory of nucleation in bistable reaction systems. Rostocker Phys. Manuskripte **2**, 93–103 (1977)
11. J. Frantzen, Disease epidemics and plant competition: control of *Senecio vulgaris* with *Puccinia lagenophorae*. Basic Appl. Ecol. **1**, 141–148 (2000)
12. H.I. Freedman, A model of predator-prey dynamics as modified by the action of a parasite. Math. Biosci. **99**, 143–155 (1990)
13. J.S. Fulda, The logistic equation and population decline. J. Theor. Biol. **91**, 255–259 (1981)
14. L.Q. Gao, H.W. Hethcote, Disease transmission models with density dependent demographics. J. Math. Biol. **30**, 717–731 (1992)
15. K.P. Hadeler, H.I. Freedman, Predator-prey populations with parasitic infection. J. Math. Biol. **27**, 609–631 (1989)
16. K. Harley, I.W. Forno, *Biological Control of Weeds: A Handbook for Practitioners and Students* (Inkata Press, Melbourne, 1992)
17. R. Hengeveld (ed.), *Dynamics of Biological Invasions* (Chapman and Hall, London, 1989)
18. F.M. Hilker, *Spatiotemporal Patterns in Models of Biological Invasion and Epidemic Spread* (Logos, Berlin, 2005)
19. F.M. Hilker, H. Malchow, Strange periodic attractors in a prey-predator system with infected prey. Math. Popul. Stud. **13**(3), 119–134 (2006)
20. M. Julien, G. White, Biological control of weeds: theory and practical application, in *ACIAR Monograph Series*, vol. 49 (Australian Centre for International Agricultural Research, Bruce ACT, 1997)
21. R.M. Keane, M.J. Crawley, Exotic plant invasions and the enemy release hypothesis. Trends Ecol. Evol. **17**(4), 164–170 (2002)
22. B.W. Kooi, M.P. Boer, S.A.L.M. Kooijman, On the use of the logistic equation in models of food chains. Bull. Math. Biol. **60**, 231–246 (1998)
23. E. Kuno, Some strange properties of the logistic equation defined with r and K: inherent effects or artefacts? Res. Popul. Ecol. **33**, 33–39 (1991)
24. H. Malchow, L. Schimansky-Geier, Noise and diffusion in bistable nonequilibrium systems, in *Teubner-Texte zur Physik*, vol. 5 (Teubner, Leipzig, 1985)
25. H. Malchow, B. Radtke, M. Kallache, A.B. Medvinsky, D.A. Tikhonov, S.V. Petrovskii, Spatiotemporal pattern formation in coupled models of plankton dynamics and fish school motion. Nonlinear Anal. Real World Appl. **1**, 53–67 (2000)
26. H. Malchow, S.V. Petrovskii, E. Venturino, Spatiotemporal patterns in ecology and epidemiology: theory, models, simulations, in *CRC Mathematical and Computational Biology Series* (CRC Press, Boca Raton, 2008)
27. H. Malchow, A. James, R. Brown, Competitive and diffusive invasion in a noisy environment. Math. Med. Biol. **28**, 153–163 (2011)
28. M. Matsumoto, T. Nishimura, Mersenne Twister: a 623-dimensionally equidistributed uniform pseudorandom number generator. ACM Trans. Model. Comput. Simul. **8**(1), 3–30 (1998)
29. H. McCallum, N. Barlow, J. Hone, How should pathogen transmission be modelled? Trends Ecol. Evol. **16**(6), 295–300 (2001)
30. P.B. McEvoy, E.M. Coombs, Biological control of plant invaders: regional patterns, field experiments, and structured population models. Ecol. Appl. **9**(2), 387–401 (1999)
31. D. Mollison, Modelling biological invasions: chance, explanation, prediction. Phil. Trans. R. Soc. Lond. B **314**, 675–693 (1986)
32. D. Mollison (ed.), Epidemic models. Their structure and relation to data, in *Publications of the Newton Institute*, vol. 5 (Cambridge University Press, Cambridge, 1995)
33. R. Nathan, F.M. Schurr, O. Spiegel, O. Steinitz, A. Trakhtenbrot, A. Tsoar, Mechanisms of long-distance seed dispersal. Trends Ecol. Evol. **23**, 638–647 (2008)
34. A. Nitzan, P. Ortoleva, J. Ross, Nucleation in systems with multiple stationary states. Faraday Symp. Chem. Soc. **9**, 241–253 (1974)
35. S.V. Petrovskii, B.L. Li, Exactly solvable models of biological invasion, in *CRC Mathematical Biology and Medicine Series* (CRC Press, Boca Raton, 2006)

36. D. Pimentel (ed.), Biological invasions, in *Economic and Environmental Costs of Alien Plant, Animal, and Microbe Species* (CRC Press, Boca Raton, 2002)
37. E.N. Rosskopf, R. Charudattan, J.B. Kadir, Use of plant pathogens in weed control, in *Handbook of Biological Control. Principles and Applications of Biological Control*, ed. by T.S. Bellows, T.W. Fisher, L.E. Caltagirone, D.L. Dahlsten, G. Gordh, C.B. Huffaker (Academic, San Diego, 1999), pp. 891–918
38. L. Sattenspiel, The geographic spread of infectious diseases, in *Princeton Series in Theoretical and Computational Biology* (Princeton University Press, Princeton, 2009)
39. D.F. Sax, J.J. Stachowicz, S.D. Gaines (eds.), Species invasions, in *Insights into Ecology, Evolution, and Biogeography* (Sinauer, Sunderland, 2005)
40. J. Segarra, M.J. Jeger, F. van den Bosch, Epidemic dynamics and patterns of plant diseases. Phytopathology **91**(10), 1001–1010 (2001)
41. N. Shigesada, K. Kawasaki, *Biological Invasions: Theory and Practice* (Oxford University Press, Oxford, 1997)
42. N. Shigesada, K. Kawasaki, Y. Takeda, Modeling stratified diffusion in biological invasions. Am. Nat. **146**(2), 229–251 (1995)
43. I. Siekmann, *Mathematical Modelling of Pathogen-Prey-Predator Interactions* (Verlag Dr. Hut, München, 2009)
44. J. Suszkiw, Fungus unleashed to combat yellow starthistle. Agric. Res. Mag. **52**(8), 20–22 (2004)
45. P. van den Driessche, M.L. Zeeman, Disease induced sscillations between two competing species. SIAM J. Appl. Dynam. Syst. **3**(4), 601–619 (2004)
46. J.E. van der Plank, Host-pathogen interactions in plant disease (Academic, New York, 1982)
47. E. Venturino, The influence of diseases on Lotka-Volterra systems. Rocky Mt. J. Math. **24**, 381–402 (1994)
48. E. Venturino, The effect of diseases on competing species. Math. Biosci. **174**, 111–131 (2001)
49. L.J. Wiles, G.W. Oliver, A.C. York, H.C. Gold, G.G. Wilkerson, Spatial distribution of broadleaf weeds in North Carolina soybean (*Glycine max*) field. Weed Sci. **40**, 554–557 (1992)
50. M. Williamson, Biological invasions, in *Population and Community Biology Series*, vol. 15 (Chapman & Hall, London, 1996)
51. C.L. Wilson, Use of plant pathogens in weed control. Annu. Rev. Phytopathol. **7**(1), 411–434 (1969)
52. L.M. Wilson, C. Jette, J. Connett, J.P. McCaffrey, Biology and biological control of yellow starthistle, in *FHTET Technology Transfer Series 17-1998* (Forest Health Technology Enterprise Team, Morgantown WV, 2003)
53. D.M. Woods, W.L. Bruckart III, M. Pitcairn, V. Popescu, J. O'Brien, Susceptibility of yellow starthistle to *Puccinia jaceae* var. *solstitialis* and greenhouse production of inoculum for classical biological control programs. Biol. Contr. **50**, 275–280 (2009)
54. M. Yokozawa, Y. Kubota, T. Hara, Effects of competition mode on spatial pattern dynamics in plant communities. Ecol. Model. **106**, 1–16 (1998)
55. R.L. Zimdahl, *Weed-Crop Competition. A review* (Blackwell Publishing, Oxford, 2004)

Destruction and Diversity: Effects of Habitat Loss on Ecological Communities

Nick F. Britton

Abstract In many parts of the world habitat is being destroyed at an alarming rate. Many major ecosystems have lost more than half of their original area, and some much more than this [Millennium Ecosystem Assessment, Ecosystems and Human Well-Being: Synthesis (2005); World Wildlife Fund, Insight into Europe's Forest Protection (2001)]. At the same time biodiversity is fast declining [Butchart et al., Science 328:1164–1168, 2010]. Habitat loss is a major threat to biodiversity [Brooks et al., Conserv. Biol. 16:909–923, 2002; Baillie et al., 2004 IUCN Red List of Threatened Species: A Global Assessment (2004)], but the effects of the destruction are sometimes difficult to predict [Debinski and Holt, Conserv. Biol. 14:342–355, 2000; Prugh et al., Proc. Natl. Acad. Sci. USA 105:20770–20775, 2008; Hanski, AMBIO 40:248–255, 2011], and the effect of habitat loss and fragmentation on predator–prey interactions in particular is unclear [Ryall and Fahrig, Ecology 87:1086–1093, 2006]. One reason for the lack of clarity may be that the species-occupancy patterns that underlie diversity patterns in fragmented landscapes have often been overlooked [Prugh et al., Proc. Natl. Acad. Sci. USA 105:20770–20775, 2008; Ovaskainen and Hanski, Ecol. Lett. 6:903–909, 2003]. The patch-occupancy metapopulation paradigm, despite its simplicity, has proved successful in developing some understanding of how habitat destruction affects the local flora and fauna. We shall review and derive some results arising from this approach for single species, for competitive and mutualistic communities, for predator-prey systems and food chains, and finally for a simple food web, a predator interacting with two competing prey. We show that although the outcome of habitat destruction in terms of species extinctions may be straightforward for the simplest models and communities, it may be subtly parameter-dependent and counter-intuitive with even a small increase in complexity. Progress towards a theory

N.F. Britton (✉)
Department of Mathematical Sciences and Centre for Mathematical Biology,
University of Bath, Bath BA2 7AY, UK
e-mail: n.f.britton@bath.ac.uk

M.A. Lewis et al. (eds.), *Dispersal, Individual Movement and Spatial Ecology*,
Lecture Notes in Mathematics 2071, DOI 10.1007/978-3-642-35497-7_11,
© Springer-Verlag Berlin Heidelberg 2013

for more complex food webs [Leibold et al., Ecol. Lett. 7:601–613, 2004; Leibold and Miller, From metapopulations to metacommunities (2004); Pillai et al., Theor. Ecol. 3:223–237, 2010] will be difficult until we have a thorough understanding of these basic building blocks.

1 Introduction

Landscapes are rarely spatially homogeneous but consist of patches with different ecological characteristics, and endogenous processes such as ecological succession may lead to spatial heterogeneity even in an otherwise homogeneous landscape [23]. Landscape ecology is concerned with this spatial structure and how it affects the abundance of organisms at the landscape level. One obvious example of exogenous heterogeneity is an archipelago, consisting (for land plants and animals) of patches of inhabitable land in an uninhabitable ocean. The theory of island biogeography [39] considers biodiversity on islands as the result of competing forces of colonisation from a mainland stock and local stochastic extinction. Examples of endogenous heterogeneity include rocky intertidal zones, "an everchanging mosaic of many species which inhabit wave-generated patches or gaps" [52] and tropical rainforests [11], where the fall of large canopy trees followed by ecological succession creates a mosaic landscape.

In 1969 Levins [36] set up an abstract model of such patchy environments, providing biologists with a new paradigm, the metapopulation, with which to analyse population viability in fragmented or mosaic habitats [19, 21, 22]. In the simplest and archetypal case [36] there is only one species of interest, each patch is assumed identical, and space is modelled implicitly, in the sense that any patch is equally easily colonised from any other one. It is a *patch-occupancy* model, in the sense that each patch is in one of two states, either occupied or unoccupied by the species of interest, although according to the particular biological situation an occupied patch may contain either a single individuals or a local population of the species. Each occupied patch is subject to mortality or local extinction, and each unoccupied one to colonisation from an occupied patch. The state of the system as a whole is given by the fraction of habitat patches that is occupied. In 1987 Lande extended the model to include the effect of partial habitat destruction [33], the focus of this chapter.

Many authors [24–26, 29, 37, 41, 47–49, 59, 62] have extended patch-occupancy metapopulation models with partial habitat destruction to systems of more than one interacting species. In particular Nee and May [47] and Tilman and co-authors [62] investigated the effect of habitat removal on the trade-off between an aspect of competitiveness and reproduction and colonisation capabilities, and concluded the following.

- Rule 1: for two competitors in the absence of predators, and with a competition–colonisation trade-off, habitat removal is detrimental to the good competitor, but is *beneficial* to the good coloniser while the good competitor persists.

May [41] investigated the predator-prey case, and concluded the following.

- Rule 2: with one prey and one predator, habitat removal leads to extinction of the predator, followed by extinction of the prey.

We shall review the derivation of these rules and make use of them as benchmarks for slightly more complex ecological communities, where they may no longer be valid.

2 The Ecological Model

2.1 The Single-Species Model

The archetypal Levins model [36] is given by

$$\frac{dP}{dt} = \lambda(1 - P) - eP = cP(1 - P) - eP, \tag{1}$$

where $\lambda = cP$ is the *force of colonisation*, c is a colonisation parameter and e a local extinction or mortality parameter (see Fig. 1a), and P is the fraction of patches in the environment that is occupied by the species of interest. If the metapopulation persists, it is assumed to do so through a balance between local extinction of occupied patches and successful colonisation of unoccupied patches. The population on each occupied patch is assumed to be at constant risk of local extinction or mortality, at rate e, in other words with probability $e\delta t$ in a short period of time δt. It is assumed to produce propagules at a rate that would be sufficient to colonise $c\delta t$ new patches in time δt in a virgin (completely unoccupied) environment. In an environment with a fraction P of occupied patches, only $c(1 - P)\delta t$ of these are new occupations, so that the *per-capita* colonisation rate is $c(1 - P)$. (The model does not preclude propagules from becoming established in patches that are already occupied, leaving the state of the patch unchanged, but we only count as colonisers those propagules that colonise and therefore change the state of a vacant patch.) The force of colonisation λ is defined in analogy with the force of infection in epidemiology, as the rate at which unoccupied patches become occupied, and is given in the archetypal model by $\lambda = cP$. (The model $\lambda = cP^2/(P + a)$, incorporating a weak Allee effect [1] into the colonisation term, has also been used [10, 32, 66], and leads to a strong Allee effect in the population, with a stable trivial solution and stable and unstable positive steady states $P_2^* > P_1^*$ for sufficiently small a.)

The basic reproduction number R_0 is defined to be the expected number of new patches colonised during its lifetime by an occupied patch introduced into a virgin environment. Its lifetime lasts until it goes locally extinct, and it suffers a constant risk of local extinction at rate e, so its expected lifetime is $1/e$. During its lifetime

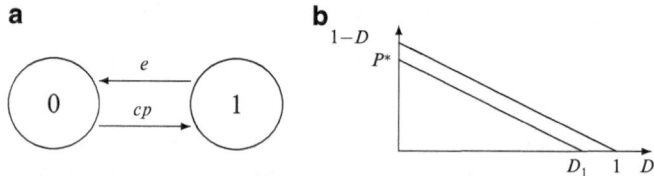

Fig. 1 (**a**) The transitions between the vacant and the occupied state in the archetypal Levins metapopulation model (1). Extinction for a given occupied site is assumed to be independent of how many other sites are occupied, whereas colonisation of a given vacant site is linearly dependent on the number of sites available to provide colonists. (**b**) The bifurcation diagram plotting non-negative steady state solutions p^* of (2) as a function of the fraction D of habitat removed. Also shown is the fraction $1 - D$ of patches remaining. The difference between these is the number of empty patches remaining, which stays constant as D increases until the species becomes extinct at $D_1 = P^*$, the fraction of patches occupied at steady state when $D = 0$

it produces successful propagules at a rate c, so that the basic reproduction number is $R_0 = c/e$. It is clear that the trivial steady state of (1) is stable whenever $R_0 < 1$, while the non-trivial steady state $P^* = 1 - 1/R_0$ is stable whenever $R_0 > 1$, or whenever it is biologically realistic, as one might expect. The species will only be able to invade a virgin environment if each occupied patch colonises at least one other (on average) before it dies.

It is now assumed [33, 47] that a fraction D of the patches in the habitat is permanently destroyed, and we define p to be the fraction of the original patches that is occupied. Propagules can only be successful when they land on the habitat that remains, so that patches now produce successful propagules at a rate $c(1 - D - p)$. The model becomes

$$\frac{dp}{dt} = cp(1 - D - p) - ep, \tag{2}$$

with effective basic reproduction number $R'_0 = (1 - D)R_0 = c(1 - D)/e$. The trivial steady state of (2) is stable whenever $R'_0 < 1$, while the non-trivial steady state $p^* = 1 - D - 1/R_0 = (1 - D)(1 - 1/R'_0)$ is stable whenever $R'_0 > 1$, or whenever it is biologically realistic. The bifurcation diagram with parameter D for (2) is shown in Fig. 1b. Note that the critical value D_1 at which the population goes extinct is given by $D_1 = P^* = 1 - 1/R_0$. In other words, whenever the fraction of patches removed (at random) is at least as great as the fraction of patches originally occupied, then the population goes to (deterministic) extinction.

Note that, as one would expect, (2) is essentially identical to (1) with c replaced by $c(1 - D)$, if we work in terms of the fraction q of the *remaining* habitat that is occupied. For, substituting $q = p/(1 - D)$ into (2), it becomes

$$\frac{dq}{dt} = c(1 - D)q(1 - q) - eq. \tag{3}$$

This transformation is often useful in analysing (2) and the multispecies extensions of it that will appear later.

2.2   Two Independent Populations

The patch-occupancy metapopulation model (2) for a single population with habitat destruction may be extended to one for two independent populations by considering that remnant patches may be in one of four states, empty (p_0) or occupied by one (p_1) or other (p_2) or both (p_{12}) of the populations. If neither population has any effect on the colonisation or extinction of the other, the equations for the system are given by

$$\frac{dp_0}{dt} = -\lambda_1 p_0 - \lambda_2 p_0 + e_1 p_1 + e_2 p_2,$$

$$\frac{dp_1}{dt} = \lambda_1 p_0 - \lambda_2 p_1 - e_1 p_1 + e_2 p_{12},$$

$$\frac{dp_2}{dt} = \lambda_2 p_0 - \lambda_1 p_2 - e_2 p_2 + e_1 p_{12}, \tag{4}$$

$$\frac{dp_{12}}{dt} = \lambda_1 p_2 + \lambda_2 p_1 - e_1 p_{12} - e_2 p_{12},$$

where λ_i is the force of colonisation of species i and e_i its rate of local extinction. Let us assume that there is no habitat destruction, $p_0 = 1 - p_1 - p_2 - p_{12}$, so that *all* patches not occupied by species i experience the same force of colonisation from it. Let us define $P_i = p_i + p_{12}$, the total fraction of patches containing species i, and let $Q = p_{12} - P_1 P_2$. Then, as in [31],

$$\frac{dQ}{dt} = -(\lambda_1 + \lambda_2 + e_1 + e_2)Q, \tag{5}$$

so that $Q(t) \to 0$ as $t \to \infty$. Occupation by species 1 is (at least for large time) statistically independent of occupation by species 2, as one might expect. A similar result may be obtained in the case $D > 0$ if we work in terms of fractions of the *remaining* habitat that is occupied, as in Sect. 2.1 above. Note that there is no restriction on each force of colonisation λ_i, except that it must be the same for all patches not occupied by species i. In particular, it may depend on species $j \neq i$.

2.3   Competition: Two Competing Species

The patch-occupancy metapopulation model (4) may now be extended to one for two competing populations. Klausmeier [31], following Slatkin [59], distinguishes three ways in which two species in a patchy environment may compete with each other locally.

- Extinction competition.
- Establishment competition.
- Propagule production competition.

Fig. 2 State transitions for
patches in the metapopulation
model (6)

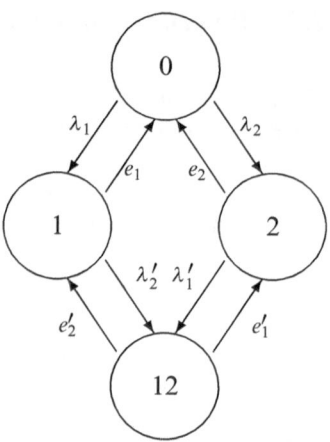

In extinction competition [24, 47, 62], one species increases the extinction rate of the other in patches in which they both occur; in establishment (pre-emptive) competition [8], one species decreases the probability that the other can colonise a patch that it occupies; and in propagule production competition, one species decreases the rate at which the other produces propagules in patches in which they both occur. Yu and Wilson [65] also identify colonisation competition, in which the colonisation coefficient c_1 of species 1 is a decreasing function of the occupancy p_2 of species 2, and vice versa, a phenomenological model for the local competition between propagules to colonise patches, depending on the global occupancy of the adults and thus the expected number of propagules of each species that arrive at each unoccupied patch.

With all three of Klausmeier's forms of competition, the equations with habitat destruction become (Fig. 2)

$$\frac{dp_1}{dt} = \lambda_1 p_0 - \lambda_2' p_1 - e_1 p_1 + e_2' p_{12},$$

$$\frac{dp_2}{dt} = \lambda_2 p_0 - \lambda_1' p_2 - e_2 p_2 + e_1' p_{12}, \qquad (6)$$

$$\frac{dp_{12}}{dt} = \lambda_2' p_1 + \lambda_1' p_2 - e_1' p_{12} - e_2' p_{12},$$

where

$$\lambda_1 = c_1 p_1 + f_{12} c_1 p_{12}, \quad \lambda_1' = c_1' p_1 + f_{12} c_1' p_{12},$$
$$\lambda_2 = c_2 p_2 + f_{21} c_2 p_{12}, \quad \lambda_2' = c_2' p_2 + f_{21} c_2' p_{12}, \qquad (7)$$

and $p_0 = 1 - D - p_1 - p_2 - p_{12}$. Extinction competition occurs if $e_1' > e_1$, $e_2' > e_2$, establishment competition if $c_1' < c_1$, $c_2' < c_2$, and propagule production competition if $f_{12} < 1$, $f_{21} < 1$. It is usual to simplify this system in some way before analysing it. The species are not in general independently distributed [59], (given in the case $D = 0$ by $p_{12} = P_1 P_2$, where $P_i = p_i + p_{12}$), although it is sometimes (implicitly) assumed that they are [37]. However, in the case of pure propagation-production competition, $e_i' = e_i$ and $c_i' = c_i$, then the analysis of Sect. 2.2 applies, and the species are independently distributed on the remnant habitat patches [31]. This reduces the number of equations from three to two. Alternatively, in particular if patches are occupied by only one individual, it is often assumed that no patch may be occupied by more than one population at a time. This may be under conditions of strong establishment competition or strong extinction competition. Under conditions of strong establishment competition, with $c_1' = c_2' = 0$, the equations become

$$\frac{dp_1}{dt} = c_1 p_0 p_1 - e_1 p_1, \quad \frac{dp_2}{dt} = c_2 p_0 p_2 - e_2 p_2, \tag{8}$$

where $p_0 = 1 - D - p_1 - p_2$. Formally, this is a Lotka–Volterra competition system. Let $R_i = c_i/e_i$ be the basic reproduction number of species i. We shall refer to the species with greater R_i as the (patch-wise) better reproducer. In this case there is no coexistence steady state (unless $R_1 = R_2$), the worse reproducer is competitively excluded, and the system reduces to the single-species case. Under conditions of strong extinction competition, propagules may (with some probability) immediately displace the original occupants from a patch. In terms of model (6) above, this is equivalent to taking a limit $e_1' + e_2' \to \infty$. With $e_1' + e_2' = O(1/\varepsilon)$, where ε is a small positive parameter, the p_{12} equation can only be asymptotically balanced if $p_{12} = O(\varepsilon)$, and then gives $(e_1' + e_2')p_{12} = (c_1' + c_2')p_1 p_2 + O(\varepsilon)$. Let $\theta_{12} = e_2'/(e_1' + e_2')$ be the probability that species 1 succeeds in occupying a contested patch, and $\theta_{21} = e_1'/(e_1' + e_2') = 1 - \theta_{21}$ the probability that species 2 succeeds. With no establishment competition ($c_1' = c_1, c_2' = c_2$), the equations with habitat destruction become

$$\frac{dp_1}{dt} = c_1 p_1 (1 - D - p_1 - p_2) - c_2 p_1 p_2 - e_1 p_1 + \theta_{12}(c_1 + c_2) p_1 p_2$$

$$= c_1 p_1 (1 - D - p_1) - \theta_{21}(c_1 + c_2) p_1 p_2 - e_1 p_1, \tag{9}$$

$$\frac{dp_2}{dt} = c_2 p_2 (1 - D - p_1 - p_2) - c_1 p_1 p_2 - e_2 p_2 + \theta_{21}(c_1 + c_2) p_1 p_2$$

$$= c_2 p_2 (1 - D - p_2) - \theta_{12}(c_1 + c_2) p_1 p_2 - e_2 p_2,$$

again to leading order in ε. We shall refer to species i as the superior (extinction) competitor if $\theta_{ij} > \theta_{ji}$, or $\theta_{ij} > \frac{1}{2}$.

A limiting case occurs when $\theta_{12} = 1$, $\theta_{21} = 0$, [24, 61], so that the superior competitor species 1 can always immediately displace the inferior species 2 in a patch, while the inferior can never displace the superior (Fig. 3). In this case we shall refer to species 1 as the predominant (extinction) competitor. This case has been widely studied and extended since its introduction, and we shall treat it in detail. Equations (9) become

Fig. 3 State transitions for
patches in the metapopulation
model with two competitors,
species 1 and 2, where
species 1 immediately
displaces species 2 from a
contested patch

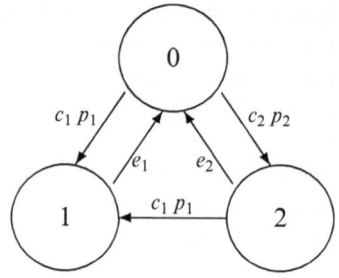

$$\frac{dp_1}{dt} = c_1 p_1 (1 - D - p_1) - e_1 p_1, \tag{10}$$

$$\frac{dp_2}{dt} = c_2 p_2 (1 - D - p_1 - p_2) - c_1 p_1 p_2 - e_2 p_2.$$

Species 2 has no impact whatever on species 1. In contrast, species 2 can only colonise empty sites (the term $c_2 p_2 (1 - D - p_1 - p_2)$), and loses sites when species 1 colonises them (the term $-c_1 p_1 p_2$). Trivial and semi-trivial steady states $S = (p_1^*, p_2^*)$ of this model are given by $S_0 = (0, 0)$, $S_1 = (1 - D - 1/R_1, 0)$, and $S_2 = (0, 1 - D - 1/R_2)$, where $R_i = c_i/e_i$ is the basic reproduction number of species i. There is a coexistence steady state $S_{12}^* = (p_1^*, p_2^*)$, with

$$p_1^* = 1 - D - \frac{1}{R_1}, \quad p_2^* = 1 - D - \frac{1}{R_2} - \left(1 + \frac{c_1}{c_2}\right) p_1^*, \tag{11}$$

in the positive quadrant whenever

$$1 - D - \frac{1}{R_2} > \left(1 + \frac{c_1}{c_2}\right) \left(1 - D - \frac{1}{R_1}\right) > 0. \tag{12}$$

in which case $R_1 > 1$, $R_2 > 1$, $D < 1 - 1/R_1$, and $D < 1 - 1/R_2$. Let us define new parameters $Q_1 = 1 - 1/R_1$, $Q_2 = 1 - 1/R_2$, so that Q_i is the steady-state population of species i in the absence of its competitor in an undisturbed environment. Then $S_1 = (Q_1 - D, 0)$, $S_2 = (0, Q_2 - D)$, and $S_{12} = (p_1^*, p_2^*)$, with

$$p_1^* = Q_1 - D, \quad p_2^* = Q_2 - D - \left(1 + \frac{c_1}{c_2}\right)(Q_1 - D), \tag{13}$$

in the positive quadrant whenever

$$Q_2 - D > \left(1 + \frac{c_1}{c_2}\right)(Q_1 - D) > 0. \tag{14}$$

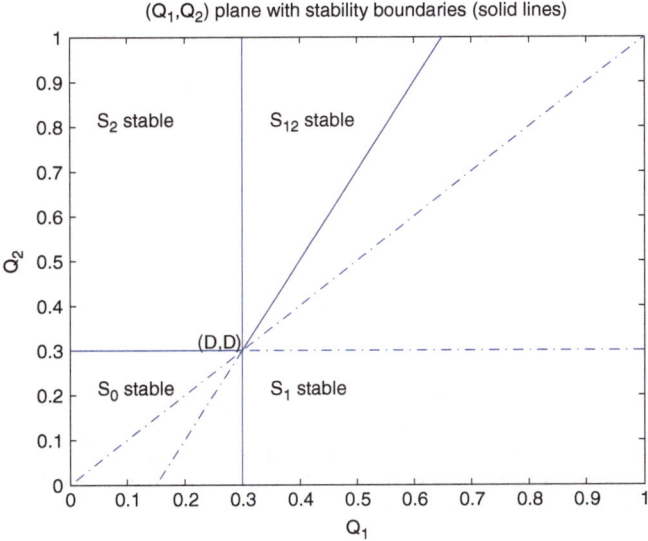

Fig. 4 Stability boundaries for (9). As D increases, the point (D, D) moves along the diagonal from $(0, 0)$ to $(1, 1)$, leading to one of the bifurcation scenarios described in the text

The Jacobian matrix for this system is lower triangular, since species 1 is completely unaffected by species 2, so that we can immediately identify the eigenvalues at the steady states as $\lambda_1 = c_1(1-D-2p_1^*)-e_1$ and $\lambda_2 = c_2(1-D-p_1^*-2p_2^*)-c_1 p_1^*-e_2$. Therefore

$$\lambda_1 = \begin{cases} c_1(Q_1 - D) & \text{at } S_0 \text{ and } S_2, \\ -c_1 p_1^* & \text{at } S_1 \text{ and } S_{12}, \end{cases} \tag{15}$$

and

$$\lambda_2 = \begin{cases} c_2(Q_2 - D) & \text{at } S_0, \\ c_2(Q_2 - D) - (c_1 + c_2)(Q_1 - D) & \text{at } S_1, \\ -c_2 p_2^* & \text{at } S_2, \\ -(c_1 + c_2)p_1^* - c_2 p_2^* & \text{at } S_{12}. \end{cases} \tag{16}$$

Let us assume henceforth that $R_i > 1$, $Q_i > 0$ for each i, so that neither species goes extinct in the absence of its competitor in an undisturbed environment. Then S_0 is stable whenever both $Q_1 < D$ and $Q_2 < D$, S_1 whenever both $Q_1 > D$ and $Q_2 - D > (1 + c_1/c_2)(Q_1 - D)$, S_2 whenever both $Q_1 < D$ and $Q_2 > D$, and S_{12} whenever both $Q_1 > D$ and $Q_2 - D < (1 + c_1/c_2)(Q_1 - D)$, as shown in Fig. 4. It is clear that as D increases with Q_1 and Q_2 fixed, there are three possible bifurcation scenarios, as follows, with all bifurcations transcritical.

- If $Q_1 > Q_2$, then species 2 cannot survive for any D; S_1 is stable for D small, but loses stability to S_0 as D increases past Q_1.

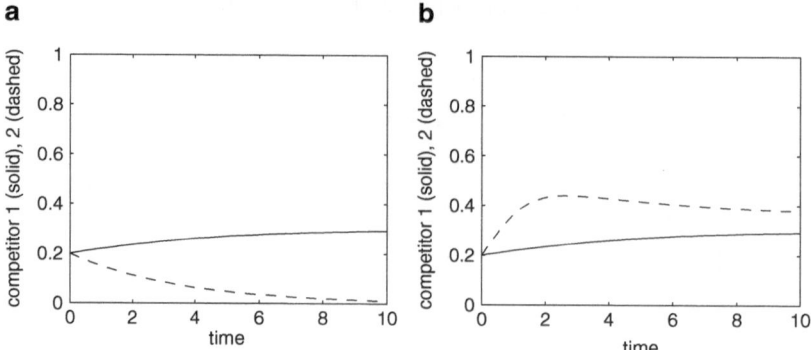

Fig. 5 Numerical results for a metapopulation model with competition. Parameter values are $c_1 = 1, e_1 = 0.7, e_2 = 0.7$, and (**a**) $c_2 = 1.2$ (no coexistence steady state), (**b**) $c_2 = 3$ (coexistence steady state). In both cases the inferior competitor is the better disperser, $R_2 > R_1$, or $Q_2 > Q_1$, but in (**b**) the better dispersal characteristics of the inferior competitor lead to sufficient niche differentiation for it to coexist with the superior one, $Q_2 > (1 + c_1/c_2)Q_1$

- If $Q_2 > (1 + c_1/c_2)Q_1 > Q_1$, S_{12} is stable for D small, but loses stability to S_2 as D increases past Q_1 and then to S_0 as D increases past Q_2.
- If $Q_1 < Q_2 < (1 + c_1/c_2)Q_1$, S_1 is stable for D small, but loses stability to S_{12} as $c_1 D/c_2$ increases past $(1 + c_1/c_2)Q_1 - Q_2$, then to S_2 as D increases past Q_1, then to S_0 as D increases past Q_2.

In the first two scenarios destruction causes a loss of biodiversity, while in the third biodiversity unexpectedly but temporarily increases at the bifurcation from S_1 to S_{12}.

Note that the competitor species 1 and species 2 may stably coexist. In the absence of habitat destruction, this occurs if $Q_2 > (1 + c_1/c_2)Q_1 > Q_1 > 0$. Species 2 must be a better reproducer than species 1, in the sense that $R_2 > R_1$, equivalent to $Q_2 > Q_1$. Coexistence depends on patches becoming free through local extinction, and the worse competitor being better placed to exploit these free patches, a mechanism known as *fugitive coexistence*. It relies on the mosaic structure of a metapopulation, and cannot be described in the Lotka–Volterra approach to modelling competition, although the equations are formally identical [38]. Species 2 must not only be a better reproducer than species 1, $R_2 > R_1$ or $Q_2 > Q_1$, but it must exceed some threshold of reproduction superiority, $Q_2 > (1 + c_1/c_2)Q_1$. In other words, species 2 must exceed a threshold of niche differentiation over species 1 [61] (Fig. 5).

Now consider the effect of habitat destruction. From (11) or (13), $dp_1^*/dD = -1 < 0$ but $dp_2^*/dD = c_1/c_2 > 0$, so that habitat removal is detrimental to species 1 but beneficial to species 2 in the coexistence regime. We have derived Rule 1 of the Introduction. It states that habitat destruction shifts the balance away from the better competitor, species 1, towards the better coloniser, species 2. Let us consider the third bifurcation scenario: $Q_2 < (1 + c_1/c_2)Q_1$, so that 1 drives

2 to extinction in the undisturbed habitat, but $Q_2 > Q_1$, so that 2 is the better reproducer. Let R'_2 be the effective basic reproduction number of a patch of species 2 introduced into an environment consisting of species 1 at equilibrium, equal to the rate of colonisation divided by the rate of loss of patches of species 2 at S_1. It is instructive to consider how R'_2 changes with D. At S_1 the patches accessible to species 2 are those undestroyed patches that are not occupied by species 1, a fraction $1 - D - p_1^* = 1 - Q_1$, so the per capita colonisation rate of species 2 at S_1 is $c_2(1 - Q_1)$, independent of D. Patches of species 2 die with constant per capita local extinction rate e_2 and are lost to competitors at per capita rate $c_1 p_1^* = c_1(Q_1 - D)$, a decreasing function of D. As D increases species 2 has less competition for the same number of patches, and $R'_2 = c_2(1 - Q_1)/(e_2 + c_1(Q_1 - D))$ is an increasing function of D. As D increases past D_1, where $D_1 = (c_2/c_1)((1 + c_1/c_2)Q_1 - Q_2)$, and $0 < D_1 < 1$ if $Q_1 < Q_2 < (1 + c_1/c_2)Q_1$, then R'_2 increases past unity and invasion of species 2 is possible. Under these conditions on the parameters, habitat destruction always leads to a potential temporary increase in species richness before eventual extinction of both species. In general, if there is no trade-off between competitiveness and reproductive ability, so that $R_1 > R_2$ or $Q_1 > Q_2$, species 2 will not persist for any D. If there is such a trade-off, so that $R_2 > R_1$ or $Q_2 > Q_1$, reproductive ability is at a premium when habitat destruction is taking place, and the best competitor will always go extinct first as D increases.

2.4   Competition: More Than Two Competing Species

The general model (9) may be extended in the obvious way [8, 24, 30, 62] to the case of n competitors, to give

$$\frac{dp_i}{dt} = c_i p_i \left(1 - D - \sum_{j=1}^{n} p_j\right) + \sum_{j=1}^{n} c_i \theta_{ij} p_i p_j - \sum_{j=1}^{n} c_j \theta_{ji} p_i p_j - e_i p_i, \quad (17)$$

where θ_{ij} is the probability that species i wins a patch in a contest with species j, and of course $\theta_{ji} = 1 - \theta_{ij}$. In the limiting case of a predominance hierarchy, with species i immediately displacing species j whenever $i < j$, this becomes

$$\frac{dp_i}{dt} = c_i p_i \left(1 - D - \sum_{j=1}^{i} p_j\right) - \sum_{j=1}^{i-1} c_j p_i p_j - e p_i. \quad (18)$$

For simplicity, following previous authors, we have fixed the extinction parameter to be identical and equal to e for all species i. The steady state $S = (p_1^*, p_2^*, \ldots, p_n^*)$ with $p_i^* \neq 0$ for all i is given by

$$p_1^* = 1 - D - \frac{e}{c_1}, \quad p_2^* = \frac{e}{c_1} - \frac{c_1}{c_2}(1 - D), \quad p_3^* = \frac{c_1}{c_2}(1 - D) - \frac{ec_2}{c_1 c_3}, \quad (19)$$

and, in general,

$$p_{2i-1}^* = \frac{c_1 c_3 \dots c_{2i-3}}{c_2 c_4 \dots c_{2i-2}}(1 - D) - e\frac{c_2 c_4 \dots c_{2i-2}}{c_1 c_3 \dots c_{2i-1}}, \tag{20}$$

$$p_{2i}^* = e\frac{c_2 c_4 \dots c_{2i-2}}{c_1 c_3 \dots c_{2i-1}} - \frac{c_1 c_3 \dots c_{2i-1}}{c_2 c_4 \dots c_{2i}}(1 - D).$$

The Jacobian matrix of the system (18) is lower triangular, so the stability of its steady states is easy to determine. The steady state S above is stable whenever it is positive, $p_i^* > 0$ for all i, so that the metapopulation paradigm allows the fugitive coexistence of any number of competing species. Let us assume that S is positive for $D = 0$. As D increases, and while S remains positive, the patch occupancy of the odd-numbered species decreases, while that of the even-numbered species increases. The first species to go extinct must be an odd-numbered one, but its identity depends on the exact values of the parameters. The system is said to be "noninteractive" in the sense of Hastings [24] if $c_i/c_{i-1} > c_{i-1}/c_{i-2}$ for all i. In this case the first species to go extinct is species 1, the best competitor. In [62], the authors chose the colonisation parameters to give a geometric abundance series $p_i^* = P_i^* = q(1 - q)^i$ at steady state in the intact environment. This requires $c_i = e/(1 - q)^{2i-1}$, so $c_i/c_{i-1} = c_{i-1}/c_{i-2}$, on the borderline between interactive and noninteractive in the sense of Hastings. The basic reproduction number of species i is then given by $R_i = c_i/e_i = c_i/e = 1/(1 - q)^{2i-1}$, so that $R_i > R_j$ if $i > j$, and worse competitors are therefore assumed to be better reproducers. Then it may be shown that, as D increases, species go deterministically extinct in the order of their competitive ability, with the best competitor first, just as in the two-competitor case. This conclusion depends on the detailed assumptions made, and other authors look in more generality at changes in abundance rankings as D increases [44].

2.5 Predation

In 1958 Huffaker [28] published the results of some classic experiments on a prey-predator system in a patchy environment. The system consisted of predatory mites preying on phytophagous mites, which fed on oranges. The oranges were arranged in a lattice, and the colonisation rates between them were experimentally manipulated. With careful control of the colonisation rates, Huffaker managed to achieve coexistence between the prey and the predators, with a tendency towards oscillations.

We shall set up a patch-occupancy model for a prey–predator system with one prey and one predator species. Each patch may therefore be in one of four states, so that we expect a system of three independent equations. This may be reduced to two on the assumption that the predator is a specialist on the prey, and that predator-only patches therefore do not exist, or are so short-lived that they may be neglected

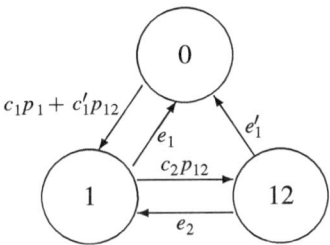

Fig. 6 State transitions for patches in the metapopulation model for predation, with species 1 the prey and species 2 the predator. A patch in state 12 contains prey and predators, but if the prey in such a patch go extinct then it is assumed that the predators immediately do so as well, so that there are no pure predator patches

[4, 14, 25, 26, 41, 55]. Others have considered the "ignorant predator" case, where predators may colonise patches that are not occupied by prey but are subject to a greater extinction rate if they do so, but have then made an unjustified independence assumption to reduce the number of equations to two [60]. We shall use the subscript 2 to denote the predator. Of the sites in the system, a fraction p_1 is assumed to be in state 1, occupied by prey only, a fraction p_{12} in state 12, occupied by both prey and predators, and a fraction $p_0 = 1 - p_1 - p_{12}$ in state 0, empty. The possible moves that a site can make from state to state, and the rate of movement from one state to another, are shown in Fig. 6. Colonisation by the prey is assumed to take place from sites in state 1 at rate $c_1 p_1$, and from sites in state 12 at rate $c'_1 p_{12}$. For sites in state 1 the extinction rate is e_1, the local extinction rate of the prey in the absence of predators. Sites in state 12 may suffer local extinction of the predators, to move to state 1 at rate e_2, or local extinction of the prey, assumed to be followed immediately by the local extinction of the predators themselves, so that these sites move directly to state 0 at rate e'_1. If the predator has negligible effect on local prey dynamics, the case of "donor control" [12], then $c'_1 = c_1$ and $e'_1 = e_1$. Otherwise it is usual to assume that $c_1 > c'_1$, so that predator-free prey populations produce colonisers more readily, and $e'_1 > e_1$, so that the rate of local extinction of the prey is greater in the presence of the predators.

The equations for the fraction of sites in each state are given by

$$\frac{dp_1}{dt} = (1 - D - p_1 - p_{12})(c_1 p_1 + c'_1 p_{12}) - c_2 p_1 p_{12} - e_1 p_1 + e_2 p_{12}, \quad (21)$$

$$\frac{dp_{12}}{dt} = c_2 p_1 p_{12} - (e'_1 + e_2) p_{12},$$

The limiting case $c'_1 = 0$, $e_2 = 0$ was introduced by May [41] and is the case we shall analyse here. The equations for the fraction of sites in each state are then given by

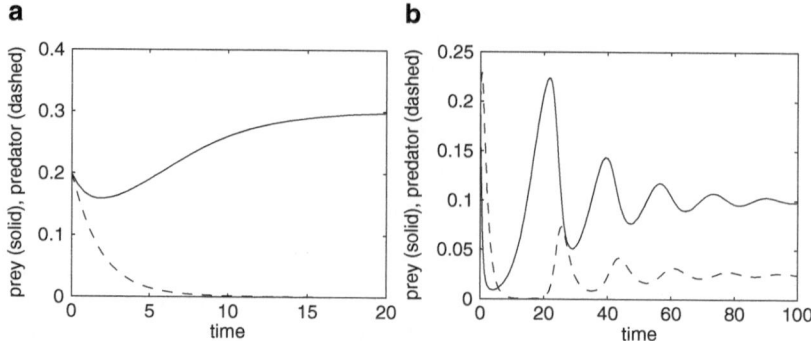

Fig. 7 Numerical results for a metapopulation model of a predator-prey interaction in the cases (a) $e_1'/c_2 > Q_1 > 0$ (no coexistence steady state) and (b) $0 < e_1'/c_2 < Q_1$, where $Q_1 = 1 - 1/R_1$ (coexistence steady state)

$$\frac{dp_1}{dt} = c_1(1 - D - p_1 - p_{12})p_1 - c_2 p_1 p_{12} - e_1 p_1, \tag{22}$$

$$\frac{dp_{12}}{dt} = c_2 p_1 p_{12} - e_1' p_{12}.$$

Steady states $S = (p_1^*, p_{12}^*)$ are given by $S_0 = (0, 0)$, $S_1 = (Q_1 - D, 0)$, biologically realistic if $Q_1 > D$, where $Q_1 = 1 - 1/R_1 = 1 - e_1/c_1$ as before, and $S_{12} = (p_1^*, p_{12}^*)$, where

$$p_1^* = \frac{e_1'}{c_2}, \quad p_{12}^* = \frac{c_1}{c_1 + c_2}\left(Q_1 - D - p_1^*\right), \tag{23}$$

biologically realistic if $Q_1 - D > p_1^*$. Of course, S_{12} cannot be biologically realistic unless S_1 is. It is easy to show that S_0 is stable if S_1 is not biologically realistic, S_1 is stable if it is biologically realistic but S_{12} is not, and S_{12} is stable if it is biologically realistic (Fig. 7). Let us assume that $Q_1 > p_1^* = e_1'/c_2 > 0$, so that S_{12} is biologically realistic for $D = 0$, since otherwise predators can never persist. Then, as D increases, the predator goes extinct as S_{12} ceases to be biologically realistic, at $D = D_{12} = Q_1 - p_1^* = Q_1 - e_1'/c_2$, and then the prey goes extinct as S_1 ceases to be biologically realistic, at $D = D_1 = Q_1 > D_{12}$. Within the coexistence regime, habitat destruction has a potentially negative effect on both prey and predator populations by reducing the space available to them. A reduction in the prey population would have a negative effect on the predator population, while a reduction in the predator population would have a positive effect on the prey population. The effect of habitat destruction on the predators must be negative, while in this simple model its effect on the prey is neutral, with the loss of predators exactly compensating for the loss of habitat. This is essentially Volterra's principle in the metapopulation context, and Rule 2 follows immediately.

2.6   Food Chains

Models for food chains with similar assumptions lead to similar results. Let us consider a food chain of length three, with 1 a basal prey species, 2 an intermediate predator, and 3 a top predator. Let each patch be in one of the states 0, 1, 12, or 123, assuming as above that predators do not persist in patches where they have no prey. The Holt [26] equations for the fraction of sites in each state are given by

$$
\frac{dp_1}{dt} = (1 - D - p_1 - p_{12} - p_{123})(c_1 p_1 + c_1' p_{12} + c_1'' p_{123})
$$

$$
- p_1(c_2 p_{12} + c_2' p_{123}) - e_1 p_1 + e_2 p_{12} + e_2' p_{123}, \tag{24}
$$

$$
\frac{dp_{12}}{dt} = p_1(c_2 p_{12} + c_2' p_{123}) - c_3 p_{12} p_{123} + e_3 p_{123} - (e_1' + e_2) p_{12},
$$

$$
\frac{dp_{123}}{dt} = c_3 p_{12} p_{123} - (e_1'' + e_2' + e_3) p_{123},
$$

In the limiting case $c_1' = c_1'' = c_2' = 0$, $e_2 = e_2' = e_3 = 0$, as in [41], the equations become

$$
\frac{dp_1}{dt} = c_1(1 - D - p_1 - p_{12} - p_{123})p_1 - c_2 p_1 p_{12} - e_1 p_1,
$$

$$
\frac{dp_{12}}{dt} = c_2 p_1 p_{12} - c_3 p_{12} p_{123} - e_2 p_{12}, \tag{25}
$$

$$
\frac{dp_{123}}{dt} = c_3 p_{12} p_{123} - e_3 p_{123}.
$$

If there is a coexistence steady state for $D = 0$ it is stable, and the species go extinct in the order 3, 2, 1 as D increases, as one would expect.

Holt's [26] basic conclusion was that there is a fundamental constraint of spatial inefficiency, and that "metapopulation dynamics can constrain the length of specialist food chains, particularly in heterogeneous landscapes where the basal species is specialised to a rare habitat". Calcagno and co-authors [9] considered a model similar to (24) but extended to food chains of indefinite length, in the absence of habitat removal. They made alternative assumptions on the colonisation and extinction parameters, and included a patch selection mechanism for colonisers. They also found constraints on food chain length arising from the metapopulation dynamics.

2.7   Mutualism

Mutualism in patch-occupancy metapopulations has been discussed in [31, 48, 56]. Mutualism between species i and j may act in the following ways.

- Decrease the extinction rate of species i in patches that species j occupies, an extinction mutualism.
- Increase the probability that species i can colonise a patch if species j already occupies that patch, an establishment mutualism.
- Increase the number of propagules produced by species i in patches in which both species occur, a propagation-production mutualism.

Mutualism may therefore be modelled in the same way as competition, using (6) in the two-species case, but with $e_i' \leq e_i$, $c_i' \geq c_i$, $f_{ij} \geq 1$. If $R_1 > R_2 > 1$, the mutualism is facultative for both species. Extinction and establishment mutualisms are discussed in [56]. In the case of a pure propagation-production mutualism, $e_i' = e_i$, $c_i' = c_i$, we have already seen in Sect. 2.2 that as $t \to \infty$ then $p_{12}/(1-D) \to P_1 P_2/(1 - D)^2$, where $P_i = p_i + p_{12}$, so that the species are independently distributed on the remnant habitat patches. As habitat destruction increases in this case, there are two possible bifurcation scenarios [31]. In the first scenario, the mutualism becomes obligate for species 2, then species 2 goes extinct, then species 1 does. In the second, the mutualism becomes obligate for species 2, then for species 1 as well, and then both species go extinct simultaneously.

More complex mutualistic communities have also been modelled using a patch-occupancy approach [15]. Such communities are often bipartite, consisting for example of several species of flowering plants and their insect pollinators, or of fruiting plants and frugivorous animals. A patch occupancy model with m species of flowering plant and n species of insect pollinator consists of $2^{m+n} - 1$ equations, and soon becomes intractable even for modest values of m and n. For example, a community of two species of flowering plant and one species of insect pollinator that can pollinate either of them consists of seven equations, or six if one assumes that the insect cannot survive on a patch that is not occupied by a plant. Let P_i be the fraction of patches occupied by plant species i, whether or not they are also occupied by the other plant species or by the insect pollinator, and P_{12} the fraction occupied by both plant species, similarly. In the case that there is no interaction between the plant species except through the pollinator, and no competition between them for the services of the pollinator, then the argument of Sect. 2.2 applies to show asymptotic independence of the plant species on undestroyed patches, $P_{12}/(1-D) \to P_1 P_2/(1-D)^2$ as $t \to \infty$. This may be used to reduce the number of equations to five. Other assumptions have been used to reduce the number from five to three [15], one for each of the species concerned, but these have not been justified.

2.8 A Simple Food Web

As an example of the increase in complexity that may arise from slightly less simple networks of interactions between species, we shall now consider a system of one predator preying on two competing species. Such a system was also considered

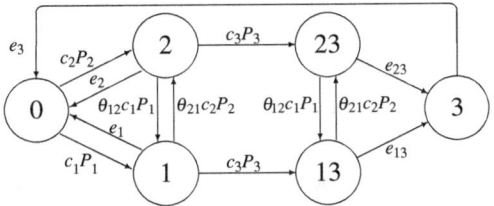

Fig. 8 The transitions of patches between states in the model with two competing prey, species 1 and 2, and one predator, species 3. Prey-predator patches include species 3 and either species 1 (type 13) or species 2 (type 23). Upper case P_i denotes *all* patches including species i, whether or not they include any other species, so $P_1 = p_1 + p_{13}$, $P_2 = p_2 + p_{23}$, $P_3 = p_3 + p_{13} + p_{23}$

by [42], but they made an unjustified independence assumption to reduce the number of equations in their model. A model of the transitions of the system is shown diagrammatically in Fig. 8. The parameters θ_{12} and θ_{21} are the competition parameters introduced in Sect. 2.3 above. Patches containing either prey species may be colonised by predators from other predator or prey-predator patches, but in these prey-predator patches extinction of prey is followed by extinction of predators and return to the empty state.

With a fraction D of the patches removed, and with species 1 competitively predominant over species 2, $\theta_{12} = 1$, $\theta_{21} = 0$, the equations are

$$\frac{dp_1}{dt} = c_1 P_1 (p_0 + p_2) - c_3 P_3 p_1 - e_1 p_1,$$

$$\frac{dp_{13}}{dt} = c_3 P_3 p_1 + c_1 P_1 p_{23} - e_{13} p_{13},$$

$$\frac{dp_2}{dt} = c_2 P_2 p_0 - c_1 P_1 p_2 - c_3 P_3 p_2 - e_2 p_2, \qquad (26)$$

$$\frac{dp_{23}}{dt} = c_3 P_3 p_2 - c_1 P_1 p_{23} - e_{23} p_{23},$$

$$\frac{dp_3}{dt} = e_{13} p_{13} + e_{23} p_{23} - e_3 p_3,$$

$$p_0 = 1 - D - p_1 - p_2 - p_3 - p_{13} - p_{23},$$

where p_0 is the fraction of empty undisturbed patches.

This competitor–predator model was investigated numerically to find its steady state behaviour for different values of D. As habitat is progressively destroyed, how does the composition of the ecosystem change? Possible transitions between stable steady states that occur as D increases are shown in Fig. 9, and corresponding numerical results in Fig. 10. Some of these are well understood. Transition A, $S_1 \to S_0$ (or $S_2 \to S_0$), is the classical result of Nee and May [47] discussed in Sect. 2.1 above, and is analogous to herd immunity. Transition B, $S_{12} \to S_2$, was

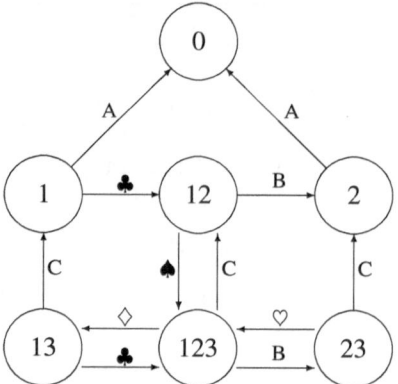

Fig. 9 Possible transitions between stable steady states as the removed fraction D increases, depending on parameter values. Note that the *circles* no longer represent the states of individual patches but state of the landscape as a whole; thus the *circle* 123 represents a landscape in which all three species stably coexist, at a coexistence steady state S_{123}. The transitions marked with *letters* (or close analogues) have been discussed before, A in [47], B in [61], and C in [41]

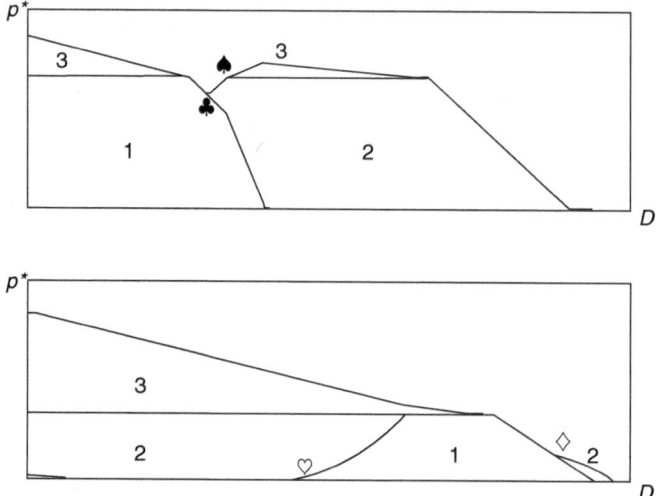

Fig. 10 Numerical results showing the new transitions, represented by *clubsuit, diamondsuit, heartsuit* and *spadesuit* symbols. In the *upper panel*, where $c_1 = 2$, $c_2 = 1$, $c_3 = 10$, $e_1 = 1$, $e_2 = 0.1$, $e_3 = 2$, $e_{13} = 1$, $e_{23} = 1$, the transitions are $S_{13} \to S_1 \to S_{12} \to S_{123} \to S_{23} \to S_2 \to S_0$; in the *lower*, where $c_1 = 2$, $c_2 = 4$, $c_3 = 6$, $e_1 = 0.12$, $e_2 = 0.1$, $e_3 = 2$, $e_{13} = 2$, $e_{23} = 2$, they are $S_{123} \to S_{23} \to S_{123} \to S_{13} \to S_1 \to S_{12} \to S_2 \to S_0$

discussed in Sect. 2.3 above, and reflects Rule 1. The good competitor is completely unaffected by the poor one, and follows the graph of Fig. 1b. The good reproducer therefore has just as many patches to play with as before, *and* is less likely to be displaced by the good competitor, so its population rises. Hence if we start with stable co-existence of two competitors in the absence of predators and then remove habitat, the good competitor will go extinct first [62]. Transition B, $S_{123} \to S_{23}$, shows that Rule 1 may still hold in the presence of a predator. Transition C, $S_{13} \to S_1$ (or $S_{23} \to S_2$), was discussed in Sect. 2.5 above, and reflects Rule 2. Because habitat removal reduces predator population size, it has positive as well as negative effects on prey (which cancel out in this simple situation), but it has entirely negative effects on predators. Hence if we start with stable co-existence of one predator and one prey and then remove habitat, the predator will go extinct first [41]. Transition $S_{123} \to S_{12}$ shows that Rule 2 may still hold in the presence of a second prey species. We have seen transition ♣, $S_1 \to S_{12}$, in Sect. 2.3 above. It occurs whenever 1 drives 2 to extinction in the undisturbed habitat, but 2 is the better reproducer. Transition $S_{13} \to S_{123}$ is similar.

But Rules 1 and 2 do not necessarily hold in our new system. The other transitions appear not to have been described before. Consider transition ◇, $S_{123} \to S_{13}$. It may be species 2, the good reproducer, that goes extinct first. Thus Rule 1 does not necessarily hold in the presence of predators. Why is this? A clue is obtained when we note that predator-mediated co-existence may occur in this system [5]. For certain parameter values the three species may co-exist, where if no predators are present the good competitor drives the poor to extinction. The mechanism is fugitive co-existence, whereby the predators provide the empty patches that the good reproducers exploit so efficiently that it ensures their persistence. The predator, despite being non-selective, has positive as well as negative effects on the good reproducer, but only negative effects on the good competitor. As habitat is removed, then, the predator population falls, empty patches become scarce, and this can lead to extinction of the good reproducer. In the presence of predators Rule 1 may be reversed, and habitat removal may shift the balance from the good reproducer to the good competitor.

Transition ♡, $S_{23} \to S_{123}$, increases species richness as D increases. This is another manifestation of the reversal of Rule 1 in the presence of predators. We may understand it by starting in state S_{123}, where predator-mediated coexistence again allows competitors 1 and 2 to coexist with predator 3, and then considering the effect of habitat addition rather habitat removal. Habitat addition shifts the balance from the good competitor to the good reproducer, so that competitor 1 goes extinct. The predator is crucial, of course, by Rule 1.

Even more surprising is the transition ♠, $S_{12} \to S_{123}$, where habitat removal favours predators. Rule 2 can only be relied upon if there is a single prey species. This is because, in certain circumstances, habitat removal can increase total prey populations. The mechanism is as follows. Consider a landscape in state S_{12} as habitat is removed. As always, the good competitor population falls and the good reproducer population rises, but how does the total prey population change? The easiest way to decide this is to compare the per capita colonisation rate $c_2 p_0^*$ of species 2 with the per capita extinction rate $c_1 p_1^* + e_2$, which since we are at steady state remain equal to each other whatever value D takes, and recall that $1 - D - p_1^* = p_0^* + p_2^*$ is independent of D, as in Fig. 1b. It follows that the slope of p_2^* as a function of D is c_1/c_2. Hence, if $c_1 > c_2$, the good reproducer population rises faster than the good competitor population falls, and the total prey population rises. This favours predators, and may lead to a situation where predators can invade the steady state.

2.9 Heterogeneity

In the models above all habitable patches are identical, and space is modelled implicitly, so that any patch is equally easily colonised from any other one. In reality this is not true even with no habitat destruction, and habitat destruction not only changes available habitats but alters the connectivity between patches. Habitat loss in models with explicit spatial structure leads to habitat fragmentation and percolation effects, but explicit spatial models are outside the scope of this chapter. A review is given in [51].

Heterogeneous patches may be modelled in the implicit spatial context. For example, Holt [26] considers habitable patches of two different types, with equations in the case of two independent species given by

$$\frac{dp_1}{dt} = (c_{11}p_1 + c_{12}p_2)(h_1 - p_1) - e_1 p_1, \tag{27}$$

$$\frac{dp_2}{dt} = (c_{21}p_1 + c_{22}p_2)(h_2 - p_2) - e_2 p_2.$$

Here h_i is the fractions of habitable patches of type i, $i = 1, 2$, and $h_1 + h_2 = 1 - D$. This allows him to consider habitat specialists and generalists, and source–sink metapopulations. The generalisation to any number of patch types is straightforward.

Heterogeneity may also be modelled using a metacommunity approach. In its original sense [63] a metacommunity is a collection of ecological communities connected by a global dispersal pool, each one occupying a different habitable patch, and with population dynamics at the level of the individual patch. The term has since been used to denote a metapopulation with several species, as in Sect. 2.8 above. We shall define it here as a collection of communities connected by dispersal, where the communities themselves are metapopulations, as in [2,27,35,45,46]. For example, Mouquet and co-authors [46] consider a metacommunity approach to competition. Within each community k there is a competitive metapopulation, satisfying (17) with no displacement, $\theta_{ij} = 0$ for all i and j, but the colonisation and extinction parameters c_i and e_i depend also on the community, and are given by c_{ik} and e_{ik} in community k. If there is a single community the species with greatest R_0 excludes all its competitors, but if there are several communities linked by a dispersal pool then a community acting as a source for species i may allow it to persist globally. Habitat destruction, interpreted as destruction of some of the communities, may cause extinctions directly, if one of the communities destroyed is a source for the focal species, or indirectly, if it shifts the regional balance to a different species that may now out-compete it.

The original metacommunity of Wilson [63] may be thought of alternatively and perhaps less ambiguously as a structured metapopulation, which is in general a metapopulation where the state of a patch is not merely a list of the species occupying that patch, but contains more detailed information, such as the size of the occupying populations. This is more realistic and allows more detailed modelling of the processes of colonisation and extinction to take place [18], but it has an adverse effect on tractability.

3 Discussion

In this chapter we reviewed patch-occupancy metapopulation models for habitat destruction and its effect on species extinctions, for single species, competitive and mutualistic communities, predator–prey systems and food chains. The reviewed results show in particular that under a competition–colonisation trade-off habitat removal favours a good reproducer at the expense of a good competitor (in the absence of a predator, Rule 1) or a prey at the expense of a predator (in the absence of a second prey, Rule 2). We then investigated their predictions for a simple food web consisting of a predator with two competing prey, and showed that rules 1 and 2 cannot be extended to this slightly more complex case. Starting from co-existence of all three species, any one may go extinct first, depending on the choice of parameters. Moreover, any of these transitions may proceed in the opposite direction, so that habitat removal may lead to an *increase* in species richness. Parameter sets where such an increase occurs are not difficult to find, so that this does not seem to be abnormal behaviour in such systems, although it should be emphasised that the overall trend of habitat loss is towards decreased richness. It is clear that detailed biological knowledge will be required, not only of a species itself but also of the species it interacts with, to predict which species are most vulnerable to extinction as a result of habitat destruction in a real and far more complex system. Intuition is not a reliable guide even in the simple system we have considered here.

For more complicated food webs with trophic interactions, Pillai and co-authors [54], in a comprehensive treatment of the problem, advocate defining variables $p_{(i,j)}$ to be the fraction of patches occupied by a consumer species j and its resource species i, for each resource–consumer pair (i, j) in the web, and then writing a set of equations for these variables. The implicit assumption is that a consumer species goes immediately locally extinct if it inhabits a patch with no appropriate resource. The approach may involve transforming food-web graphs to allow vertices to represent more than one species at a time, and it may be more transparent to use a state-transition model and simplify it by making explicit assumptions.

Holt's [26] constraint of spatial inefficiency (Sect. 2.6) applies to food chains but not necessarily to food webs. It is shown in [17], using simulations of empirical and randomly created food webs, that persistence increases with diversity and connectance in trophic metacommunities, but the effect of habitat destruction is not explicitly considered.

An ecosystem perspective may be added to spatially structured communities subject to colonisation–extinction dynamics, by including nutrient flows. It is shown in [16] that integrating ecosystem and spatial dynamics can lead to various indirect interactions that contribute significantly to community organisation.

Most studies of the effect of habitat destruction on metapopulations consider its consequences in terms of species richness, and do not take into account other aspects of biodiversity, such as the evenness of the distribution of individuals between species. An exception is [44], which shows numerically that the Shannon index H [40, 53] typically increases after an extinction.

What happens as habitat is removed in a real system? Debinski and Holt [13] give an overview of habitat fragmentation experiments. Expectations that species richness should decrease with decreasing area were supported in only six out of 14 studies. It is shown in this chapter that theory can make much more complex predictions than has hitherto been suspected. These might have a rôle to play in explaining some counter-intuitive observations.

References

1. W.C. Allee, *Animal Aggregations: A Study in General Sociology* (Chicago University Press, Chicago, 1931)
2. P. Amarasekare, Spatial dynamics of foodwebs. Annu. Rev. Ecol. Evol. Syst. **39**, 479–500 (2008)
3. J. Baillie, C. Hilton-Taylor, S.N. Stuart, *2004 IUCN Red List of Threatened Species: A Global Assessment* (International Union for Conservation of Nature, Cambridge, 2004)
4. J. Bascompte, R.V. Solé, Effects of habitat destruction in a prey–predator metapopulation model. J. Theor. Biol. **195**, 383–393 (1998)
5. N.F. Britton, N.R. Franks, G.P. Boswell, Dispersal and conservation in heterogeneous landscapes, in *Insect Movement: Mechanisms and Consequences*, ed. by I.P. Woiwod, D.R. Reynolds, C.D. Thomas (CABI Publishing, Wallingford, 2001), pp. 299–320
6. T.M. Brooks et al., Habitat loss and extinction in the hotspots of biodiversity. Conserv. Biol. **16**, 909–923 (2002)
7. S.H.M. Butchart et al., Global biodiversity: indicators of recent decline. Science **328**, 1164–1168 (2010)
8. V. Calcagno, N. Mouquet, P. Jarne, P. David, Coexistence in a metacommunity: the competition–colonization trade-off is not dead. Ecol. Lett. **9**, 897–907 (2006)
9. V. Calcagno, F. Massol, N. Mouquet, P. Jarne, P. David, Constraints on food chain length arising from regional metacommunity dynamics. Proc. R. Soc. B **278**, 3042–3049 (2011)
10. L.-L. Chen, Z.-S. Lin, The effect of habitat destruction on metapopulations with an Allee-like effect: a study case of Yancheng in Jiangsu Province, China. Ecol. Model. **213**, 356–364 (2008)
11. J.H. Connell, Diversity in tropical rainforests and coral reefs. Science, **199**, 1302–1310 (1978)
12. D.L. DeAngelis, *Dynamics of Nutrient Cycling and Food Webs* (Chapman and Hall, London, 1992)
13. D.M. Debinski, R.D. Holt, A survey and overview of habitat fragmentation experiments. Conserv. Biol. **14**, 342–355 (2000)

14. F.A.S. Dos Santos, M.I.S. Costa, A correct formulation for a spatially implicit predator–prey metacommunity model. Math. Biosci. **223**, 79–82 (2010)
15. M.A. Fortuna, J. Bascompte, Habitat loss and the structure of plant–animal mutualistic networks. Ecol. Lett. **9**, 281–286 (2006)
16. D. Gravel, N. Mouquet, M. Loreau, F. Guichard, Patch dynamics, persistence, and species coexistence in metaecosystems. Am. Nat. **176**, 289–302 (2010)
17. D. Gravel, E. Canard, F. Guichard, N. Mouquet, Persistence increases with diversity and connectance in trophic metacommunities. PLoS ONE **6**(5), e19374 (2011). doi:10:1371/journal.pone.0019374
18. I. Hanski, Coexistence of competitors in patchy environment. Ecology **64**, 493–500 (1983)
19. I. Hanski, *Metapopulation Ecology* (Oxford University Press, Oxford, 1999)
20. I. Hanski, Habitat loss, the dynamics of biodiversity, and a perspective on conservation. AMBIO **40**, 248–255 (2011)
21. I. Hanski, O.E. Gaggiotti (eds.), *Ecology, Genetics and Evolution of Metapopulations* (Elsevier Academic, San Diego, 2004)
22. I. Hanski, M.E. Gilpin, *Metapopulation Biology: Ecology, Genetics and Evolution* (Academic, San Diego, 1997)
23. L. Hansson, L. Fahrig, G. Merriam (eds.), *Mosaic Landscapes and Ecological Processes* (Springer, New York, 1995)
24. A. Hastings, Disturbance, coexistence, history, and competition for space. Theor. Popul. Biol. **18**, 363–373 (1980)
25. A. Hastings, Population dynamics in patchy environments, in *Modeling and Differential Equations in Biology*, ed. by T.A. Burton (Dekker, New York, 1980), pp. 217–223
26. R.D. Holt, Consequences of spatial heterogeneity, in *Metapopulation Biology: Ecology, Genetics and Evolution*, ed. by I. Hanski, M.E. Gilpin (Academic, San Diego, 1997), pp. 149–164
27. M. Holyoak, M.A. Leibold, R.D. Holt (eds.), *Metacommunities: Spatial Dynamics and Ecological Communities* (University of Chicago Press, Chicago, 2005)
28. C.B. Huffaker, Experimental studies on predation: dispersion factors and predator-prey oscillations. Hilgardia **27**, 343–383 (1958)
29. P. Kareiva, V. Wennergren, Connecting landscape patterns to ecosystem and population process. Nature **373**, 299–302 (1995)
30. C.A. Klausmeier, Extinction in multispecies and spatially explicit models of habitat destruction. Am. Nat. **152**, 303–310 (1998)
31. C.A. Klausmeier, Habitat destruction and extinction in competitive and mutualistic communities. Ecol. Lett. **4**, 57–63 (2001)
32. M.J. Labrum, Allee effects and extinction debt. Ecol. Model. **222**, 1205–1207 (2011)
33. R. Lande, Extinction thresholds in demographic models of territorial populations. Am. Nat. **130**, 624–635 (1987)
34. M.A. Leibold, T.E. Miller, From metapopulations to metacommunities, in *Ecology, Genetics and Evolution of Metapopulations*, ed. by I. Hanski, O.E. Gaggiotti (Elsevier Academic, San Diego, 2004), pp. 133–150
35. M.A. Leibold et al., The metacommunity concept: a framework for multi-scale community ecology. Ecol. Lett. **7**, 601–613 (2004)
36. R. Levins, Some demographic and genetic consequences of environmental heterogeneity for biological control. Bull. Entomol. Soc. Am. **15**, 237–240 (1969)
37. R. Levins, D. Culver, Regional coexistence of species and competition between rare species. Proc. Natl. Acad. Sci. USA **68**, 1246–1248 (1971)
38. M. Loreau, Does functional redundancy exist? Oikos **104**, 606–611 (2004)
39. R.H. MacArthur, E.O. Wilson, *The Theory of Island Biogeography* (Princeton University Press, Princeton, 1967)
40. A.E. Magurran, *Ecological Diversity and its Measurement* (Princeton University Press, Princeton, 1988)

41. R.M. May, The effects of spatial scale on ecological questions and answers, in *Large-Scale Ecology and Conservation Biology*, ed. by P.J. Edwards, R.M. May, N.R. Webb (Blackwell Scientific Press, Oxford, 1994), pp. 1–18
42. C.J. Melian, J. Bascompte, Food web structure and habitat loss. Ecol. Lett. **5**, 37–46 (2002)
43. Millennium Ecosystem Assessment, *Ecosystems and Human Well-Being: Synthesis* (Island Press, Washington, 2005)
44. A. Morozov, B.-L. Li, Abundance patterns in multi-species communities exposed to habitat destruction. J. Theor. Biol. **251**, 593–605 (2008); **253**, 628 (2008)
45. N. Mouquet, M. Loreau, Coexistence in metacommunities: the regional similarity hypothesis. Am. Nat. **159**, 420–426 (2002)
46. N. Mouquet, B. Matthiessen, T. Miller, A. Gonzalez, Extinction debt in source–sink metacommunities. PLoS One **6**(3), e17567 (2011). doi:10.1371/journal.pone.0017567
47. S. Nee, R.M. May, Dynamics of metapopulations: habitat destruction and competitive coexistence. J. Anim. Ecol. **61**, 37–40 (1992)
48. S. Nee, R.M. May, M.P. Hassell, Two-species metapopulation models, in *Metapopulation Biology: Ecology, Genetics and Evolution*, ed. by I. Hanski, M.E. Gilpin (Academic, San Diego, 1997), pp. 123–147
49. R.M. Nisbet, W.S.C. Gurney, *Modelling Fluctuating Populations* (Wiley, Chichester, 1982)
50. O. Ovaskainen, I. Hanski, The species–area relationship derived from species-specific incidence functions. Ecol. Lett. **6**, 903–909 (2003)
51. O. Ovaskainen, I. Hanski, Metapopulation dynamics in highly fragmented habitats, in *Ecology, Genetics and Evolution of Metapopulations*, ed. by I. Hanski, O.E. Gaggiotti (Elsevier Academic, San Diego, 2004), pp. 73–103
52. R.T. Paine, S.A. Levin, Intertidal landscapes: disturbances and the dynamics of pattern. Ecol. Monogr. **51**, 145–178 (1981)
53. E.C. Pielou, *Ecological Diversity* (Wiley Interscience, New York, 1975)
54. P. Pillai, M. Loreau, A. Gonzalez, A patch-dynamic framework for food web metacommunities. Theor. Ecol. **3**, 223–237 (2010)
55. S. Prakash, A.M. de Roos, Habitat destruction in a simple predator–prey patch model: how predators enhance prey persistence and abundance. Theor. Popul. Biol. **65**, 153–163 (2004)
56. S. Prakash, A.M. de Roos, Habitat destruction in mutualistic metacommunities. Theor. Popul. Biol. **65**, 153–163 (2004)
57. L.R. Prugh, K.E. Hodges, A.R.E. Sinclair, J.S. Brashares, Effect of habitat area and isolation on fragmented animal populations. Proc. Natl. Acad. Sci. USA **105**, 20770–20775 (2008)
58. K.L. Ryall, L. Fahrig, Response of predators to loss and fragmentation of prey habitat: a review of theory. Ecology **87**, 1086–1093 (2006)
59. M. Slatkin, Competition and regional coexistence. Ecology **55**, 128–134 (1974)
60. R.K. Swihart, Z. Feng, N.A. Slade, D.M. Mason, T.M. Gehring, Effects of habitat destruction and resource supplementation in a predator–prey metapopulation model. J. Theor. Biol. **210**, 287–303 (2001)
61. D. Tilman, Competition and biodiversity in spatially structured habitats. Ecology **75**, 2–16 (1994)
62. D. Tilman, R.M. May, C.L. Lehman, M.A. Nowak, Habitat destruction and the extinction debt. Nature **371**, 65–66 (1994)
63. D.S. Wilson, Complex interactions in metacommunities, with implications for biodiversity and higher levels of selection. Ecology **73**, 1984–2000 (1992)
64. World Wildlife Fund, *Insight into Europe's Forest Protection* (WWF, Gland, 2001)
65. D.W. Yu, H.B. Wilson, The competition–colonization trade-off is dead: long live the competition–colonization trade-off. Am. Nat. **158**, 49–63 (2001)
66. S.-R. Zhou, C.-Z. Liu, G. Wang, The competitive dynamics of metapopulations subject to the Allee-like effect. Theor. Popul. Biol. **65**, 29–37 (2004)

Emergence and Propagation of Patterns in Nonlocal Reaction-Diffusion Equations Arising in the Theory of Speciation

Vitaly Volpert and Vitali Vougalter

Abstract Emergence and propagation of patterns in population dynamics is related to the process of speciation, appearance of new biological species. This process will be studied with a nonlocal reaction-diffusion equation where the integral term describes nonlocal consumption of resources. This equation can have several stationary points and, as it is already well known, a travelling wave solution which provides a transition between them. It is also possible that one of these stationary points loses its stability resulting in appearance of a stationary periodic in space structure. In this case, we can expect a possible transition between a stationary point and a periodic structure. The main goal of this work is to study such transitions. The loss of stability of the stationary point signifies that the essential spectrum of the operator linearized about the wave intersects the imaginary axis. Contrary to the usual Hopf bifurcation where a pair of isolated complex conjugate eigenvalues crosses the imaginary axis, here a periodic solution may not necessarily emerge. To describe dynamics of solutions, we need to consider two transitions: a steady wave with a constant speed between two stationary points, and a periodic wave between the stationary point which loses its stability and the periodic structure which appears around it. Both of these waves propagate in space, each one with its own speed. If the speed of the steady wave is greater, then it runs away from the periodic wave, and they propagate independently one after another.

V. Volpert (✉)
Institute Camille Jordan, UMR 5208 CNRS, University Lyon 1, Villeurbanne, 69622 France
e-mail: volpert@math.univ-lyon1.fr

V. Vougalter
Department of Mathematics and Applied Mathematics, University of Cape Town Private Bag,
Rondebosch 7701, South Africa
e-mail: Vitali.Vougalter@uct.ac.za

M.A. Lewis et al. (eds.), *Dispersal, Individual Movement and Spatial Ecology*,
Lecture Notes in Mathematics 2071, DOI 10.1007/978-3-642-35497-7_12,
© Springer-Verlag Berlin Heidelberg 2013

1 Speciation Theory and Nonlocal Reaction-Diffusion Equations

In their recent book "Speciation" [10], J.A. Coyne and H.A. Orr wrote that "one of the most striking development in evolutionary biology during the last 20 years has been a resurgence of interest in the origin of species". The theory of speciation began with Darwin's "On the origin of species" [11]. It was continued by two schools, naturalists and mutationists, both critical to Darwin's theory. The first one insisted on the role of geographic isolation in speciation (allopatric speciation), the second on nonadaptive and macromutational leaps. This critics was related to the inability to understand how a continuous process in a homogeneous population (sympatric speciation) can result in the emergence of discontinuous entities. In the 1930–1940s, Dobzhansky stressed the importance of reproductive isolation and Mayr defined species as group of interbreeding populations. It was also the period of intensive development of theoretical population genetics due to the works by Fisher, Haldane, Wright. More recent development of the theory of speciation is related to molecular analysis, ecology and some other topics [10, 18]

From a more general point of view, speciation is related to the emergence of discrete clusters in biological populations. The question about clustering was addressed by many authors (see, e.g., [10], page 49). Dawkins considered it as a general property of living matter while Coynne and Orr regarded it "as one of the most important questions in evolutionary biology—perhaps the most important question about speciation".

We will study the question about the emergence and propagation of patterns in population dynamics with nonlocal reaction-diffusion equations taking into account three main properties: reproduction, small variations, competition for resources. Each of them should be specified, and biological relevance of the assumptions should be discussed. We will return to this discussion below.

The approach based on nonlocal reaction-diffusion equations can give an explicit and easily derived condition of the emergence of patterns [8, 19–21, 24, 25]. From the mathematical point of view, it is a linear stability analysis of a homogeneous in space solution. Once the conditions of the emergence of patterns are known, we need to determine how they appear and evolve in time. This depends on the initial population distribution. If it is localized in space, then it can begin to spread or remain spatially localized [2, 3, 31]. Level lines of the population density in the first case are shown in Fig. 1 [3]. This is a result of numerical simulation of a nonlocal reaction-diffusion equation in the bistable case which corresponds to a model of sexual reproduction with a nonlocal consumption of resources. In the beginning, the population is localized at the center of the interval. It grows, then splits into two sub-populations, then splits again and so on. From the mathematical point of view, it is a periodic travelling wave propagating from the center of the interval to the left and to the right.

These results admit two important biological interpretations. If the space variable corresponds to a morphological parameter, for example the size of some animals,

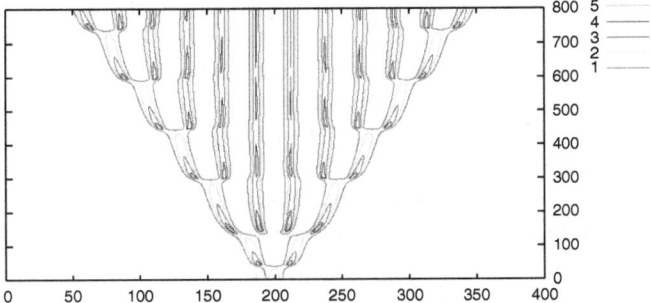

Fig. 1 Level lines of the population density described by a nonlocal reaction-diffusion equation (adapted from [3]). The horizontal axis is the space variable, vertical axis is time. Initially, the population is localized in the center of the interval. It splits into sub-populations and spreads in space (cf. Fig. 2)

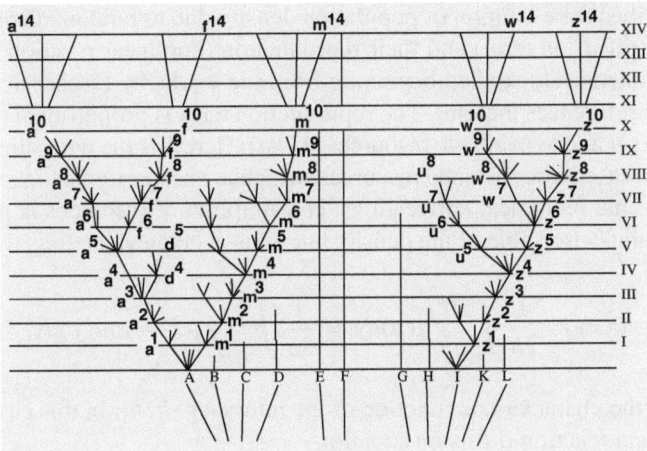

Fig. 2 Darwin's diagram illustrating emergence of species in the process of evolution (adapted from [11]). Horizontal axis corresponds to a morphological parameter, vertical axis to time measured in generations

then each sub-population can be interpreted as a separate species. With this interpretation, we are close to Darwin's representation of the emergence of species (Fig. 2). It is interesting to note that he showed some species which do not split (vertical lines). Such solutions, standing waves or pulses, can be obtained with nonlocal reaction-diffusion equations [3, 31]. This first interpretation of clustering described by nonlocal reaction-diffusion equations is related to sympatric speciation (see Discussion).

In the second interpretation, the space variable corresponds to the physical distance. Biological species can spread in space forming either a uniform density distribution or nonuniform clusters. The first situation is well known since the works by Fisher [17] and Kolmogorov–Petrovskii–Piskunov (KPP) [27] on propagation of dominant gene. It can be described by conventional reaction-diffusion equations. The second case requires introduction of nonlocal reaction-diffusion equations. Emerging clusters can be at the origin of geographic isolations resulting in the allopatric speciation.

Thus, both types of speciation can be related to the emergence of clusters in population density. Propagation of clusters can be considered as a periodic travelling wave described by nonlocal reaction-diffusion equations. In this work, we will study the structure of such waves and will show that there can exist different modes of propagation. In order to introduce the model, let us first consider the classical logistic equation in population dynamics

$$\frac{\partial u}{\partial t} = d \, \frac{\partial^2 u}{\partial x^2} + ku(1 - u), \tag{1}$$

which describes the evolution of population density due to random displacement of individuals (diffusion term) and their reproduction (nonlinear reaction term). This equation is intensively studied beginning from the works by Fischer and KPP (see [32] and the references therein). The reproduction term is proportional to the population density u and to available resources $(1 - u)$. Here 1 is the normalized carrying capacity decreased by consumed resources, which are proportional to the population density. In some biological applications, consumption of resources is proportional not to the point-wise value of the density but to its average value (e.g., [20]):

$$\bar{u}(x) = \frac{1}{2h} \int_{x-h}^{x+h} u(y)dy = \frac{1}{2h} \int_{-\infty}^{\infty} \chi_h(x - y)u(y)dy,$$

where χ_h is the characteristic function of the interval $[-h, h]$. In this case we arrive to the nonlocal reaction-diffusion equation

$$\frac{\partial u}{\partial t} = d \, \frac{\partial^2 u}{\partial x^2} + ku(1 - \bar{u}). \tag{2}$$

In a more general setting, we consider the integro-differential equation

$$\frac{\partial u}{\partial t} = d \, \frac{\partial^2 u}{\partial x^2} + F(u, J(u)), \tag{3}$$

where

$$J(u) = \int_{-\infty}^{\infty} \phi(x - y)u(y, t)dy,$$

$F(u, v)$ is a sufficiently smooth function of its two arguments. In what follows it is convenient to consider it in the form

$$F(u, J(u)) = f(u)(1 - J(u)),$$

where $f(u)$ is a sufficiently smooth function, $f(u) > 0$ for $u > 0$, $f(0) = 0$, $f(1) = 1$. We will assume everywhere below that ϕ is a bounded even function with a compact support, and $\int_{-\infty}^{\infty} \phi(y)dy = 1$. Hence $u = 0$ and $u = 1$ are stationary solutions of (3).

If we look for travelling wave solutions of (3), it is convenient to substitute $u(x, t) = v(x - ct, t)$. Then

$$\frac{\partial v}{\partial t} = d \frac{\partial^2 v}{\partial x^2} + c \frac{\partial v}{\partial x} + F(v, J(v)). \tag{4}$$

Travelling wave is a stationary nonhomogeneous in space solution of this equation. Generalized travelling wave can be time dependent. The existence of travelling waves was studied in [1–7, 16].

An important property of (3) is that its homogeneous in space stationary solution $u = 1$ can lose its stability resulting in appearance of periodic in space stationary solutions (see, e.g., [8, 20, 25]). If we consider this problem in a bounded interval, then this solution bifurcates due to a real eigenvalue which crosses the origin. The situation is more complex if we consider this equation on the whole axis. In this case, not an isolated eigenvalue crosses the origin but the essential spectrum. Conventional bifurcation analysis does not allow us to study bifurcations of nonhomogeneous in space solutions.

We will study behavior of solutions in this case by the combination of numerical simulations and stability analysis. Linear stability analysis allows us to determine the stability boundary (Sect. 2). Numerical simulations show that the solution with a localized initial condition propagates in space. Stability analysis in a weighted space, where the whole spectrum lies in the left half-plane, gives an estimate of the speed of propagation of this solution. Though the spectrum is in the left half-plane, conventional results on nonlinear stability of the solution appear to be inapplicable because the operator does not satisfy the required conditions. We prove a weaker nonlinear stability result of the homogeneous in space solution in a properly chosen weighted space (Sect. 3). This is the stability on a half-axis in the coordinate frame moving faster than the propagation of the periodic in space solution. Let us note that related questions for reaction-diffusion equations are studied in [6, 22].

Next, we study travelling waves connecting the points $u = 0$ and $u = 1$. If the essential spectrum of the operator linearized about a wave crosses the imaginary axis, we cannot use the existing results on the wave stability. We prove the wave stability on a half-axis in some weighted spaces. This result can be interpreted as follows. The wave connecting the points $u = 0$ and $u = 1$ propagates with some speed c_0. The solution which provides the transition from $u = 1$ to a periodic in space solution propagates with some speed c_1. If $c_0 > c_1$, then the wave moves faster and the distance between them increases. This implies the wave stability on the half-axis. Considered on the whole axis, the wave is not stable because of the periodic perturbation, which grows since the essential spectrum is partially in the right-half plane. We will illustrate these results by numerical simulations in 1D and 2D cases.

2 Spectrum of the Operator Linearized about a Stationary Solution

2.1 Equation (3)

We analyze the stability of the solution $u = 1$ of (3). Linearizing this equation about this stationary solution, we obtain the eigenvalue problem:

$$du'' - \int_{-\infty}^{\infty} \phi(x - y)u(y)dy = \lambda u. \tag{5}$$

Applying the Fourier transform, we have

$$\lambda_d(\xi) = -d\xi^2 - \tilde{\phi}(\xi),$$

where $\tilde{\phi}(\xi)$ is the Fourier transform of the function $\phi(x)$. We will assume that it is a real-valued, even, bounded and continuous function, which is not everywhere positive. An example where these conditions are satisfied is given by the following function ϕ:

$$\phi(x) = \begin{cases} 1/(2N), & -N \le x \le N \\ 0, & |x| > N \end{cases}$$

Thus, $\tilde{\phi}(0) = 1$, and there exist one or more intervals where $\tilde{\phi}(\xi)$ is negative. Hence we can make some conclusions about the structure of the function $\lambda_d(\xi)$. There exists $d = d_c$ such that

$$\lambda_d(\xi) < 0, \quad \xi \in R, \quad d > d_c$$

and

$$\lambda_{d_c}(\xi) \le 0, \quad \xi \in R, \quad \lambda_{d_c}(\pm \xi_0) = 0$$

for some $\xi_0 > 0$. Finally, $\lambda_d(\xi) > 0$ in some intervals of ξ for $0 < d < d_c$.

These assumptions signify that the essential spectrum of the operator

$$L_0 u = du'' - \int_{-\infty}^{\infty} \phi(x - y)u(y)dy$$

is in the left-half plane for $d > d_c$ and it is partially in the right-half plane for $d < d_c$.

Fig. 3 Schematic representation of the part $\lambda_+(\xi)$ of the essential spectrum for $d = d_c$

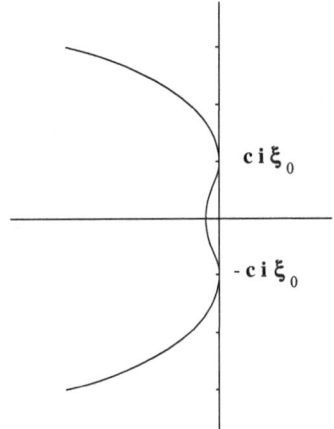

2.2 Equation (4)

Consider next equation (4). As above, we linearize it about the solution $u = 1$ and obtain the eigenvalue problem

$$du'' + cu' - \int_{-\infty}^{\infty} \phi(x - y)u(y)dy = \Lambda u. \tag{6}$$

Applying the Fourier transform, we get the expression

$$\Lambda_d(\xi) = -d\xi^2 + ci\xi - \tilde{\phi}(\xi)$$

which describes the essential spectrum [2,3,30]. From the properties of the function $\lambda_d(\xi)$ it follows that $\Lambda_d(\xi)$ is in the left-half plane of the complex plane for $d > d_c$. For $d = d_c$, it has two values, $\pm ci\xi_0$ at the imaginary axis (Fig. 3), and it is partially in the right-half plane for $d < d_c$. Thus, the essential spectrum passes to the right-half plane when the parameter d decreases and crosses the critical value $d = d_c$.

Let us introduce the function $v(x) = u(x)\exp(-\sigma x)$, where $\sigma > 0$ is a constant. We substitute the function $u(x) = v(x)\exp(\sigma x)$ into (6):

$$dv'' + (c + 2d\sigma)v' + (d\sigma^2 + c\sigma)v - e^{-\sigma x}\int_{-\infty}^{\infty}\phi(x - y)v(y)e^{\sigma y}dy = \Lambda v.$$

Denote

$$\psi(x) = \phi(x)e^{-\sigma x}.$$

Then

$$\Lambda_{d,\sigma}(\xi) = -d\xi^2 + (c + 2d\sigma)i\xi + d\sigma^2 + c\sigma - \tilde{\psi}(\xi), \quad \xi \in R,$$

where $\tilde{\psi}(\xi)$ is the Fourier transform of the function $\psi(x)$.

Put $\sigma = -c/(2d)$. Then

$$\Lambda_{d,\sigma}(\xi) = -d\xi^2 - \tilde{\psi}(\xi) - \frac{c^2}{4d}, \quad \xi \in R.$$

If

$$\frac{c^2}{4d} > \sup_{\xi} \mathrm{Re}\, (-d\xi^2 - \tilde{\psi}(\xi)), \tag{7}$$

then the spectrum is in the left-half plane and the solution of the linear equation

$$\frac{\partial v}{\partial t} = d\,\frac{\partial^2 v}{\partial x^2} + (c+2d\sigma)\,\frac{\partial v}{\partial x} + (d\sigma^2 + c\sigma)v - e^{-\sigma x} \int_{-\infty}^{\infty} \phi(x-y)v(y,t)e^{\sigma y}\,dy \tag{8}$$

will converge to zero as $t \to \infty$ uniformly in x. We have proved the following lemma.

Lemma 1. *If condition (7) is satisfied, then the solution $v(x,t)$ of (8) with a bounded initial condition converges to zero in the uniform norm as $t \to \infty$.*

This lemma gives an estimate of the speed of propagation of the perturbation for the equation

$$\frac{\partial z}{\partial t} = d\,\frac{\partial^2 z}{\partial x^2} - \int_{-\infty}^{\infty} \phi(x-y)z(y,t)\,dy. \tag{9}$$

Indeed, $z(x-ct,t) = u(x,t) = v(x,t)e^{\sigma x}$ $(\sigma > 0)$. For c sufficiently large, $z(x-ct,t)$ converges to 0 uniformly in x on every negative half-axis.

Figure 4 shows an example of numerical simulations of a propagating perturbation.

2.3 Essential Spectrum of the Operator Linearized about a Wave

Suppose that (4) has a stationary solution $w(x)$ with the limits $w(-\infty) = 0$, $w(\infty) = 1$. Suppose that the wave propagates from the right to the left, that is $c < 0$ and the solution $u(x,t) = w(x-ct)$ converges to 1 uniformly on every bounded set.

We linearize (4) about $w(x)$ and obtain the eigenvalue problem

$$du'' + cu' + a(x)u - b(x)J(u) = \lambda u, \tag{10}$$

where

$$a(x) = f'(w(x))(1 - J(w)), \quad b(x) = f(w(x)).$$

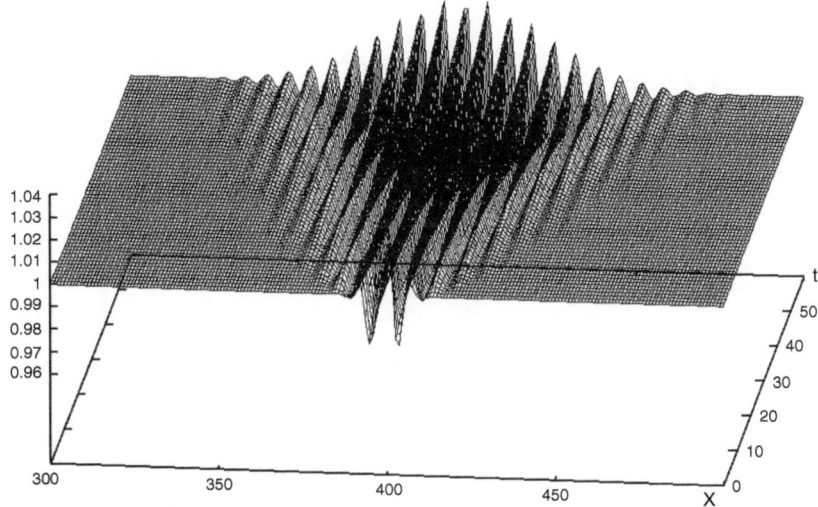

Fig. 4 Solution $u(x, t)$ of (9). The perturbation spreads to the left and to the right with a constant speed. Dimensionless units are used for x and t

The essential spectrum of the operator

$$Lu = du'' + cu' + a(x)u - b(x)J(u)$$

is given by two curves on the complex plane:

$$\lambda_-(\xi) = -d\xi^2 + ci\xi + f'(0), \quad \lambda_+(\xi) = -d\xi^2 + ci\xi - \tilde{\phi}(\xi), \quad \xi \in R.$$

The curve $\lambda_-(\xi)$ is a parabola. It lies in the left-half plane if $f'(0) < 0$ and it is partially in the right-half plane if $f'(0) > 0$. The second curve $\lambda_+(\xi)$ coincides with $\Lambda_d(\xi)$ considered in Sect. 2.2. It is located completely in the left-half plane for $d > d_c$ and it is partially in the right-half plane for $d < d_c$. It is shown schematically in Fig. 3 for the critical value $d = d_c$.

Let us now introduce a weight function $g(x)$. We assume that it is positive, sufficiently smooth and such that

$$g(x) = \begin{cases} e^{\sigma_+ x}, & x \geq 1 \\ e^{\sigma_- x}, & x \leq -1 \end{cases}, \tag{11}$$

where the exponents $\sigma_\pm$ will be specified below. We substitute $u(x) = v(x)g(x)$ into (10):

$$d(gv'' + 2v'g' + vg'') + c(v'g + vg') + avg - b\int_{-\infty}^{\infty} \phi(x - y)v(y)g(y)du = \lambda vg$$

or

$$dv'' + (c + 2dg_1)v' + (dg_2 + cg_1)v - b \int_{-\infty}^{\infty} \phi(x - y)\gamma(x, y)v(y)dy = \lambda v,$$

where

$$g_1(x) = \frac{g'(x)}{g(x)}, \quad g_2(x) = \frac{g''(x)}{g(x)}, \quad \gamma(x, y) = \frac{g(y)}{g(x)}. \tag{12}$$

The essential spectrum of the operator

$$Mv = dv'' + (c + 2dg_1)v' + (dg_2 + cg_1 + a)v - b \int_{-\infty}^{\infty} \phi(x - y)\gamma(x, y)v(y)dy \tag{13}$$

is given by the following expressions:

$$\lambda_\sigma^-(\xi) = -d\xi^2 + (c + 2d\sigma_-)i\xi + (d\sigma_-^2 + c\sigma_- + f'(0)),$$
$$\lambda_\sigma^+(\xi) = -d\xi^2 + (c + 2d\sigma_+)i\xi + d\sigma_+^2 + c\sigma_+ - \tilde{\psi}(\xi), \quad \xi \in R. \tag{14}$$

The second expression coincides with $\Lambda_{d,\sigma}(\xi)$ (Sect. 2.2). When we study stability of waves, we will choose, when it is possible, $\sigma_\pm$ in such a way that the essential spectrum lies in the left-half plane.

3 Nonlinear Stability

Consider the equation (cf. (4))

$$\frac{\partial u}{\partial t} = d \frac{\partial^2 u}{\partial x^2} + c \frac{\partial u}{\partial x} + f(u)\left(1 - \int_{-\infty}^{\infty} \phi(x - y)u(y, t)dy\right). \tag{15}$$

Let us look for its solution in the form $u(x, t) = w(x) + v(x, t)g(x)$, where $w(x)$ is a stationary solution and $g(x)$ is a weight function. Then

$$\frac{\partial v}{\partial t} = d \frac{\partial^2 v}{\partial x^2} + (c + 2dg_1) \frac{\partial u}{\partial x} + (dg_2 + cg_1)v + B(v), \tag{16}$$

where the nonlinear operator $B(v)$ writes

$$B(v) = g^{-1} f(w + vg)\left(1 - \int_{-\infty}^{\infty} \phi(x - y)g(y)v(y, t)dy - \int_{-\infty}^{\infty} \phi(x - y)w(y)dy\right)$$
$$- g^{-1} f(w)\left(1 - \int_{-\infty}^{\infty} \phi(x - y)w(y)dy\right)$$

$$= -f(w + vg) \int_{-\infty}^{\infty} \phi(x - y)\gamma(x, y)v(y, t)dy$$

$$+ \frac{f(w + vg) - f(w)}{vg} v \left(1 - \int_{-\infty}^{\infty} \phi(x - y)w(y)dy\right)$$

$$= -f(w + vg) \int_{-\infty}^{\infty} \phi(x - y)\gamma(x, y)v(y, t)dy$$

$$+ f'(w + \theta vg) v \left(1 - \int_{-\infty}^{\infty} \phi(x - y)w(y)dy\right),$$

where $\theta(x) \in (0, 1)$ is some function. Then

$$B(v) = B_0(v) + B_1(v),$$

where

$$B_0(v) = -f(w) \int_{-\infty}^{\infty} \phi(x - y)\gamma(x, y)v(y, t)dy + f'(w) v \left(1 - \int_{-\infty}^{\infty} \phi(x - y)w(y)dy\right)$$

is a linear and

$$B_1(v) = -(f(w + vg) - f(w)) \int_{-\infty}^{\infty} \phi(x - y)\gamma(x, y)v(y, t)dy$$

$$+ (f'(w + \theta vg) - f'(w)) v \left(1 - \int_{-\infty}^{\infty} \phi(x - y)w(y)dy\right) \qquad (17)$$

a nonlinear operators.

Hence we can write (16) as

$$\frac{\partial v}{\partial t} = M_0 v + B_0 v + B_1(v), \qquad (18)$$

where

$$M_0 v = d \frac{\partial^2 v}{\partial x^2} + (c + 2dg_1) \frac{\partial u}{\partial x} + (dg_2 + cg_1)v.$$

Note that $M = M_0 + B_0$, where M is the operator introduced in Sect. 2.3.

We note that the function $\gamma(x, y)$ under the integral is bounded since $(x - y)$ is bounded in the support of the function ϕ. However, if the weight function $g(x)$ is unbounded, then the nonlinear operator $B_1(v)$ does not satisfy a Lipschitz condition because $f(w + vg)$ and $f'(w + \theta vg)$ do not satisfy it (considered as operators acting on v). Therefore we cannot apply conventional results on stability of solutions [26, 32]. We will use a weaker result presented in the next section.

3.1   An Abstract Theorem on Stability of Stationary Solutions

Consider the evolution equation

$$\frac{\partial v}{\partial t} = Av + T(v),$$ (19)

where A is a linear operator acting in a Banach space E, $T(v)$ is a nonlinear operator acting in the same space. Suppose that A is a sectorial operator and its spectrum lies in the half-plane Re $\lambda < -\beta$, where β is a positive number, and the operator T satisfies the estimate

$$\|T(v)\| \leq K\|v\|,$$ (20)

where K is a constant independent of v. Suppose next that there exists a mild solution of (19), that is a function $v(t) \in E$ which satisfies the equation

$$v(t) = e^{A(t-t_0)}v(t_0) + \int_{t_0}^{t} e^{A(t-s)}T(v(s))ds.$$ (21)

Since the operator A is sectorial, then we have the estimate

$$\|e^{At}v\| \leq Re^{-\beta t}\|v\|$$ (22)

with some constant $R \geq 1$. Let $0 < \alpha < \beta$. Suppose that

$$2RK < \beta - \alpha.$$ (23)

We will show that $\|v\|$ converges to zero. Similar to the estimates in the proof of Theorem 5.1.1 in [26],[1] we obtain from (21):

$$\|v(t)\| \leq R\|v(t_0)\| + KR \sup_{s\in(t_0,t)} \|v(s)\| \int_{t_0}^{t} e^{-\beta(t-s)}ds.$$

Hence

$$\sup_{s\in(t_0,t)} \|v(s)\| \leq R\|v(t_0)\| + \frac{1}{2} \sup_{s\in(t_0,t)} \|v(s)\|.$$

Therefore $\|v(t)\| \leq 2R\|v(t_0)\|$ for all $t \geq t_0$, that is the norm of the solution remains uniformly bounded.

---

[1]We cannot directly use the theorem because the nonlinear operator does not satisfy a Lipschitz condition. This is the reason why we suppose in addition the existence of a mild solution.

On the other hand, from the same equation we obtain

$$\|v(t)\| \leq Re^{-\alpha(t-t_0)}\|v(t_0)\| + KR \int_{t_0}^{t} e^{-\beta(t-s)}\|v(s)\|ds.$$

Let

$$\omega(t) = \sup_{s \in (t_0, t)} \|v(s)\| e^{\alpha(s-t_0)}.$$

Then from the last inequality

$$\|v(t)\| e^{\alpha(t-t_0)} \leq R\|v(t_0)\| + KR \int_{t_0}^{t} e^{-\beta(t-s)} e^{\alpha(t-t_0)}\|v(s)\|ds$$

$$\leq R\|v(t_0)\| + KR\,\omega(t) \int_{t_0}^{t} e^{-(\beta-\alpha)(t-s)}ds$$

$$\leq R\|v(t_0)\| + \frac{1}{2}\,\omega(t).$$

Hence $\omega(t) \leq 2R\|v(t_0)\|$. Thus the norm of the solution exponentially converges to zero. We have proved the following theorem.

Theorem 1. *Suppose that a linear operator A and a nonlinear operator T satisfy estimates (22) and (20), respectively. If there exists a mild solution (21) and condition (23) is verified, then the solution of equation (19) exponentially converges to 0 as $t \to \infty$.*

Existence of a Mild Solution

Suppose that $f(0) = 0$ and the initial condition of the Cauchy problem for (15) is non-negative. Then the solution is also non-negative, and the solution of this equation can be estimated from above by the solution of the equation

$$\frac{\partial z}{\partial t} = d\,\frac{\partial^2 z}{\partial x^2} + c\,\frac{\partial z}{\partial x} + f(z). \tag{24}$$

If $f(z) \leq m$ for all $z \geq 0$, then the supremum of the solution of the last equation can be estimated by $\exp(mt) \sup_x |z(x, 0)|$. Therefore, there exists the classical solution $u(x, t)$ of (15) in every bounded time interval. Then the operator $B_1(v)$ can be written as

$$B_1(v, t) = -(f(u) - f(w)) \int_{-\infty}^{\infty} \phi(x - y)\gamma(x, y)v(y, t)dy$$

$$+(f'((1 - \theta)w + \theta u) - f'(w))\,v\left(1 - \int_{-\infty}^{\infty} \phi(x - y)w(y)dy\right),$$

where $u(x, t)$ is considered as a given function. Taking into account that $\gamma(x, y)$ is uniformly bounded, this operator satisfies a Lipschitz condition with respect to v and a Hölder condition with respect to t. Then we can affirm the existence of a mild solution of equation (18) in $L^2(R)$.

We note that if $f(z_0) = 0$ for some $z_0 > 0$, and the initial condition $z(x, 0)$ is such that $0 < z(x, 0) < z_0$ for all $x \in R$, then the solution also satisfies these inequalities for all $t: 0 < z(x, t) < z_0$. This provides the boundedness of the solution $u(x, t)$ of (15).

3.2   Stability of the Homogeneous Solution

Consider the homogeneous solution $w(x) = 1$. Let $g(x) = e^{\sigma x}$ and $\sigma = -c/(2d)$. Then

$$Mv = d \frac{\partial^2 v}{\partial x^2} - \frac{c^2}{4d} v - f(1) \int_{-\infty}^{\infty} \phi(x - y) e^{-\sigma(x-y)} v(y, t) dy,$$

$$B_1(v) = -(f(1 + vg) - f(1)) \int_{-\infty}^{\infty} \phi(x - y) e^{-\sigma(x-y)} v(y, t) dy.$$

We recall that $f(1) = 1$ but we keep it here for convenience. We multiply the equation

$$\frac{dv}{dt} = Mv$$

by v and integrate over R:

$$\frac{d\|v\|^2}{dt} = 2 \int_{-\infty}^{\infty} \Psi(\xi)(\tilde{v}(\xi))^2 d\xi,$$

where

$$\Psi(\xi) = -d\xi^2 - \frac{c^2}{4d} - f(1)\tilde{\psi}(\xi),$$

$\tilde{\psi}(\xi)$ is the Fourier transform of the function $\psi(x) = \phi(x)e^{-\sigma x}$. Suppose that

$$\sup_{\xi} \Psi(\xi) < -\beta, \tag{25}$$

where β is a positive constant. Then

$$\frac{d\|v\|^2}{dt} \leq -2\beta\|v\|^2,$$

where $\|v\|$ is the norm in $L^2(R)$. Hence $\|v(t)\| \leq e^{-\beta(t-t_0)}\|v(t_0)\|$.

We next estimate the operator $B_1(v)$. Denote by $V(f)$ the maximal variation of the function f,

$$V(f) = \sup_{x,y} |f(x) - f(y)|.$$

Then

$$\|B_1(v)\|^2 \leq (V(f))^2 \int_{-\infty}^{\infty} \left(\int_{-\infty}^{\infty} \psi(x - y)v(y,t)dy \right)^2 dx$$

$$= (V(f))^2 \int_{-\infty}^{\infty} \left| \tilde{\psi}(\xi)\tilde{v}(\xi,t) \right|^2 d\xi$$

$$\leq \left(V(f) \sup_{\xi} |\tilde{\psi}(\xi)| \right)^2 \|v\|^2.$$

Thus, we have determined the values of the constants in estimate (23):

$$R = 1, \quad K = V(f) \sup_{\xi} |\tilde{\psi}(\xi)|$$

and β is determined by (25). We have proved the following theorem.

Theorem 2. *Let the maximum of the function*

$$\Psi_0(\xi) = -d\xi^2 - f(1)\text{Re}\,\tilde{\psi}(\xi)$$

be attained at $\xi = \xi_0$ and $\Psi_0(\xi_0) > 0$. Suppose that $c < 0$ and $c^2 > 4d\,\Psi_0(\xi_0)$. If for some α such that

$$0 < \alpha < \frac{c^2}{4d} - \Psi_0(\xi_0) \tag{26}$$

the estimate

$$2V(f) \sup_{\xi} |\tilde{\psi}(\xi)| < \frac{c^2}{4d} - \Psi_0(\xi_0) - \alpha \tag{27}$$

holds, then the stationary solution $u = 1$ of (15) is asymptotically stable with weight and the following convergence occurs:

$$\|(u(x,t) - 1)e^{(c/2d)x}\| \leq 2Re^{-\alpha t} \|(u(x,0) - 1)e^{(c/2d)x}\|, \quad t \geq 0.$$

We note that condition (26) provides linear stability of the stationary solution in the sense that the spectrum of the linearized problem lies in the left-half plane. However, it may not be sufficient for nonlinear stability understood as convergence of the solution of nonlinear problem to the stationary solution. This implies the additional condition (27).

3.3   Stability of Waves

Consider the equation

$$\frac{\partial u}{\partial t} = d \frac{\partial^2 u}{\partial x^2} + c \frac{\partial u}{\partial x} + f(u)(1 - J(u)) \tag{28}$$

in R. Assume that it has a stationary solution $w(x)$ with the limits $w(\pm\infty) = w_\pm$ at infinity. This function satisfies the equation

$$dw'' + cw' + f(w)(1 - J(w)) = 0, \quad w(\pm\infty) = w_\pm. \tag{29}$$

If $f'(0) \neq 0$, then it decays exponentially at $-\infty$ with the exponent μ which can be found from the equation

$$d\mu^2 + c\mu + f'(0) = 0. \tag{30}$$

If $f'(0) < 0$, then there is a unique positive value of μ which provides a bounded solution at $-\infty$:

$$\mu_1 = -\frac{c}{2d} - \sqrt{\frac{c^2}{4d^2} - \frac{f'(0)}{d}}.$$

If $f'(0) > 0$, then there are two positive values:

$$\mu_{1,2} = -\frac{c}{2d} \pm \sqrt{\frac{c^2}{4d^2} - \frac{f'(0)}{d}}, \quad \mu_1 > \mu_2 > 0$$

($c < 0$). We will restrict ourselves to the latter. It corresponds to the monostable case where the waves exist for all positive speeds greater or equal to some minimal speed [4, 32].

We recall the operator linearized about the wave:

$$Lu = du'' + cu' + a(x)u - b(x)J(u),$$

where

$$a(x) = f'(w)(1 - J(w)), \quad b(x) = f(w).$$

The part of its essential spectrum corresponding to $-\infty$ is given by the curve

$$\lambda_-(\xi) = -d\xi^2 + ci\xi + f'(0), \quad x \in R.$$

Since we assume that $f'(0) > 0$, then it is partially in the right-half plane. We recall that it is obtained as a set of all complex λ for which the equation

$$L^- u \equiv du'' + cu' + f'(0)u = \lambda u$$

has a bounded solution in R. Let us substitute $u = e^{\sigma - x}v$ in this equation. Then

$$dv'' + (c + 2d\sigma_-)v' + (d\sigma_-^2 + c\sigma_- + f'(0))v = \lambda v.$$

Therefore the essential spectrum

$$\lambda_\sigma^-(\xi) = -d\xi^2 + (c + 2d\sigma_-)i\xi + d\sigma_-^2 + c\sigma_- + f'(0), \quad \xi \in R$$

is completely in the left-half plane if

$$d\sigma_-^2 + c\sigma_- + f'(0) < 0.$$

There exists a value of σ_- such that this condition is satisfied if $c^2 > 4df'(0)$. We assume that it holds and

$$-\frac{c}{2d} - \sqrt{\frac{c^2}{4d^2} - \frac{f'(0)}{d}} < \sigma_- < -\frac{c}{2d} + \sqrt{\frac{c^2}{4d^2} - \frac{f'(0)}{d}}. \tag{31}$$

Consider the weighted norm

$$\|w\|_{C_{\sigma_-}(R)} = \sup_x |(1 + e^{-\sigma - x})w(x)|.$$

The wave $w(x)$ is bounded in this norm if (a) $w(x) - w_+ \sim e^{\mu_1 x}$ as $x \to -\infty$ and unbounded if (b) $w(x) - w_+ \sim e^{\mu_2 x}$ as $x \to -\infty$.

In the case of the reaction-diffusion equation [32] and for some nonlocal reaction-diffusion equations [4] the wave with the minimal speed behaves as (a) and all waves with greater speeds behave as (b). In what follows we will be interested in waves with sufficiently large speeds. Therefore we will consider the case (b) where the wave does not belong to the weighted space.

We look for the solution of equation (28) (the same as (15)) in the form $u = w + vg$, where the weight function g is given by (11). Then v satisfies the equation

$$\frac{\partial v}{\partial t} = Mv + B_1(v), \tag{32}$$

where the operator M is given by (13) and the operator $B_1(v)$ by (17). The essential spectrum of the operator M is given by the curves (14). If condition (31) is satisfied, then the curve $\lambda_\sigma^-(\xi)$ lies in the left-half plane. This is also true for $\lambda_\sigma^+(\xi)$ if

$$-d\xi^2 + d\sigma_+^2 + c\sigma_+ - \mathrm{Re}\,\tilde{\psi}(\xi) < 0, \quad \xi \in R.$$

We recall that $f(1) = 1$ and $\tilde{\psi}(\xi)$ is the Fourier transform of the function $\phi(x)e^{-\sigma + x}$. In order to show its dependence on σ_+, we will denote it by $\tilde{\psi}(\xi; \sigma_+)$. Set

$$\Psi(\xi; \sigma_+) = -d\xi^2 + d\sigma_+^2 + c\sigma_+ - \mathrm{Re}\,\tilde{\psi}(\xi; \sigma_+).$$

Assumption 1. The maximum of the function $\Psi(\xi; 0)$ is positive and there exists such σ_+ that the maximum of the function $\Psi(\xi; \sigma_+)$ is negative.

This assumption means that the essential spectrum of the operator L is partially located in the right-half plane and that for some σ_+ the essential spectrum of the operator M lies completely in the left-half plane.

Discrete Spectrum

Let us now discuss the structure of the discrete spectrum of the operator M. It can be directly verified that the operator L has a zero eigenvalue with the corresponding eigenfunction w'. By virtue of the assumptions on the wave $w(x)$, its derivative does not belong to the weighted space $C_g(R)$ with the norm

$$\|v\|_g = \sup_x \left| \frac{v}{g} \right|.$$

Hence if the zero eigenvalue of the operator L is simple, then the operator M does not have zero eigenvalues. This observation justifies the following assumption.

Assumption 2. All eigenvalues of the operator M lie in the half-plane $\text{Re } \lambda < -\beta'$ with some positive β'.

Nonlinear Operator

The operator $B_1(v)$ admits the estimate

$$\|B_1(v)\| \leq K\|v\| \tag{33}$$

in the $C(R)$ norm where

$$K = V(f) \sup_{x,y} |\phi(x - y)\gamma(x, y)| + V(f') \sup_x |1 - J(w)|. \tag{34}$$

We can now formulate the main theorem of this section. Its proof follows from the result of Sect. 3.1.

Theorem 3. *Suppose that there exists a solution of problem (29) such that $w(x) - w_+ \sim e^{\mu_2 x}$ as $x \to -\infty$. Let Assumptions 1 and 2 be satisfied, and the spectrum of the operator M lie in the half-plane $\text{Re } \lambda < -\beta$ with some positive β. Suppose that $2RK < \beta - \alpha$, where α is a positive constant, $0 < \alpha < \beta$, K is given by (34) and R is the constant in the estimate*

$$\|e^{Mt}v(t)\| \leq Re^{-\beta t}\|v(0)\|.$$

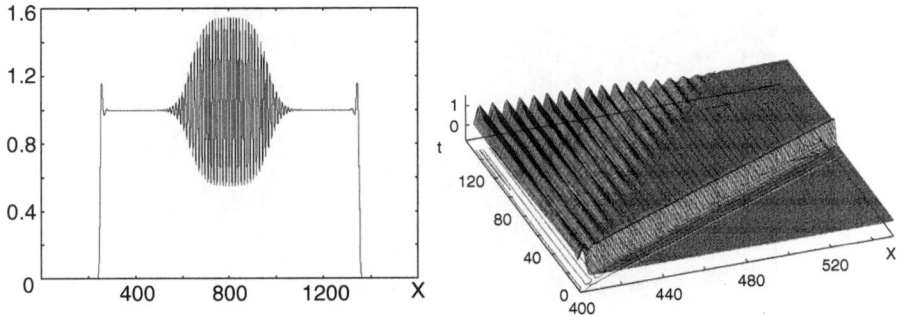

Fig. 5 Periodic wave moves slower than the wave between the stationary points. Solution $u(x, t)$ as a function of x for a fixed t (*left*). The same solution as a function of two variables (*right*)

Then the following estimate holds:

$$\|(u(x, t) - w(x))g(x)\| \leq 2Re^{-\alpha t} \|(u(x, 0) - w(x))g(x)\|, \quad t \geq 0.$$

3.4 Numerical Examples

Figure 5 shows an example of numerical simulations of wave propagation. The wave between 0 and 1 propagates faster than the wave between 1 and the periodic in space stationary solution. The distance between them grows. The first wave is stable on a half-axis. This is in agreement with the result of the previous section. At the same time, this travelling wave is unstable in the uniform norm on the whole axis.

We present here some example of two-dimensional numerical simulations. Qualitatively, they are similar to the 1D case. However, in comparison with the 1D case, there is an additional parameter, the form of the support of the function $\phi(x, y)$. Consider an initial condition with a bounded support in the center of the computational domain. Then we observe a circular travelling wave propagating from the center outside. A spatial structure emerges behind the wave. The peaks of the density form a regular square grid in the case of the square support of ϕ (Fig. 6, left). In the case of the circular support of the function ϕ, emerging structures are also circular (Fig. 6, right). Depending on the parameters, these can be just circles or peaks forming circles.

4 Discussion

Let us recall that allopatric speciation implies the existence of geographic or genetic isolation where there is no gene exchange between different taxa, parapatric speciation admits partial exchange and sympatric speciation occurs without geographic

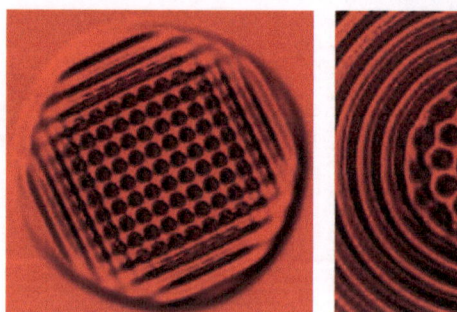

Fig. 6 Propagation of circular wave and formation of a structure behind the wave. The maxima of the density form a square grid in the case of a square support of ϕ (*left*) and a circular structure in the case of a circular support (*right*)

or genetic barriers. It is generally accepted that allopatric speciation is biologically realistic though the mechanism which leads to the appearance of isolating barriers may be sometimes unclear. Sympatric speciation continues to instigate intensive discussions. In spite of big body of experimental data, observations in nature and theoretical models, it is difficult to make definite conclusions about its existence in nature because of the complexity of these phenomena and variety of possible mechanisms [10, 18].

In this work we use nonlocal reaction-diffusion equations in order to describe emergence of patterns in population density. In the morphological space, this model describes sympatric speciation, in the physical space, the appearance of geographical isolation which can lead to allopatric speciation. Thus, the question about the emergence and propagation of patterns in population density is more general than the question about speciation, and it may be simpler from the point of view of biological interpretation since speciation implies certain biological properties including genetic difference.

The model takes into account competition for resources, reproduction and small variations. Let us briefly discuss these assumptions. The reproduction term is proportional to the population density u^k with the first power for the asexual reproduction and the second power for the sexual reproduction, and to available resources $(1 - J(u))$. Here 1 is normalized carrying capacity and $J(u)$ describes consumption of resources. In the case of nonlocal consumption of resources, which corresponds to intra-specific competition, this is an integral term. The diffusion term describes random motion of individuals in the physical space or mutation resulting in small variation in the phenotype in the morphological space.

In this work we considered the case $k = 1$. Emergence and propagation of patterns can also occur for $k = 2$ [2, 3]. An interesting difference between these two cases is the existence of standing waves or pulses which is observed for the latter but not for the former [19–21]. This may signify that asexual populations are necessarily invasive while the sexual reproduction can create stable stationary population distributions.

The population density $u(x, t)$ is a function of the space variable x and of time t. The reproduction term shows the rate of growth of the density at the space point x. Hence, consumption of resources is nonlocal while reproduction is local. This assumption is natural for the asexual reproduction but it requires some comments for the sexual reproduction. In the case of a physical space, this means that the two parents have their average location at the same space point, in the case of a morphological space that they have the same phenotype. The last assumption is often discussed in the context of sympatric speciation. What happens if it is not satisfied and the reproduction is also nonlocal, that is two parents can have different phenotypes? The nonlinear term becomes proportional to $uJ_1(1 - J)$, where J_1 is now an integral which shows some distribution of mating partners. Moreover, we need to specify how offsprings phenotype depends on parents phenotype. In this case, we can obtain a nonlocal operator instead of the diffusion term. It can be verified that the change in the nonlinear term does not influence the result of the linear stability analysis of the solution $u = 1$, but the nonlocal diffusion operator will certainly influence it, possibly changing the conditions of the emergence of localized patterns. The detailed analysis of this question is beyond the scope of this paper.

Let us note that sympatric speciation is studied in [5, 14] by individual based modelling and in [9, 12] by probabilistic methods. Similar to the assumptions discussed above, competition for resources and similar parents phenotype are supposed. The assumption about nonhomogeneous resource distribution, suggested in [14], is not necessary to get splitting of the population. Integro-differential equations with a nonlocal production term but without diffusion are studied in [13, 15, 29] and, in a more general context, in [23]. The Hamilton–Jacobi equation is derived and used to study evolution of population clusters in [28].

Formation of spatial patterns in population density may be a precursor of speciation but may not necessarily lead to it. There are many examples of such structuring, like anthills, human settlements or animal herds. Applicability of nonlocal reaction-diffusion models in these cases should be justified. In a more general framework of Universal Darwinism, such models can be potentially applicable to many other ecological, sociological or economical processes but this will require a careful investigation in each particular case.

Acknowledgements The first author was supported by the grant no. 14.740.11.0877 of the Ministry of Education and Research of Russian Federation, "Investigation of Spatial and Temporal Structures in Fluids with Applications to Mathematical Biology".

References

1. S. Ai, Traveling wave fronts for generalized Fisher equations with spatio-temporal delays. J. Differ. Equat. **232**, 104–133 (2007)
2. A. Apreutesei, A. Ducrot, V. Volpert, Competition of species with intra-specific competition. Math. Model. Nat. Phenom. **3**, 1–27 (2008)

3. A. Apreutesei, A. Ducrot, V. Volpert, Travelling waves for integro-differential equations in population dynamics. Discrete Cont. Dyn. Syst. Ser. B **11**, 541–561 (2009)
4. A. Apreutesei, N. Bessonov, V. Volpert, V. Vougalter, Spatial structures and generalized travelling waves for an integro-differential equation. DCDS B **13**(3), 537–557 (2010)
5. S. Atamas, Self-organization in computer simulated selective systems. Biosystems **39**, 143–151 (1996)
6. M. Beck, A. Ghazaryan, B. Sandstede, Nonlinear convective stability of travelling fronts near Turing and Hopf instabilities. J. Differ. Equat. **246**, 4371–4390 (2009)
7. H. Berestycki, G. Nadin, B. Perthame, L. Ryzkik, The non-local Fisher-KPP equation: traveling waves and steady states. Nonlinearity **22**(12), 2813–2844 (2009)
8. N.F. Britton, Spatial structures and periodic travelling waves in an integro-differential reaction-diffusion population model. SIAM J. Appl. Math. **6**, 1663–1688 (1990)
9. N. Champagnat, R. Ferriere, S. Meleard, Unifying evolutionary dynamics: from individual stochastic processes to macroscopic models. Theor. Popul. Biol. **69**(3), 297–321 (2006)
10. J.A. Coyne, H.A. Orr, *Speciation* (Sinauer Associates, Sunderland, 2004)
11. C. Darwin, *On the Origin of Species by means of Natural Selection* (John Murray, London, 1859)
12. L. Desvillettes, C. Prevost, R. Ferriere, Infinite dimensional reaction-diffusion for population dynamics. Preprint no. 2003–04 du CMLA, ENS Cachan
13. L. Desvillettes, P.E. Jabin, S. Mischler, G. Raoul. On selection dynamics for continuous structured populations. Commun. Math. Sci. **6**(3), 729–747 (2008)
14. U. Dieckmann, M. Doebeli, On the origin of species by sympatric speciation. Nature **400**, 354–357 (1999)
15. M. Doebeli, H.J. Blok, O. Leimar, U. Dieckmann, Multimodal pattern formation in phenotype distributions of sexual populations. Proc. R. Soc. B **274**, 347–357 (2007)
16. A. Ducrot, Travelling wave solutions for a scalar age-structured equation. Discrete Contin. Dyn. Syst. Ser. B **7**, 251–273 (2007)
17. R.A. Fisher, The wave of advance of advantageous genes. Ann. Eug. **7**, 355–369 (1937)
18. S. Gavrilets, *Fitness Landscape and the Origin of Species* (Princeton University Press, Princeton, 2004)
19. S. Genieys, V. Volpert, P. Auger, Adaptive dynamics: modelling Darwin's divergence principle. Comptes Rendus Biologies **329**(11), 876–879 (2006)
20. S. Genieys, V. Volpert, P. Auger, Pattern and waves for a model in population dynamics with nonlocal consumption of resources. Math. Model. Nat. Phenom. **1**, 63–80 (2006)
21. S. Genieys, N. Bessonov, V. Volpert, Mathematical model of evolutionary branching. Math. Comput. Modell. (2008). doi: 10/1016/j.mcm.2008.07.023
22. A. Ghazaryan, B. Sandstede, Nonlinear convective instability of turing-unstable fronts near onset: a case study. SIAM J. Appl. Dyn. Syst. **6**(2), 319–347 (2007)
23. A.N. Gorban, Selection theorem for systems with inheritance. Math. Model. Nat. Phenom. **2**(4), 1–45 (2007)
24. S.A. Gourley, Travelling front solutions of a nonlocal Fisher equation. J. Math. Biol. **41**, 272–284 (2000)
25. S.A. Gourley, M.A.J. Chaplain, F.A. Davidson, Spatio-temporal pattern formation in a nonlocal reaction-diffusion equation. Dyn. Syst. **16**(2), 173–192 (2001)
26. D. Henry, Geometric theory of semilinear parabolic equations. *Lecture Notes in Mathematics*, vol. 840 (Springer, New York, 1981)
27. A.N. Kolmogorov, I.G. Petrovskii, N.S. Piskunov, L'étude de l'équation de la chaleur avec croissance de la quantité de matière et son application à un problème biologique. Bull. Moskov. Gos. Univ. Mat. Mekh. **1**(6), 1–25 (1937)
28. B. Perthame, S. Genieys, Concentration in the nonlocal Fisher equation: the Hamilton-Jacobi limit. Math. Model. Nat. Phenom. **2**(4), 135–151 (2007)
29. S. Pigolotti, C. Lopez, E. Hernandez-Garcia, Species clustering in competitive Lotka-Volterra models. Phys. Rev. Lett. **98**, 258101 (2007)

30. V. Volpert, Elliptic partial differential equations. *Fredholm Theory of Elliptic Problems in Unbounded Domains*, vol. 1 (Birkhauser, Basel, 2011)
31. V. Volpert, S. Petrovskii, Reaction-diffusion waves in biology. Phys. Life Rev. **6**, 267–310 (2009)
32. A.I. Volpert, Vit. Volpert, Vl. Volpert, Traveling wave solutions of parabolic systems. *Translation of Mathematical Monographs*, vol. 140 (American Mathematical Society, Providence, 1994)

Numerical Study of Pest Population Size at Various Diffusion Rates

Natalia Petrovskaya, Nina Embleton, and Sergei V. Petrovskii

Abstract Estimating population size from spatially discrete sampling data is a routine task of ecological monitoring. This task may however become challenging in the case that the spatial data are sparse. The latter often happens in nationwide pest monitoring programs where the number of samples per field or area can be reduced to just a few due to resource limitation and other reasons. In this rather typical situation, the standard (statistical) approaches may become unreliable. Here we consider an alternative approach to evaluate the population size from sparse spatial data. Specifically, we consider numerical integration of the population density over a coarse grid, i.e. a grid where the asymptotical estimates of numerical integration accuracy do not apply because the number of nodes is not large enough. We first show that the species diffusivity is a controlling parameter that directly affects the complexity of the density distribution. We then obtain the conditions on the grid step size (i.e. the distance between two neighboring samples) allowing for the integration with a given accuracy at different diffusion rates. We consider how the accuracy of the population size estimate may change if the sampling positions are spaced non-uniformly. Finally, we discuss the implications of our findings for pest monitoring and control.

N. Petrovskaya (✉) · N. Embleton
School of Mathematics, University of Birmingham, Birmingham, B15 2TT, UK
e-mail: n.b.petrovskaya@bham.ac.uk; embleton@for.mat.bham.ac.uk

S.V. Petrovskii
Department of Mathematics, University of Leicester, Leicester, LE1 7RH, UK
e-mail: sp237@le.ac.uk

M.A. Lewis et al. (eds.), *Dispersal, Individual Movement and Spatial Ecology*,
Lecture Notes in Mathematics 2071, DOI 10.1007/978-3-642-35497-7_13,
© Springer-Verlag Berlin Heidelberg 2013

1 Introduction

Theoretical ecologists routinely operate with quantities like average population densities and/or population sizes, apparently assuming that they can be measured in the field with sufficient accuracy. Indeed, there are a variety of approaches to estimate the population size depending on the species taxonomy and biological traits [33]. However, it is almost never measured directly by counting all the animals (e.g. insects) in a given field or forest. Much more typically, an estimate is obtained through collecting samples and their subsequent analysis, e.g. by using statistical methods [28]. The accuracy of the estimate then depends significantly on the number of samples. This has long been a focus of applied statistical analysis, yet there are some issues that remain unresolved. The focus of statistical methods has been more on calculating the variance in the sampling data (and on the relation between the variance and the mean [34]) rather than on the mean density itself.

The essence of the standard approach can be readily seen from the following outline. Let $u_0, \ldots, u_{N-1}$ be the values of the population density of a given species obtained at the location of the samples $\mathbf{r}_0, \ldots, \mathbf{r}_{N-1}$, respectively, where N is thus the number of samples. In order to obtain the average population density $\bar{u}$ and/or the (total) population size I in an area A, this information must somehow be 'integrated' over the area. A commonly used statistical method to estimate the population size is based on the arithmetic average [31]:

$$I \approx \tilde{I} = A\hat{u}, \quad \text{where} \quad \hat{u} = \frac{1}{N} \sum_{n=0}^{N-1} u_n \approx \bar{u}. \tag{1}$$

This approach works well when N is sufficiently large because the theory predicts that $\hat{u}$ converges to $\bar{u}$ when N tends to infinity. However, if N is not large, the application of (1) become questionable, especially when the density distribution is not spatially homogeneous but exhibits some form of aggregation.

A few questions arise here. Firstly, it is not clear what it actually means for N to be "large" or "small." Mathematically rigorous criteria assessing the minimum number of samples required to obtain a robust estimate of the population size are largely missing, and in field studies the decision about the optimum number of sampling locations is often made based on intuition [5]. A quantitative look at the applicability of (1) together with some of its alternatives (based on different ideas, see below) shows that the answer to the above question depends on the peculiarities of spatial density distributions [21, 23]. A value of N that can be regarded as large (i.e. sufficient to provide a good estimate of the population density) in one case may appear to be small (i.e. insufficient) in another case. An intuitively clear conclusion is that the more aggregative the population distribution, the larger the value of N that should be used [21, 23]. Also, the location of samples can be important.

In order to illustrate the problem, we consider an example. Consider a population density distribution over a one-dimensional domain of length L as $u(x) = U_0 \sin^2 \left(\frac{M \pi x}{L} \right)$. Obviously, it consists of M peaks. Assume that samples are collected on a regular sampling grid defined as $x_n = \frac{nL}{M}$, $n = 0, \ldots, M$. Then (1) will give $I = 0$ which has very little to do with the reality. The reason for this failure is that the population density in the peaks is not taken into account at all. One can then expect that the result might be better if we have at least one sample taken around the location of each population maximum. Indeed, if we almost double the number of sampling points, i.e. by adding a sampling point at the middle of each interval (x_n, x_{n+1}), it is readily seen that (1) gives a much better approximation of the population size! Although the exact value here is of course an artifact of the special form of $u(x)$, an increase in the accuracy of the population size estimate with an increase in the number of sampling points is a general tendency. The rule of thumb is that the 'minimum sufficient' number of traps should be about double the expected number of the maxima in the population distribution. (Interestingly, this heuristic estimate is in good agreement with results of a more quantitative analysis, see [21].) In its turn, a rough estimate for the number of peaks can be made if we know which environmental or biological factors control the width of the peaks.

Note that the estimate based on (1) is sensitive not only to the number of sampling points but also to their location. In the case that samples are taken at $\tilde{x}_n = (n + \frac{1}{2})\frac{L}{M}$, (1) gives the value $U_0 L$ that significantly overestimate the true value $\frac{1}{2}U_0 L$ of the population size. The situation improves in a hypothetical case when the number N of sampling points per unit area can be made arbitrary large. Theory predicts that, for an asymptotical case of large N, on a regular sampling grid the error of the estimate decays as a power law of the inter-sampling distance and the estimate is not sensitive to the location of sampling points. We call a grid with these properties a *fine grid*. In the case that the number of sampling points is not large enough and the asymptotical properties do no hold, we call it a *coarse grid*. The spatial structure of the population distribution is therefore properly resolved on a fine grid but is not resolved on a coarse grid. We will discuss the definition of the coarse grid in a more quantitative way in Sect. 3.1.

Secondly—coming back to the discussion of issues arising with (1)—in the case that N is in the intermediate range, i.e. neither too small nor sufficiently large, can the accuracy of the population density estimation be improved by replacing formula (1) by something else, perhaps slightly more complicated? Note that (1) is space-implicit because it does not contain any information about the geometry of the sampling grid $\{r_0, \ldots, r_{N-1}\}$ such as sample location or inter-sample distance. Could a spatially-explicit approach be possibly more accurate, e.g. if the contribution from each sample is somehow weighted by the corresponding sampling area?

And thirdly, in the case that the number N of samples is apparently too small to provide an estimate with reasonable accuracy, can some quantitative information still be obtained from the sampling data? The existing practice is to treat the sparse sampling data in a comparative way (i.e. a larger number of insects in the sample means larger population in the area around), but can we make any inference about the absolute values of the density?

For many ecological applications, these questions are of particular interest. One example is given by pest control. Dangerous pest species (e.g. pest insects) are usually the subject of nationwide or regional monitoring programs. This implies that information is collected (i.e. samples are taken) simultaneously across a large region. Due to resource limitation, it means that the number of samples per field then may become as small as just one or a few [13,17]. Moreover, even under an idealized assumption of unlimited resources, a large number of samples in an agricultural field would hardly be possible anyway. Sampling introduces a disturbance to agricultural procedures and pest monitoring specialists would never be allowed to make this disturbance large as it can significantly damage the agricultural product.

The accuracy of a population size estimate from the data collected on a given sampling grid may also depend on the properties of the density distribution. Indeed, while for the N-peak population distribution considered above the minimum required number of samples is $2N$, for a uniform distribution the exact value of the average population density is correctly estimated from just one sample. In our recent work [21, 23], we identified the case of "extreme population aggregation" (when most of the population is located within a relatively small area) as being the most difficult one, for which obtaining a reliable estimate is challenging. This case is going to be the main focus of our analysis here. Extreme aggregation means that a peak of the population density may fall in between two neighboring sampling points so that much of the information is lost. Correspondingly, we are going to provide further quantitative insight into the analysis of sparse sampling data. In particular, we will show that the interpretation of such data may become completely different: on a very coarse grid, due to increased uncertainty, the estimate should be treated probabilistically rather than deterministically.

Note that the properties of the population spatial distribution of the monitored species are usually not known in advance (although in some cases a priori information may indeed be available such as, for instance, the so-called edge effect when the pest population density tends to increase towards the field edge [1]). It is well known, however, that ecological populations often exhibit significant spatial heterogeneity [18, 35], so that the cases of high aggregation are not rare. Peaks in the population density frequently appears at the beginning of the breeding season and their early detection is important for successful implementation of the pest monitoring programs.

In a domain of a given size, the answers to the questions discussed above depend on the typical width of the peak or peaks in the population density distribution [21, 23]. It is therefore important to identify the biological factor that can be responsible for the scale of spatial heterogeneity. In the next section, we will show that in many cases the peak's width may be linked to species diffusivity, so that the controlling parameter is the diffusion coefficient.

1.1   Spatial Heterogeneity and the Effect of Diffusion

Population distribution over space appears as a result of the movement of its individuals. The individual movement can be of different types, the two extreme cases are given by random movement and ordered (ballistic) movement. The movement type is affected by a variety of environmental and biological factors [35]. For instance, search for food can be random in an idealized, perfectly homogeneous environment but it turns into a more ordered movement once information about the location of food item(s) becomes available, e.g. through the gradient in scent or odour concentration, the latter being usually referred to as chemotaxis. Another ecologically important example of chemotaxis commonly observed in insects is given by the effect of female pheromones on the movement of males.

In the presence of a directional bias, the actual individual movement is usually a combination of different movement modes, one of them being the random search. In fact, it is through this random search that the individuals detect the gradient in the environmental conditions [2, 29]. The random component of individual movement is therefore common.

Note that the randomness mentioned here is not a simple issue as it may depend on the spatial and temporal scale of the movement. On a short temporal scale, animal movement is hardly random as the direction of the next "step" along the movement path is likely to be correlated with the direction of the previous step, cf. "correlated random walk" [12]. On a long time scale, however, the correlated walk becomes completely random because of the tumbling effect of turning angles [4].

Below we consider diffusion as a paradigm of the random movement. It implicitly assumes that individuals perform Brownian motion. However, this is not a principle restriction and there is no loss of generality. In Sect. 6 we discuss how the results of our analysis can be extended onto an alternative case of Levy flight resulting in population "superdiffusion."

The purpose of this section is to reveal the link between diffusion and heterogeneity in the spatial population distribution. We begin with a simple yet illuminating example when the population distribution is described by the scalar diffusion equation, thus neglecting for the moment the impact of population multiplication and the interspecific interactions:

$$\frac{\partial u(x,t)}{\partial t} = D\frac{\partial^2 u}{\partial x^2} , \tag{2}$$

where D is the diffusion coefficient due to the random self-movement of individuals [18].

For the purposes of this section, we consider a population in the unbounded domain, $-\infty < x < \infty$. The population density distribution over space, i.e. the solution of (2), depends on the initial conditions. In the case of a point-source release of a population with initial size I at a position x_0, it is well known that

$$u(x,t) = \frac{I}{\sqrt{4\pi Dt}} \exp\left(-\frac{(x-x_0)^2}{4Dt}\right) . \tag{3}$$

It is readily seen that the characteristic width of the distribution (3), i.e. the characteristic length of the spatial heterogeneity, is given as

$$\Delta \sim \sqrt{Dt}\,, \tag{4}$$

where the sign $\sim$ means "up to a constant coefficient". The solutions of the diffusion equation obtained for some other ecologically sensible initial conditions possess similar properties (see [25], Sect. 9.3), i.e. the characteristic size of the arising spatial heterogeneity is given by (4). A more general approach based on the analysis of dimensions shows that this is, in fact, a generic property of the diffusion equation. Briefly, the matter is that the diffusion equation contains a single parameter, the diffusion coefficient D, and its dimension is distance$^2 \cdot$ time^{-1}. Therefore, for any given time t, the only quantity with the dimension of length is $\sqrt{Dt}$; see [3] for more details.

The next level of complexity is a single-species model with multiplication, i.e. a diffusion-reaction equation. Consider a particular case when reproduction is described by the logistic function:

$$\frac{\partial u(x,t)}{\partial t} = D \frac{\partial^2 u}{\partial x^2} + \alpha u \left(1 - \frac{u}{K}\right), \tag{5}$$

where α is the per capita growth rate and K is the carrying capacity. The dimension of α is time^{-1} and hence the only way to create a quantity with the dimension of length from the parameters of (5) is

$$\Delta_{fr} \sim \sqrt{D/\alpha}. \tag{6}$$

For a wide class of initial conditions, in the large-time limit (5) describes a travelling front [16] and then Δ_{fr} gives the characteristic length of the system's spatial heterogeneity, i.e. the width of the front.

In case of a multi-species system, e.g. as described by a system of diffusion-reaction equations, application of the dimensions analysis is less instructive as such systems contain more than one parameter with the dimension of time or inverse time, and often more than one diffusion coefficient. However, there are some alternative approaches. Let us assume that the diffusion coefficients for all n species in the system have approximately the same value, i.e. $D_1 \approx D_2 \approx \ldots \approx D_n \approx D$. Consider the case when the corresponding non-spatial system has a unique positive state and this state is as an unstable focus. In this case, the system is known to develop complex, chaotic spatiotemporal pattern sometimes referred to as "biological turbulence." [15]. The characteristic length Δ_g of the emerging multi-hump spatiotemporal pattern, i.e. the width of a single hump, is then given as [27]

$$\Delta_g = 2\pi c^* \left(\frac{D}{\max \operatorname{Re}(\lambda)}\right)^{1/2}, \tag{7}$$

where $\max \mathrm{Re}(\lambda)$ is the maximum real part of the eigenvalues of the linearized system and c^* is a numerical coefficient of the order of unity. Note that, since $\max \mathrm{Re}(\lambda)$ has the dimension of time^{-1}, (7) is in a good agreement with the dimensions analysis; in fact, it can be regarded as a generalization of (6).

An observation important for our analysis is that, in all three cases (4), (6) and (7) the characteristic length of the spatial heterogeneity is proportional to $\sqrt{D}$, i.e.

$$\Delta_g = \omega \sqrt{D}, \tag{8}$$

where ω is a factor that can depend on the parameters of the intra- and interspecific interactions, but not on the diffusion coefficient.

1.2 Goals and the Road Map

The main goal of this paper is to further develop a new approach to the analysis of spatial sampling data recently suggested in [21, 22]. The approach interprets the problem of population size estimation from sampling data as a problem of numerical integration on a coarse grid, and it has been shown [23] to be potentially more efficient than the standard statistical approach. The focus is on a few particular issues that have not been considered before such as (a) how the integration accuracy may depend on possible "defects" in the sampling grid (i.e. when one or a few of the grid nodes are moved away from their regular grid location), (b) how can we distinguish quantitatively between the cases of a coarse and an "ultra-coarse" grid, so that in the latter case only a probabilistic interpretation of the sampling data may be possible, and (c) how these issues are affected by the species diffusivity.

We want to emphasize that we do not claim to provide complete, practical, ready-to-use recipes as to how to calculate the population size from sparse spatial data. In fact, we do not even claim to provide a comprehensive theoretical analysis of the problem. Due to the complexity of the problem, it would hardly be possible in one paper. However, a good understanding and potential practical application of the approach will not be possible until all particular aspects are properly scrutinized. The issues that are in the focus of this paper are important milestones along the way.

The paper is organized as follows. In the next section, we describe the numerical integration method designed to evaluate the population size from discrete spatial data. In Sect. 3 we introduce a population dynamics model that we use to generate ecologically meaningful population distributions for various diffusion rates. We then check the efficiency of the numerical integration method by applying it to spatial population distributions of different complexity. In Sect. 4, we perform a detailed mathematical analysis of the impact of the diffusion rates on the accuracy of numerical integration on a coarse grid. In Sect. 5, we investigate the effect of the grid's non-uniformness on the population size estimation. Finally, in Sect. 6 we summarize our findings and discuss their potential implications for the pest monitoring practices.

2   Numerical Integration on Coarse Grids: The Problem Outline

In our recent study [21–23], we have developed, as an alternative to the statistical method (1), a novel approach to estimate the population size based on ideas of numerical integration. Since here we are mostly interested in the theoretical aspects of the approach, we restrict our consideration to a hypothetical 1D case. In terms of sampling in a real agricultural system, that may correspond to a transect; see Fig. 1, top.

We start with the case when the sampling positions x_n $(n = 0, \ldots, N - 1)$ are equidistant, i.e. $x_{n+1} = x_n + h$ where $h > 0$ is constant. Equation (1) can then be written as

$$\tilde{I} = Nh \cdot \hat{u} = \sum_{n=0}^{N-1} u_n h \approx \int_a^b u(x)dx = I, \tag{9}$$

where $h = L/(N - 1) \approx L/N$, $x_0 = a$, $x_{N-1} = b$ and $L = (b - a)$ is the size of the domain. It is readily seen that (9) coincides with the simplest method of numerical integration. This coincidence is not just by chance: a closer look at the problem shows that estimation of the population size based on the values of population density at discrete space (i.e. the position of the sampling points; see Fig. 1) is exactly the same as the general problem of numerical integration [21]. We therefore can make use of tools and methods of numerical integration accumulated in the field of numerical mathematics, e.g. see [8]. In this section, we briefly revisit (to the extent required by the goals of this paper) the main ideas of numerical integration and reveal the problems that arise when we apply these ideas to population size estimation from sparse spatial data.

A standard problem of numerical integration is to approximate the integral I by a sum $\tilde{I}$:

$$I = \int_a^b u(x)dx \approx \tilde{I}, \tag{10}$$

where the particular expression for the sum $\tilde{I}$ depends on the choice of the integration rule; one option is given by (9), some more advanced options will be considered in Sect. 2.1. In its turn, the change of integration to summation implies that, instead of the integrand $u(x)$ being defined on a continuous domain $[a, b]$, we are provided with a discrete set of values $u(x_0), u(x_1), \ldots, u(x_{N-1})$ (Fig. 1, bottom).

For any method of numerical integration, an essential requirement is that the approximation $\tilde{I} \approx I$ should be accurate enough to meet the condition $e < \epsilon$, where ϵ is the given tolerance and the integration error e is defined as

$$e = \frac{|I - \tilde{I}|}{|I|}. \tag{11}$$

Fig. 1 (*Top*) An example of field data collected in a field study on an insect pest [11], the numbers show the number of insect caught at the corresponding location in space, the *boxed numbers* show the samples along a transect; (*bottom*) a sketch of the numerical integration problem, the *diamonds* show the population density at the position $X_0, \ldots, X_{N-1}$ of the samples while the actual continuous density distribution (shown by the *dashed curve*) remains unknown

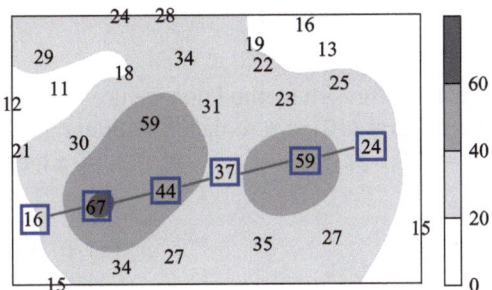

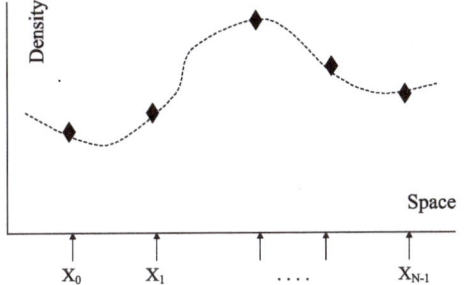

Below, we refer to the set of points x_n, $n = 0, 1, \ldots, N - 1$ in the domain $[a, b]$ as the computational grid G. The location of the grid nodes is generally defined as $x_{n+1} = x_n + h_n$, where $h_n > 0$ is the grid step size. The grid is called uniform if the grid step size is constant, $h_n \equiv h = (b-a)/(N-1)$, and non-uniform otherwise. In ecological applications, the integrand function $u(x)$ has the meaning of the density of the pest population, while the grid nodes x_n are the points where the samples are taken. Hence the density $u(x)$ becomes a discrete function available at points x_n only (see Fig. 1).

The accuracy of ecological data is usually not very high and hence the error tolerance $\epsilon \sim 0.25 - 0.5$ is regarded as acceptable [19, 30]. However, even this relatively undemanding level of required accuracy cannot always be provided when the function $\{u_n \equiv u(x_n), n = 0, \ldots, N - 1\}$ is integrated on a *coarse grid*, i.e. where the number N of nodes is small. The lack of information about the integrand function $u(x)$ may lead to an inaccurate evaluation of the integral (10) and the numerical integration of sparse data may result in a large integration error.

Meanwhile, it has been shown in [21,23] that an integral estimate $\tilde{I}$ computed on coarse grids does not necessarily lie beyond the range of accuracy required in real-life ecological problems. The results obtained in [21, 23] show that the accuracy of integration on coarse grids is defined by the spatial heterogeneity of the integrand function. For instance, the examples considered in [21, 22] demonstrate that, when coarse grids are considered, numerical integration of a monotone function gives considerably better accuracy than the integration of a function that has several "humps" or oscillates rapidly. In turn, the spatial structure of the population density $u(x)$ (i.e. the integrand) is determined by several physical/biological parameters, in particular, by diffusion.

2.1 The Method

We now discuss a method that we use for numerical integration. In order to compute the integral (10), we replace the integrand function $u(x)$ at each grid subinterval $c_n = [x_n, x_{n+1}]$, $n = 0, \ldots, N - 2$, by a local polynomial of degree K, that is,

$$p_K^n(x) = \sum_{k=0}^{K} a_{kn} x^k,\tag{12}$$

where the expansion coefficients a_{kn} are reconstructed independently at each subinterval c_n (as indicated by the subscript n in their notation). The integral (10) is then evaluated as

$$I = \int_a^b u(x) dx \approx \sum_{n=0}^{N-2} I_n,\tag{13}$$

where the integral I_n is readily computed over the grid cell c_n as $I_n = \int_{x_n}^{x_{n+1}} p_K^n(x) dx$.

The details of the implementation of the composite integration rule (13) can be found in [21, 22]. Let us note here that the numerical technique we use in the problem is the same as the Newton–Cotes family of methods of numerical integration [8] if uniform grids ($h_n \equiv h = (b - a)/(N - 1)$) are considered. In particular, the polynomial degrees $K = 0$, $K = 1$ and $K = 2$ correspond to the well-known methods of numerical integration such as the midpoint rule, the trapezoidal rule and the Simpson rule, respectively. These are the first three methods from the Newton–Cotes family. However, our approach is more flexible as it allows one to deal with non-uniform grids where the grid step size $h_n \neq const$.

One important observation about the integral evaluation is that asymptotic error estimates for the approximation (13) will depend on the polynomial degree K. It is well known [8] that the integration error (11) can be evaluated on uniform grids as

$$e = Ch^{K+1},\tag{14}$$

where C is a coefficient that does not depend on h or K but may depend on the properties of the integrand. The estimate (14), however, only holds on fine grids where the grid step size h is very small (ultimately, tends to zero). We have demonstrated in our previous work [21, 23] that the asymptotic error estimate (14) does not hold on coarse grids where, generally speaking, we cannot reduce the integration error by using higher order polynomials. Hence, other ways of controlling the accuracy of integration have to be established when one has to deal with coarse grids, where N is small.

3   Simulation Data

In order to assess the effectiveness of our approach to integrate discrete sampling data, we now need data. Note that, to make a sensible assessment, we need to know not only the values of the population density arranged along a line (e.g. see Fig. 1, top) but also the actual population size to compare with our estimate. However, field data satisfying this requirement are rarely available. Moreover, to study the effect of the grid step size (i.e. the effect of different sample spacing) on the accuracy of the estimate, we need to compare the results obtained on different grids, which is almost impossible to obtain in the field (but see [23]).

For the above reasons, instead of field data, here we use the population density distribution generated by an ecological model. Specifically, we use the spatially explicit Rosenzweig–MacArthur model which, in dimensionless variables, has the following form [16]:

$$\frac{\partial u(x,t)}{\partial t} = d \frac{\partial^2 u}{\partial x^2} + u(1-u) - \frac{uv}{u+h} \,, \tag{15}$$

$$\frac{\partial v(x,t)}{\partial t} = d \frac{\partial^2 v}{\partial x^2} + k \frac{uv}{u+h} - mv \,. \tag{16}$$

Here u and v are the dimensionless densities of prey and predator, respectively, at time t and position x where $t > 0$ and $0 < x < 1$. The distances are therefore measured in fractions of the original domain length L. (See [15, 21] for more details with regard to the choice of the dimensionless variables and parameters.) The dimensionless diffusion coefficient d quantifies the species diffusivity due to the "random" movement of the individuals. For the sake of simplicity, we assume it to be the same for both species.

It is readily seen that the relation between the dimensionless diffusion coefficient and the characteristic length of the system's spatial heterogeneity remains exactly the same as it was in dimensional units, i.e.

$$\delta_g = \frac{\Delta_g}{L} = \omega \sqrt{d} \,. \tag{17}$$

Here the coefficient ω depends on the system's parameters, cf. (7). However, an extensive numerical study performed in [26, 27] revealed that in the predator-prey system (15)–(16), its value is relatively robust to changes in the parameter values, typically being about 25.

An important feature of the system (15)–(16) is that interaction between reaction and diffusion is known [15] to result in pattern formation, e.g. see Fig. 2, where the properties of the pattern[1] depend on the value of dimensionless diffusivity d. In particular, for d being of the order of 1 or larger, the solution $u(x,t)$ will

---

[1] At least, for any t not too small, in order to avoid the effect of the initial conditions.

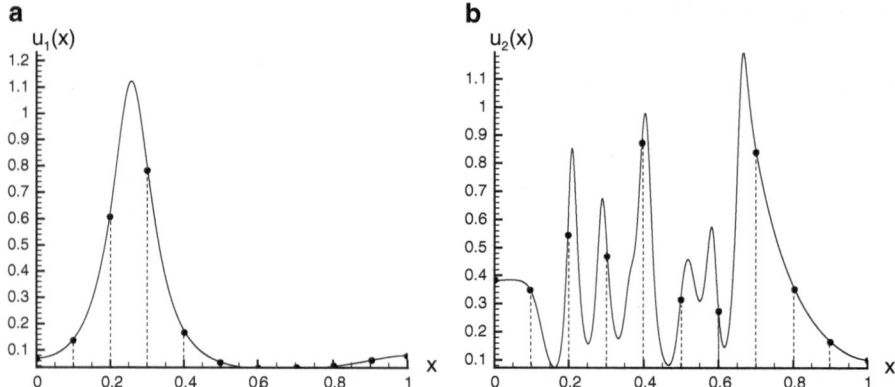

Fig. 2 Ecological test cases. Typical spatial distribution of the pest population density in the model (15)–(16) for the values of the dimensionless diffusivity $d = 10^{-4}$ (**a**) and $d = 10^{-5}$ (**b**). The continuous functions $u_1(x)$ and $u_2(x)$ are presented by *solid lines*, while the function values available for integration on a coarse grid are shown as *black filled circles*

be a monotone function of x, which means that the local population oscillations are almost synchronized over the entire domain. However, oscillations at different positions can become de-synchronized for $d \ll 1$ (see [27]). In the latter case the initial conditions $u(x, 0)$, $v(x, 0)$ evolve to an ensemble of irregular humps and hollows. For an intermediate value of d, the pattern can consist of just one or a few peaks only (see Fig. 2a). The number of humps increases for smaller values of d resulting in oscillations shown in Fig. 2b. From an ecological perspective, it means that in a domain of a given length a slowly diffusing population is more likely to form a complicated spatial pattern than a fast diffusing one.

As we showed in the introduction, the rule of thumb is that the number of nodes in the numerical grid should be at least two times larger than the number of peaks in the population distribution. It means that, when the size of a pest population is evaluated on a given grid, one may expect lower accuracy of integration for slowly diffusing species than for fast diffusing species. The complex spatial structure of the population density of a slowly diffusing pest may be not well resolved on coarse grids. Our next step is to compute the integration error for the functions shown in Fig. 2 to establish a quantitative link between the spatial heterogeneity of the integrand function and the accuracy of numerical integration, especially when the number of grid nodes is small.

3.1 Estimating Pest Population Size on Coarse Grids: Numerical Test Cases

The discussion of the accuracy of numerical integration requires us to understand how to actually compute the integration error for the ecological distributions of Fig. 2. The problem is that the definition (11) of the integration error is based

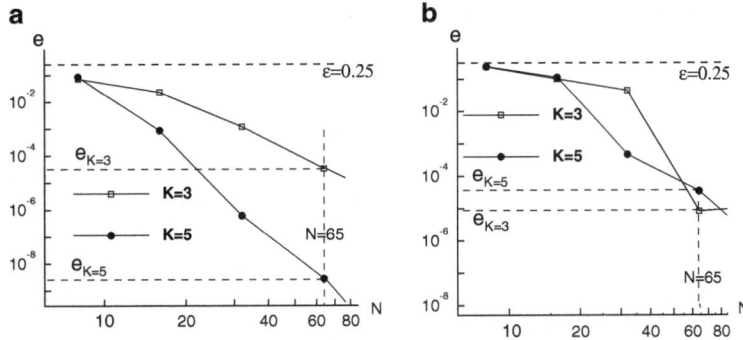

Fig. 3 The integration error (11) as a function of the number of grid nodes for ecological test cases shown in Fig. 2. (**a**) The density distribution $u_1(x)$. The error $e_{K=3}$ of the approximation by polynomials of degree $K = 3$ remains always bigger than the error $e_{K=5}$ computed for the approximation by polynomials of degree $K = 5$. (**b**) The density distribution $u_2(x)$. The approximation by high order polynomials ($K = 5$) cannot always provide better accuracy

on the knowledge of the exact answer. However, the analytical solution of the system (15)–(16) is not known and we cannot compute the integral I exactly, as required by the definition (11). Therefore, we have to define the "exact" value of the integral when the pest population density $u(x)$ is integrated numerically. For this purpose we compute a numerical solution to the system (15)–(16) on a very fine uniform grid G_f of $N_f = 2^{15} + 1 \equiv 32769$ nodes, and we consider the result as the "exact" solution to the problem. The corresponding value of the integral I is then regarded as the exact integral and is then used to estimate the accuracy of the integral $\tilde{I}$ computed on a coarse grid G_c.

For the purpose of our study, we are going to compute the integration error as a function of the number of grid nodes, $e = e(N)$, as we want to understand what happens to the approximation $\tilde{I}$ when we increase or decrease the number of grid nodes. A usual technique to generate a finer uniform grid from a coarser one is to halve each grid subinterval by inserting a new node at the subinterval midpoint. Let us denote the number of grid subintervals as $\hat{N}$, where we have $\hat{N} = N - 1$. We generate a sequence of uniform grids, where the number of subintervals on each grid is defined as $\hat{N} = s\hat{N}_0$. The number $\hat{N}_0$ of grid subintervals on the initial grid is taken $\hat{N}_0 = 8$ and the scaling coefficient s varies as $s = 2^m, m = 0, 1, 2, \ldots, 12$. The integrand function $u(x)$ is then readily available at nodes of each grid generated as above, as we simply project it from the fine grid G_f where it has originally been computed. Hence the integration error (11) can be easily defined on any grid in the sequence to obtain the convergence rate $e(N)$ of our numerical method.

Let us refer to the density distributions shown in Fig. 2a, b as $u_1(x)$ and $u_2(x)$ respectively. We integrate $u_1(x)$ and $u_2(x)$ and compute the error (11) on each uniform grid generated as above. The integration error as a function of the number N of grid nodes for the integrand function $u_1(x)$ is shown in Fig. 3a, while the error for the function $u_2(x)$ is displayed in Fig. 3b. The integration error is shown on a logarithmic scale. In both cases the error is computed for approximation by polynomials of degree $K = 3$ and $K = 5$.

It is readily seen from the figure that the behavior of the error curve depends on the integrand function. For the function $u_1(x)$ the convergence results are in good agreement with the error estimate (14). Namely, the polynomial approximation with $K = 5$ provides better accuracy if we compare it with the $K = 3$ approximation on each grid in the sequence; see Fig. 3a. The integration error is always within the required range $e < 0.25$, as we already have $e \approx 0.1$ on the initial grid of $N_0 = 9$ nodes. Thus, already the initial grid has a sufficient number of nodes to provide an accurate estimate of the integral in case that the distribution $u_1(x)$ is considered.

Meanwhile, for the function $u_2(x)$ the use of higher order polynomials to approximate the integrand does not always result in a more accurate approximation on grids with a small number of nodes. It can be seen from Fig. 3b that the integration error of a higher order polynomial approximation ($K = 5$) remains about the same as the error of the $K = 3$ approximation for $N = 9$ and $N = 17$. Moreover, the error $e_{K=5}$ can even be greater than $e_{K=3}$ as it is shown in Fig. 3b for a grid of $N = 65$ nodes. As we have already discussed, the complex multi-peak spatial pattern $u_2(x)$ may require a finer grid to resolve the function's spatial oscillations. Also, despite the initial grid of $N_0 = 9$ nodes still providing the accuracy acceptable for ecological applications, the error $e \approx 0.25$ is considerably bigger in comparison with the integration error obtained on the same grid for the integrand function $u_1(x)$.

The above examples demonstrate that, while it is sufficient to have a grid of several nodes in order to provide accurate integration results for a simple spatial distribution, the same number of grid nodes may give the accuracy beyond the acceptable level if a more complex spatial pattern is considered. The error behavior when the convergence rate does not follow its asymptotic value (14) is called a *coarse grid problem* [21, 22] and the corresponding grid is called a coarse grid. Since the error cannot be controlled based on the estimate (14), it becomes important to understand which factors determine the error on coarse grids where the spatial structure of the integrand function is not well resolved. This will be done in the next section.

4   The Impact of the Diffusion Rates on the Accuracy of Numerical Integration

In this section we derive the functional relationship between the diffusion coefficient and the grid step size required to provide good integration accuracy. Our previous discussion revealed that using higher order polynomials to approximate the integrand function does not necessarily result in a more accurate estimate of the integral on coarse grids. Hence we now reduce our attention to a technically simple yet illuminating case when the integrand function is approximated by linear polynomials ($K = 1$).

Fig. 4 Hump approximation
for the population density
distribution $u_1(x)$.
(**a**) Quadratic approximation
with $h = 0.125$.
(**b**) Quadratic approximation
with $h = 0.0625$

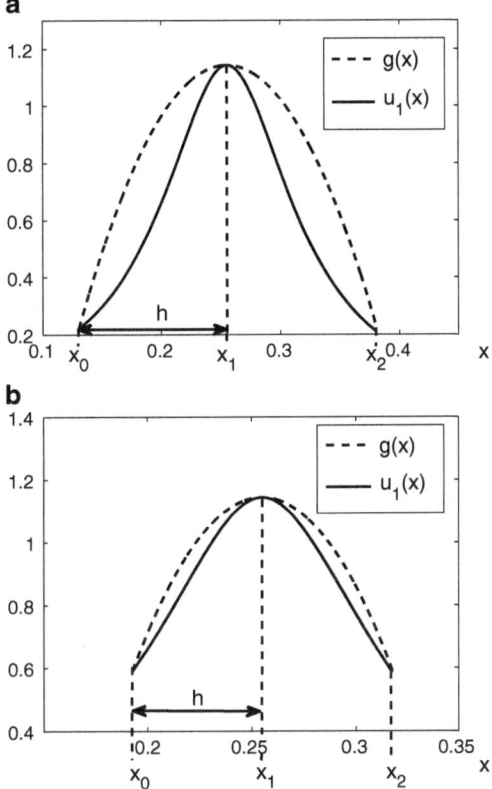

Let a nonnegative function $u(x)$ have a "hump" (i.e., a local maximum) on the interval [0, 1]. The first assumption we make for our analysis is that the hump can be handled as a quadratic function. Namely, let us introduce the subinterval $[x_0, x_2]$ of length $2h$ in the vicinity of the hump[2] (see Fig. 4). We then assume that in the vicinity of the hump the integrand $u(x)$ can be considered as

$$u(x) \approx g(x) = B - A(x - x_1)^2, \quad x \in [x_0, x_2], \tag{18}$$

where $A > 0$, $B > 0$ and the function $g(x)$ has the maximum at the interval midpoint $x_1 = x_0 + h$. We also require $g(x)$ to be a nonnegative function over the interval $[x_0, x_2]$, that is $g(x_0) = g(x_2) = B - Ah^2 \geq 0$. That gives us the following condition relating A, B and h:

$$h^2 \geq \frac{B}{A}. \tag{19}$$

---

[2]Note that the notation x_0, x_1, x_2 we use to discuss the hump approximation is not the same as the numeration of grid nodes we introduced in the previous section. In other words, the "endpoints" x_0 and x_2 are arbitrarily located interior points of the interval [0, 1].

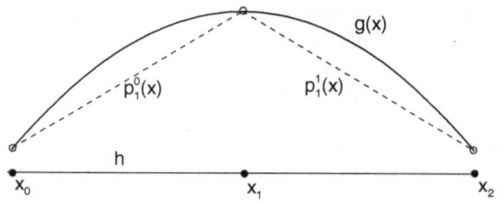

Fig. 5 Approximation of a quadratic function by linear polynomials over a uniform grid of three nodes

Examples of the approximation of a hump by a quadratic function for the pest population density $u_1(x)$ are shown in Fig. 4, where various choices of points x_0, x_1 and x_2 in the vicinity of the hump are illustrated. The details of such an approximation can be found in the Appendix.

It is obvious that we introduce an additional error to the integration problem when we tackle a hump as a quadratic function (see the discussion in the Appendix). However, as we will see below, such an approximation enables us to make correct conclusions about a grid step size that should be recommended for accurate integration of the function $u(x)$. Thus our next step is to investigate what happens when we replace the quadratic function $g(x)$ (and, therefore, the original function $u(x)$) with two linear polynomials in the vicinity of the hump, as our method of numerical integration requires us to do. We first consider a uniform grid where one of the grid nodes is located at the maximum point. We then study the case of an arbitrary location of a maximum point on a coarse uniform grid. Finally, we discuss non-uniform grids in order to understand what impact the grid distortion will make on the integration error.

4.1 Uniform Grid

Let the quadratic function $g(x)$ be integrated at the interval $[x_0, x_2]$ where we consider a local grid of two subintervals $c_0 = [x_0, x_1]$ and $c_1 = [x_1, x_2]$, the node location being $x_1 = x_0 + h$ and $x_2 = x_1 + h$. We use linear polynomials $p_1^n(x) = \sum_{k=0}^{1} a_{kn} x^k$ at each grid cell c_n, $n = 0, 1$, where we reconstruct polynomial coefficients from the condition $p_1^n(x_k) = g(x_k)$, $k = 0, 1, 2$, as $p_1^0(x) = B + Ah(x - x_1)$ and $p_1^1(x) = B + Ah(x_1 - x)$. The approximation of a quadratic function by linear polynomials over a grid of the two subintervals is illustrated in Fig. 5.

We then compute the approximate integral $\tilde{I}$ as

$$\tilde{I} = \int_{x_0}^{x_1} p_1^0(x)dx + \int_{x_1}^{x_2} p_1^1(x)dx = 2Bh - Ah^3, \qquad (20)$$

while the exact integral is

$$I = \int_{x_0}^{x_2} g(x)dx = 2Bh - \frac{2Ah^3}{3}. \tag{21}$$

Consider the error of integration (11) and let us require that $e < 0.25$. Correspondingly, we have $|I - \tilde{I}| = \frac{Ah^3}{3}$. Therefore, we obtain:

$$\frac{Ah^3}{3} < \frac{1}{4}\left|2Bh - \frac{2Ah^3}{3}\right|. \tag{22}$$

Solving (19) and (22) together and taking also into account that $I > 0$ (as $B > 0$ and $g(x) \geq 0$ for any $x \in [x_0, x_2]$), we obtain $h < h_0 = \sqrt{B/A}$.

In order to reveal the impact of diffusion, we now define the "hump width" δ_g of the quadratic function $g(x)$ as the distance between its roots, so that $\delta_g = 2\sqrt{B/A}$. Correspondingly, we obtain that the required accuracy $e < 0.25$ is ensured for $h < h_0 = \delta_g/2$. Finally, recalling that the characteristic length of the spatial heterogeneity is given by (17) and substituting it into the expression above, we arrive at

$$h < h_0 = \frac{\omega\sqrt{d}}{2}. \tag{23}$$

Whatever the value of the diffusion coefficient d, condition (23) is sufficient to integrate the "hump" with the desirable accuracy $e < 0.25$. The limiting value $e = 0.25$ is reached for h_0.

4.2 The Analysis of the Grid Step Size for Ecological Distributions

In this subsection we validate our findings—in particular, condition (23)—by considering the density distributions shown in Fig. 2. We first study the function $u_1(x)$ that has a single hump. The aim of our numerical test is to find the number N^* of grid nodes sufficient for accurate integration of the pest population density $u_1(x)$. In other words, we integrate the function $u_1(x)$ over the domain $[0, 1]$ on a sequence of uniform grids and compute the corresponding integration error (11). We then look for the grid step size h^* whose value provides us with the integration error $e < 0.25$. That should give us the number $N^* \approx 1/h^*$ of grid nodes (or the number of samples in the pest monitoring problem) required to resolve the spatial heterogeneity. The number N^* (or the grid step size h^*) obtained in this straightforward integration procedure is then compared with the estimate (23).

Table 1 The integration error (11) for the density distribution $u_1(x)$ on a sequence of uniformly refined grids with grid step size h

N	3	5	9	17
h	0.5	0.25	0.125	0.0625
e	0.6948	0.5459	0.0823	0.0036

The integrand function $u_1(x)$ is approximated by piecewise linear polynomials ($K = 1$) on each grid in the sequence

Table 2 The integration error for the density distribution $u_2(x)$ on a sequence of uniformly refined grids

N	3	5	9	17	33
h	0.5	0.25	0.125	0.0625	0.03125
e	0.1579	0.1567	0.2193	0.1304	0.0001

The integrand function $u_2(x)$ is approximated by piecewise linear polynomials, see Fig. 6

Let us note again that the hump itself is not, of course, a quadratic function and the integration error obtained for the integrand $u_1(x)$ is not the same as the integration error derived for the quadratic function. However, since a single hump can be approximated by a quadratic function with good accuracy (see Appendix), we expect that the results of our numerical experiment will be in reasonable agreement with the estimate (23) that can be obtained from the information about the diffusion coefficient only.

The estimate (23) gives us the value $h^* \sim 0.12$ for the diffusion coefficient $d = 10^{-4}$ used to generate the density distribution $u_1(x)$. The integration error when the function $u_1(x)$ is approximated by linear polynomials over a uniform grid of N nodes is shown in Table 1. We compute the error (11) on a very coarse grid of 2 subintervals, we then refine the grid by halving each grid subinterval, compute the integral error again and repeat the refinement procedure until the error is smaller than the threshold value $e = 0.25$. It can be seen from the table that the results of numerical integration are in good agreement with our estimate (23). While a very coarse grid does not provide the accuracy $e < 0.25$, the grid of $N = 9$ nodes ($h = 0.125$) gives the integration error much smaller than the required limit $e = 0.25$. Hence the number of grid nodes can be evaluated as $N^* \approx 9$.

Consider now the density distribution $u_2(x)$ shown in Fig. 2b. The diffusion coefficient used to generate the distribution $u_2(x)$ is $d = 10^{-5}$. Hence the estimate (23) of the grid step size is $h \sim 0.03$. In other words, a uniform grid should contain about 30 nodes in order to guarantee the integration error $e < 0.25$.

The values of the integration error (11) are shown in Table 2. A substantial jump in accuracy is evident when the grid is refined from $N = 17$ to $N = 33$ nodes. This is further illustrated by Fig. 6 where it can be seen that for 16 subintervals, the majority of the humps in $u_2(x)$ are approximated by a single polynomial and the spatial heterogeneity is not well resolved. When the grid is refined to 32 subintervals, all but two of the humps are approximated by two or more linear

Fig. 6 Piecewise linear
approximation for the
population density
distribution $u_2(x)$.
(**a**) Approximation using
$N = 17$ grid nodes
(**b**) Approximation using
$N = 33$ nodes

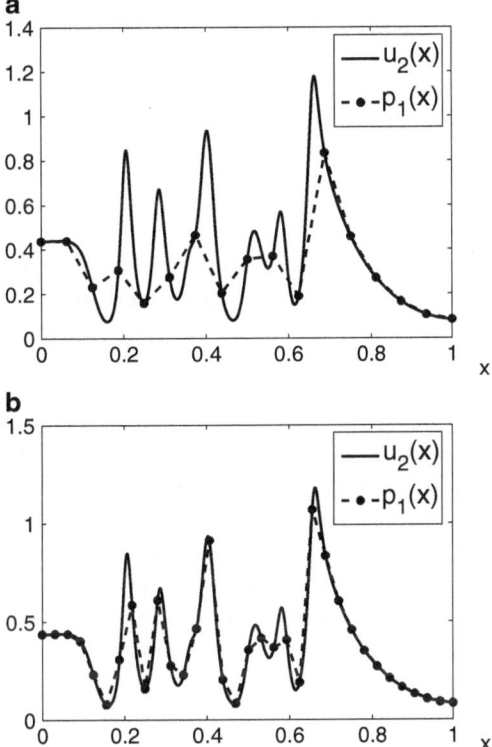

polynomials. Approximating a hump with a single linear polynomial is equivalent
to the approximation of a quadratic by linear polynomials over a local grid of two
nodes instead of considering three nodes for the approximation. That extreme case
will be discussed in more detail in the next section.

At the same time it is worth noting here that for the density distribution $u_2(x)$
the integration error is not entirely the same as expected from our analysis, as the
integration error actually remains within the required range $e < 0.25$ on any grid
that we use in our computations. We believe that this may happen because of the
"cancelation effect" that may arise when underestimated contribution of the humps
is balanced by overestimated contribution of the hollows. However, we would like to
emphasize that the error value cannot be predicted on coarse grids. In other words,
while the estimate (23) guarantees the error $e < 0.25$ on a grid of $N = 33$ nodes, it
cannot be said a priori what the error is on coarse grids with $N < 33$.

4.3 Arbitrary Location of the Peak on a Uniform Coarse Grid

In the previous subsection we assumed that there are three grid nodes in the region
of the hump and the position of the central node coincides with the position of
the maximum. Especially the last assumption is not entirely realistic because in

Fig. 7 Piecewise linear
approximation of $g(x)$

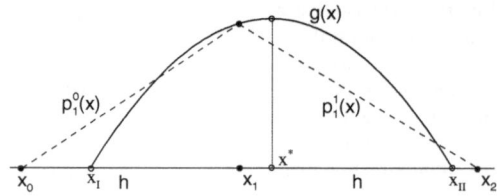

applications to pest monitoring the position of the population density hump would usually be unknown. Hence, two practically important questions that arise from our analysis above are (a) how the integration error changes when the maximum is not at the position of the node (see Fig. 7) and (b) whether we can make the grid even coarser, e.g. what will be the integration accuracy if just one grid node is used in the subdomain where the hump is located. In other words, we are now interested in the situation given by $N = 3$ and $N = 5$ in Table 1 when the entire hump is located in between two grid nodes. The error shown in Table 1 is quite large, but can we possibly make it any smaller with the same number of nodes?

Consider a regular grid consisting of three nodes, x_0, $x_1 = x_0 + h$ and $x_2 = x_0 + 2h$. Let a population density distribution have a hump within the interval $[x_0, x_2]$. We approximate the hump by a quadratic function. Let us define the approximation $g(x)$ of the hump as

$$g(x) = \begin{cases} B - A(x - x^*)^2, & \text{if } x \in [x_I, x_{II}], \\ 0, & \text{otherwise.} \end{cases} \tag{24}$$

In the approximation above x^* is the location of the maximum point, which is now different from the node x_1, and the values x_I and x_{II} are the roots of $g(x)$. We can express x^* in terms of the grid nodes as

$$x^* = x_1 + \gamma h = x_0 + h(\gamma + 1), \tag{25}$$

where $\gamma \in [0, 1/2]$. The roots x_I and x_{II} are then given by

$$x_I = x_0 + h(\gamma + 1) - \sqrt{B/A} \quad \text{and} \quad x_{II} = x_0 + h(\gamma + 1) + \sqrt{B/A}. \tag{26}$$

The exact integral of $g(x)$ in the vicinity of the hump is thus

$$I = \int_{x_0}^{x_2} g(x)\,dx \equiv \int_{x_I}^{x_{II}} g(x)\,dx = \frac{2}{3}B\delta_g, \tag{27}$$

where δ_g is the hump width as above. We now approximate $g(x)$ by two piecewise linear polynomials as follows (see Fig. 7)

$$g(x) \approx \begin{cases} p_1^0(x), \text{ if } x \in [x_0, x_1], \\[2mm] p_1^1(x), \text{ if } x \in [x_1, x_2]. \end{cases} \tag{28}$$

An approximated value $\tilde{I}$ of the integral (29) is then obtained by integrating the piecewise linear approximation of the function $g(x)$:

$$\tilde{I} = \sum_{k=0}^{k=1} \int_{x_k}^{x^{k+1}} p_1^k \, dx = h \left(B - A\gamma^2 h^2 \right). \tag{29}$$

We now require the integration error (11) to be $e < 0.25$, which means that

$$0.75I < \tilde{I} < 1.25I, \tag{30}$$

where $I > 0$. Consider the lower bound of the inequality (30) and find the values γ_{II} of parameter γ for which the equation $\tilde{I} = 0.75I$ holds. Substituting I and $\tilde{I}$ in the above we obtain

$$Bh - A\gamma^2 h^3 = \frac{1}{2} B\delta_g.$$

Hence

$$\gamma_{II}(h, \delta_g) = \frac{\delta_g}{2h} \sqrt{\frac{2h - \delta_g}{2h}}, \tag{31}$$

where we should require the grid step size $h \geq \delta_g/2$ to get γ_{II} as a real number for any fixed δ_g. That also makes our analysis consistent with our previous assumption that the grid is very coarse; see item (b) at the beginning of this section.

We then consider the upper bound of (30) and find the values γ_I that satisfy the equation $\tilde{I} = 1.25I$. The parameter γ_I as a function of the grid step size h and the hump width δ_g is given by

$$\gamma_I(h, \delta_g) = \frac{\delta_g}{2h} \sqrt{\frac{6h - 5\delta_g}{6h}}, \quad h \geq \frac{5\delta_g}{6}. \tag{32}$$

The hump width δ_g is defined by the diffusion coefficient d, so that γ in expressions (31) and (32) becomes a function of h only for a given value of d. The curves $\gamma_I(h)$ and $\gamma_{II}(h)$ are shown in Fig. 8a, b for the dimensionless diffusivity $d = 10^{-4}$ and $d = 10^{-5}$ respectively. The range of h is chosen in both cases as $h \in [\delta_g, 1]$, where δ_g is calculated from the estimate (17).

For any given value of d, the conditions (31) and (32) define the parameter range where the integral is computed with the required accuracy. Indeed, let us fix the grid step size at a certain hypothetical $h = h^*$ (see Fig. 8) and compute $\gamma_I^* = \gamma_I(h^*)$ and $\gamma_{II}^* = \gamma_{II}(h^*)$. It then follows from the inequality (30) that for any $\gamma_I^* \leq \gamma \leq \gamma_{II}^*$ the error is $e < 0.25$. Also, let us mention that, for any fixed h, there exists the value of γ for which $\tilde{I} = I$; its value is readily obtained from (27) and (29):

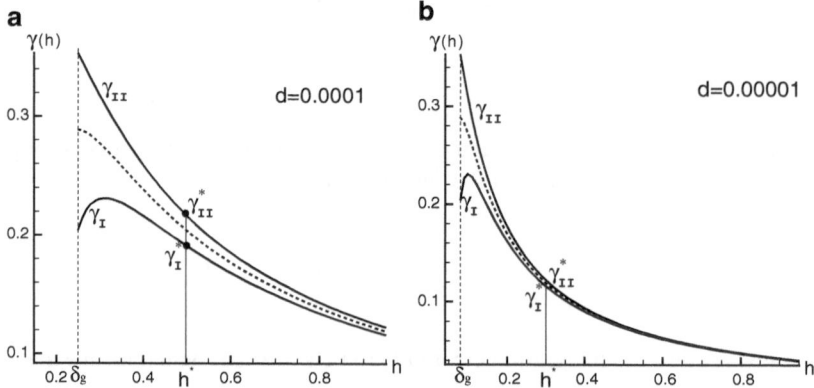

Fig. 8 The function $\gamma(h)$ for various values of the dimensionless diffusivity d. The part of the (h, γ) plane between the two *solid curves* gives the parameter range where the integration is done with the required accuracy $e < 0.25$

$$\gamma(h) = \frac{\delta_g}{2h} \sqrt{\frac{3h - 2\delta_g}{3h}}, \quad h \geq \frac{2\delta_g}{3}. \tag{33}$$

One straightforward yet important observation that can be made from Fig. 8 is that the domain where the error is $e < 0.25$ gets smaller when we decrease the diffusivity d. In other words, a narrow hump ($\delta_g \to 0$) is getting "lost" on a very coarse grid with the grid step size $h \gg \delta_g$. Another interesting observation is that installing a grid node at the location of the maximum point (which corresponds to $\gamma = 0$) does not at all result in the smallest possible integration error as (33) clearly gives the value $\gamma(h) > 0$ (see the dashed curve in Fig. 8 where $\tilde{I} = I$).

5 Nonuniform Grid

Our next task is to evaluate the integration error on a non-uniform grid, where we want to find the condition on the grid step size h that ensures the required accuracy $e < 0.25$ for a given hump width δ_g.

In order to provide insight into this issue, we use the same approach as in Sect. 4.1. We consider a single-hump distribution which we approximate with the quadratic function $g(x)$. However, the function $g(x)$ is now integrated on a grid of three nodes $\{x_0, \tilde{x}_1, x_2\}$, where the central node x_1 is now moved to the position $\tilde{x}_1$ while the maximum of the integrand remains at the midpoint x_1 of the domain $[x_0, x_2]$; see Fig. 9. In other words, the new grid is obtained from a uniform grid $\{x_0, x_1, x_2\}$ of Sect. 4.1 by the following mapping:

$$x_1 \to \tilde{x}_1 = x_1 + \beta h, \tag{34}$$

Fig. 9 Approximation of a quadratic function by linear polynomials over a non-uniform grid of three nodes

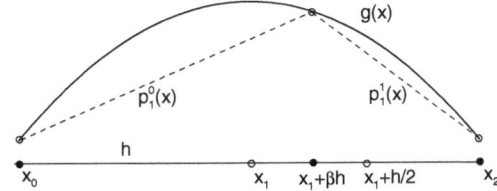

where β is a parameter quantifying the degree of the non-uniformness, $0 < \beta < 1/2$. The lower limit $\beta = 0$ thus corresponds to the original uniform grid. The upper limit $\beta = 1/2$ corresponds to the case when $\tilde{x}_1$ is the midpoint of the subinterval $[x_1, x_2]$ (see Fig. 9).

From an ecological viewpoint the transformation (34) with $0 < \beta < 1/2$ means that for some practical reason one cannot provide equidistant location of samples in the area where the measurements are made. In other words, we cannot provide sampling at the midpoint x_1 of the interval $[x_0, x_2]$ (for example, because of a natural obstacle, such as a tree) and have to install a sample somewhere in the neighborhood but still close to the point x_1. It is important to note that a hump in the density distribution still remains well resolved, as we are still allowed to use three grid points in the integration procedure.

We now apply the technique described previously in Sect. 4.1 on the non-uniform grid. The exact integral I is still given by (21). The approximate value $\tilde{I}$ of the integral is computed as

$$\tilde{I} = \int_{x_0}^{x_1+\beta h} p_1^0(x)dx + \int_{x_1+\beta h}^{x_2} p_1^1(x)dx . \tag{35}$$

The linear polynomials are now given by $p_1^0(x) = B - Ah(x_1 - \beta x_0) - Ah(\beta - 1)x$ and $p_1^1(x) = B + Ah(x_1 + \beta x_2) - Ah(1 + \beta)x$. Substituting $p_1^0(x)$ and $p_1^1(x)$ in the integrals above, we obtain

$$\tilde{I} = 2Bh - Ah^3 - A\beta^2 h^3. \tag{36}$$

Again we require that the error (11) should be $e < 0.25$. Substituting the expressions for I and $\tilde{I}$ in the condition $|I - \tilde{I}| < 0.25I$ and taking into account the condition (19), we arrive at

$$h^2 < \frac{B}{A(1 + 2\beta^2)}. \tag{37}$$

Recall that $B/A = \delta_g^2/4$ where δ_g is the hump width. Making use of (17) that relates the hump width to the diffusion rate, we obtain:

$$h < \frac{\omega\sqrt{d}}{2\sqrt{1 + 2\beta^2}}. \tag{38}$$

Therefore, the upper bound for h is a monotonously decreasing function of β. For the extreme value $\beta = 1/2$ we obtain that $h = \sqrt{2/3}h_0$, where h_0 is the restriction (23) on the grid step size obtained for the uniform grid where $\beta = 0$. Substituting (23) into (38), we arrive at

$$h < \omega \sqrt{d/6}. \tag{39}$$

Condition (39) gives us information on how to choose the grid step size if we want to have the relative error $e < 0.25$ on a non-uniform grid.

We now want to reveal how the integration error depends on the degree of the grid distortion in case the restriction (39) is ignored. Let us set $h = h_0$. For the fixed value h_0, e.g. as defined by the condition (23), the error becomes a function of β,

$$e_q(h_0, \beta) = \frac{|I(h_0) - \tilde{I}(h_0, \beta)|}{|I(h_0)|} = \frac{1 + 3\beta^2}{4}. \tag{40}$$

It is readily seen from the expression above that for $\beta = 0$ the integration error is $e_q = 0.25$, i.e. the upper limit of the required error. Since the error $e_q(h_0, \beta)$ is a monotone function of β for $0 < \beta < 1/2$, it reaches its maximum $e_q = 7/16 \approx 0.44$ at $\beta = 1/2$. Hence, moving a node away from the maximum point on a grid with the fixed grid step size $h = h_0$ can increase the error of integration almost twice.

In conclusion, let us consider the extreme case when $\beta \to 1$ in the transformation (34), i.e. when $\tilde{x}_1$ closely approaches x_2. From the integration viewpoint, the singular value $\beta = 1$ means the transition to a coarser grid where we are now allowed to use only two grid nodes instead of three. Correspondingly, we have a single linear polynomial in the vicinity of the hump instead of having two of them as considered in Sect. 4.1. It then readily follows from the restriction (38) that we should set

$$h = \frac{h_0}{\sqrt{3}} \tag{41}$$

in order to obtain a sufficiently accurate estimate of the integral.

6   Discussion and Conclusions

Estimation of pest abundance is a key topic in many ecological monitoring and control programs. The ultimate goal is to provide robust and timely recommendations on the application of pesticides, e.g. once the pest abundance exceeds a certain threshold [32].

Basic information about species presence in a given area is given by its population size, i.e. by the total number of its individuals. In practice, the information about species presence is usually obtained through collecting samples. The population size, which is an integral of the population density over the area, has to be evaluated based on the values of the population density that are known only at the position of the samples. This is a conventional problem of numerical integration. Indeed, integration of sampled data frequently arises in experimental work as well as in computational applications [9, 14, 36]. However, the situation with pest monitoring is essentially different from a standard problem of numerical integration. The matter is that the number of samples collected over an agricultural field is usually small. Evaluating pest population size becomes a problem of numerical integration of a discrete function obtained on a coarse grid. Thus the issue of integration accuracy becomes a crucial one, as we only have sparse data to deal with. Following the approach developed in our recent work [21–23], in this paper we discuss this issue in more detail.

The emphasis of this paper is on identifying the factors that can affect the accuracy of integration on coarse grids. We showed that diffusion is a factor determining the spatial heterogeneity of the integrand function and that, in turn, affects the accuracy of numerical integration. We demonstrated how the knowledge of the diffusion rate in the problem can be used to obtain an accurate estimate of the pest population size. Alternatively, this knowledge can be used to define the minimum number N of samples sufficient for accurate evaluation of the pest population size. It should be mentioned here that optimization of the number of samples required to provide robust estimates is an important issue for pest monitoring programs [5, 6, 20].

The main results of our study are bullet-pointed and discussed below.

- We showed that the problem of obtaining a robust estimate of the population size from sparse spatial data can, in principle, be solved by applying the methods and ideas of numerical integration.
- Numerical integration of sampling data has to be done on a *coarse grid*, i.e. a grid where the results of the asymptotical convergence rate for different integration methods are not applicable because the number of grid nodes is too small (correspondingly, the inter-sampling distance is too large). That makes the usual ideas about the relative accuracy of different methods irrelevant; in particular, approximation by higher order polynomials does not necessarily increase the accuracy (see Sect. 3.1).

Apparent impossibility to chose a more efficient integration method by its asymptotical properties leaves one wondering if (1), which roughly corresponds to the mid-point integration rule with linear approximation and hence the lowest convergence rate, may still be relevant. The answer to this is not straightforward. The matter is that, in the reality of ecological monitoring, the properties of the population distribution over space (e.g. whether it is almost uniform or highly aggregated, single-peaked or multi-peaked, etc.) are usually not known a priori. It means that we do not know in advance whether the given sampling grid is coarse or not: for the same number of nodes, the grid may already attain its asymptotical properties in one case (e.g. see Fig. 3a for $N = 17$ to $N = 65$) but remain coarse in another case (Fig. 3b). In this uncertainty, use of higher order polynomials is beneficial "on average," i.e. it will not make things worse on a coarse grid but may improve the accuracy on a finer grid. We also refer here to our previous work where it was shown that application of the second order polynomials may increase the estimate's robustness considerably compared to (1), especially in the case of highly aggregated population distributions [23].

- For integration on a coarse grid, we obtained condition (23) for the grid step size (i.e. the inter-sampling distance) to ensure that the estimate of the population size is obtained with a required accuracy (the error being less than 25 %) for given diffusion rates. The analytical prediction (23) is in excellent agreement with simulation results, see Tables 1 and 2.

We mention here that the analysis of the simulation results shows that, in the case that the population density has a complex multi-hump spatial structure, a reasonably accurate estimate of the population size can sometimes be obtained on an very coarse grid consisting of just three nodes; see the second column in Table 2 and the last paragraph of Sect. 4.2.

Note that the coefficient ω determining the characteristic length of the spatial pattern [see (8) and (17)] may vary depending on the parameters of intra- and interspecific interactions. Once these parameters are known, its value can be estimated theoretically, cf. (7) and (8). In ecological practice, the value of ω can be extracted from available field data (e.g. from previous studies on the given species) by fitting (8) to the characteristics of the observed spatial pattern.

- We obtained the accuracy estimates (30)–(32) in the case that the population is aggregated inside a single narrow hump and the grid is very coarse, so that the hump is "resolved" by just one node. Even in this rather extreme case, there is a parameter range where the numerical integration evaluates the population size with a required accuracy.

Interestingly, a closer look at the integration of a narrow peak on a coarse grid suggests that it may lead to a paradigm shift [23] when the integration results should be interpreted probabilistically rather than deterministically. The matter is as follows: There is a range of the peak's positions with regard to the grid nodes where the peak can be integrated with sufficient accuracy, outside of this range the accuracy becomes unacceptably low. The problem is that, especially in routine monitoring, the position of the peak would not be known in advance. Integration of the sampling data would then provide a result that could be accurate in some cases but inaccurate in other cases. This is a typical problem with uncertainty, and a standard approach to deal with it is to quantify different possible outcomes with probability. The conditions (31)–(32) can then be used to estimate the probability of accurate integration. Indeed, taking into account that $0 \leq \gamma \leq 1/2$, the probability of accurate integration with a given value of h is then double the distance between the curves $\gamma_I(h)$ and $\gamma_{II}(h)$ (see Fig. 8) along the vertical line $h = const$. For instance, for $d = 0.0001$ it is about 0.3 if $h = 0.25$ but about 0.05 if $h = 0.5$.

- We considered the effect of the grid non-uniformness, i.e. when a grid node is moved from its "regular" position, on the accuracy of our approach. This is a practically important issue because the grid of sampling positions can hardly be made precisely uniform either as a result of human error or because of peculiarities of the landscape structure. We showed that the accuracy of integration is robust with respect to a small variation in the node's position. For the case of a larger variation, we obtained conditions (38)–(40) describing what should be the average grid step size to maintain the required accuracy and/or what the accuracy is going to be should the step size be chosen irrelevantly.

Our study leaves a few open questions. First, an extension of our approach onto a 2D case should be made. The results obtained here are in good qualitative agreement with the results of the numerical study made in [23] for the 2D case. However, a modification of the analytical methods that we used in this paper will require considerable work before they can be applied to a 2D grid. Furthermore, in this paper we validated our approach using the numerical data obtained from an ecological model. Application of the methods of numerical integration to data on invertebrate sampling made in [23] led to an encouraging result. However, a further validation is necessary by applying our method to field data obtained in different environments, for different species and on different spatial grids. Finally, throughout our analysis we assumed that the population density at the location of samples was known precisely. In reality, it can of course only be known approximately. It remains unclear to what extent our approach is robust with regard to this local statistical error. A detailed consideration of these problems will become a focus of future work.

In conclusion, a more general comment should be made. In order to reveal the effect of species diffusivity on the accuracy of the population size estimation, we used the diffusion-based theoretical framework. Correspondingly, the dynamics of the population density is described by diffusion or diffusion-reaction equations and the diffusivity is quantified by the diffusion coefficient D with the dimension as distance$^2 \cdot$ time^{-1}. This description implies that the individual animal movement is the Brownian motion when the mean squared displacement $< r^2(t) >$ grows with time linearly:

$$< r^2(t) > \ \sim Dt. \tag{42}$$

The corresponding dispersal kernel is then given by a normal distribution; see (3).

This may raise a question about the generality of our results. Indeed, there has been a growing amount of evidence that some animal species perform faster dispersal (often referred to as the anomalous diffusion or "superdiffusion," or Lévy flight) when the mean squared displacement shows growth faster than linear:

$$< r^2(t) > \ \sim \ \mathscr{D}t^\nu, \tag{43}$$

where $\nu > 1$ and $\mathscr{D}$ is a coefficient similar to the diffusion coefficient in its meaning but having a different dimension, i.e. distance$^2 \cdot$ time$^{-\nu}$. The dispersal kernel in this case has a fatter tail, e.g. showing either exponential or power law rate of decay at large distances. However, the main result of the dimensions analysis still holds, i.e. there is only one quantity with the dimension of length, although its expression becomes slightly different:

$$\Delta_a \sim \sqrt{\mathscr{D} t^\nu} \,, \tag{44}$$

cf. (4).

There have been several studies concerned with the relation between the spatial heterogeneity and the "diffusivity" in a broader sense. For instance, it has been shown in [24] that the characteristic length is a power-law function of the coefficient $\mathscr{D}$ (with the exponent larger than $\frac{1}{2}$) in the case of a clearly non-Brownian motion in a turbulent environment. The dependence of the rate of decay in the population density on the combination $x/(\mathscr{D} t^\nu)$ rather than on x alone was proved in [10]. These results point out that the diffusivity rate, considered in a somewhat broader sense, still is a controlling factor that determines the characteristics of the spatial heterogeneity. Therefore, our results and conclusions about its impact on the accuracy of the population size estimation are not restricted to the case of the standard Fickian diffusion and the corresponding Brownian motion of individuals, but should remain valid in a more general case.

Appendix: Approximation of a Hump by a Quadratic Function

Let $u(x)$ be an integrand function that has a local maximum (a "hump") at point x_1, where $x_1 \in [0, 1]$. Consider points $x_0 = x_1 - h$ and $x_2 = x_1 + h$, where $h > 0$ is an arbitrary parameter defining the "hump width". For instance, the value h can be defined from the condition that $u(x_2) = 0.1 u(x_1)$. Examples of the choice of h will be given further in the text for a particular problem under consideration.

Once we know the function values $u_m \equiv u(x_m), m = 0, 1, 2$, we can approximate $u(x)$ by a quadratic polynomial. This is a well-known interpolation problem (e.g., see [7]) and below we give a brief description of this technique.

To find the equation of the quadratic $g(x)$, the function values are generally needed at three points, so that the coefficients of the function $g(x)$ can be reconstructed using the conditions $g(x_m) = u(x_m), m = 0, 1, 2$. However, in our case it is more convenient to write a quadratic function in the form

$$g(x) = B - A(x - x_1)^2, \tag{45}$$

because we require $g(x)$ to have the same maximum as the hump that it replaces. The coefficients A and B are then obtained by using just the two collocation conditions $g(x_m) = u(x_m), m = 0, 1$. Thus, the hump is replaced by a quadratic which is symmetric about the location of the maximum $x = x_1$.

Table 3 The interpolation error for the quadratic functions approximating the hump of $u_1(x)$ for various values of h

h	0.125	0.0625	0.0312	0.0156
e_{max}	0.3325	0.0922	0.0123	0.0011

Table 4 The integration error (11) when the integral is computed in the vicinity of the hump

h	0.125	0.0625	0.0312	0.0156
e_u	0.0641	0.0464	0.0279	0.0091
e_q	0.1839	0.0961	0.0341	0.0096

The integration errors are computed for the density distribution $u_1(x)$ (the row e_u), and its quadratic approximation $g(x)$ (the row e_q). The functions are approximated by piecewise linear polynomials

We now introduce the interpolation error $e_{int}(x)$ in order to evaluate what we miss when we replace a hump $u(x)$ with the function $g(x)$. The function $e_{int}(x)$ is defined at any point x of the interval $[x_0, x_2]$ as

$$e_{int}(x) = |u(x) - g(x)|. \tag{46}$$

We then consider the maximum distance between the functions $u(x)$ and $g(x)$,

$$e_{max} = \max_{x \in [x_0, x_2]} e_{int}(x). \tag{47}$$

The maximum interpolation error e_{max} depends on h, as is demonstrated by the following example. Consider the approximation of a hump by a quadratic function for the pest population density $u_1(x)$. The coefficients A and B for the quadratic function $g(x)$ are defined from the collocation conditions as

$$A = \frac{u_1(x_1) - u_1(x_0)}{h^2}, \qquad B = u_1(x_1). \tag{48}$$

The interpolation error e_{max} incurred by replacing the hump in the population distribution $u_1(x)$ by a quadratic function is shown in Table 3. As h decreases, so does the size of the interpolation error. This is further illustrated by the quadratic approximations shown in Fig. 4.

Once the integrand function $u(x)$ has been replaced by a quadratic function in the vicinity of the hump, we can integrate the function $g(x)$ by a chosen numerical method. Let us apply the method outlined in Sect. 2.1 to both functions $u_1(x)$ and $g(x)$ to integrate them in the vicinity of the hump. Consider the integration error local to the hump, i.e. on the interval $[x_0, x_2]$, where the function $u(x)$ and its corresponding quadratic replacement $g(x)$ are approximated by piecewise linear polynomials. The integration error (11) computed for the function $u_1(x)$ and for the quadratic function $g(x)$ is denoted in Table 4 as e_u and e_q, respectively.

Table 5 The integration error for the first hump in density distribution $u_2(x)$, and its quadratic approximation $g(x)$

h	0.0312	0.0156	0.0078	0.0039
e_u	0.0545	0.0532	0.0267	0.0080
e_q	0.1986	0.1024	0.0325	0.0086

The functions are approximated by piecewise linear polynomials

It can be seen from Table 4, that e_q provides a sufficiently reliable estimate for the integration error (11). This conclusion is further confirmed by the results of Table 5 where we integrate both $u_2(x)$ and $g(x)$ in the vicinity of the first hump in the multi-peak distribution $u_2(x)$ (see Fig. 2b). Thus our assumption that the density distribution $u(x)$ can be approximated by a quadratic function in the vicinity of a hump is justified by computation of the interpolation error and the integration error and such approximation can be used for further theoretical and numerical analysis.

Acknowledgements This study was partially supported by The Leverhulme Trust through grant F/00-568/X.

References

1. C.J. Alexander, J.M. Holland, L. Winder, C. Wooley, J.N. Perry, Performance of sampling strategies in the presence of known spatial patterns. Ann. Appl. Biol. **146**, 361–370 (2005)
2. W. Alt, Biased random walk models for chemotaxis and related diffusion approximations. J. Math. Biol. **9**, 147–177 (1980)
3. G.I. Barenblatt, *Scaling, Self-Similarity, and Intermediate Asymptotics* (Cambridge University Press, Cambridge, 1996)
4. F. Bartumeus, J. Catalan, G.M. Viswanathan, E.P. Raposo, M.G.E. da Luz, The influence of turning angles on the success of non-oriented animal searches. J. Theor. Biol. **252**, 43–55 (2008)
5. B. Boag, K. Mackenzie, J.W. McNicol, R. Neilson, Sampling for the New Zealand flatworm, in *Crop Protection in Northern Britain 2010*. Proceedings of the Conference held at the University of Dundee, Scotland (Page Bros, Norwich, 2010), pp. 45–50
6. G.D. Buntin, Developing a primary sampling program, in *Handbook of Sampling Methods for Arthropods in Agriculture*, ed. by L.P. Pedigo, G.D. Buntin (CRC Press, Boca Raton, 1994), pp. 99–115
7. P.J. Davis, *Interpolation and Approximation* (Dover, New York, 1975)
8. P.J. Davis, P. Rabinowitz, *Methods of Numerical Integration* (Academic, New York, 1975)
9. S.M. Dunn, A. Constantinides, P.V. Moghe, *Numerical Methods in Biomedical Engineering* (Elsevier, New York, 2006)
10. M. Giona, H.E. Roman, Fractional diffusion equation for transport phenomena in random media. Phys. A **185**, 87–97 (1992)
11. J.M. Holland, J.N. Perry, L. Winder, The within-field spatial and temporal distribution of arthropods in winter wheat. Bull. Entomol. Res. **89**, 499–513 (1999)
12. P.M. Kareiva, N. Shigesada, Analyzing insect movement as a correlated random walk. Oecologia **56**, 234–238 (1983)
13. M. Kogan, Integrated pest management: historical perspectives and contemporary developments. Annu. Rev. Entomol. **43**, 243–270 (1998)

14. S.X. Liao, M. Pawlak, On image analysis by moments. IEEE Trans. Pattern Anal. Mach. Intell. **18**(3), 254–266 (1996)
15. H. Malchow, S.V. Petrovskii, E. Venturino, *Spatiotemporal Patterns in Ecology and Epidemiology: Theory, Models, and Simulations* (Chapman & Hall/CRC Press, London, 2008)
16. J.D. Murray, *Mathematical Biology* (Springer, Berlin, 1989)
17. P. Northing, Extensive field based aphid monitoring as an information tool for the UK seed potato industry. Aspects Appl. Biol. **94**, 31–34 (2009)
18. A. Okubo, S. Levin, *Diffusion and Ecological Problems: Modern Perspectives* (Springer, Berlin, 2001)
19. M.A. Pascual, P. Kareiva, Predicting the outcome of competition using experimental data: maximum likelihood and Bayesian approaches. Ecology **77**, 337–349 (1996)
20. J.N. Perry, Simulating spatial patterns of counts in agriculture and ecology. Comput. Electron. Agric. **15**, 93–109 (1996)
21. N.B. Petrovskaya, S. Petrovskii, The coarse-grid problem in ecological monitoring. Proc. R. Soc. A **466**, 2933–2953 (2010)
22. N.B. Petrovskaya, E. Venturino, Numerical integration of sparsely sampled data. Simul. Model. Pract. Theor. **19**(9), 1860–1872 (2011)
23. N.B. Petrovskaya, S.V. Petrovskiy, A.K. Murchie, Challenges of ecological monitoring: estimating population abundance from sparse trap counts. J. R. Soc. Interface **9**, 420–435 (2012)
24. S.V. Petrovskii, On the plankton front waves accelerated by marine turbulence. J. Mar. Syst. **21**, 179–188 (1999)
25. S.V. Petrovskii, B.-L. Li, *Exactly Solvable Models of Biological Invasion* (Chapman & Hall/CRC, Boca Raton, 2006)
26. S.V. Petrovskii, H. Malchow, Spatio-temporal chaos in an ecological community as a response to unfavorable environmental changes. Adv. Complex Syst. **4**, 227–250 (2001)
27. S.V. Petrovskii, B.-L. Li, H. Malchow, Quantification of the spatial aspect of chaotic dynamics in biological and chemical systems. Bull. Math. Biol. **65**, 425–446 (2003)
28. G.A. Seber, *The Estimation of Animal Abundance and Related Parameters* (Charles Griffin, London, 1982)
29. L.A. Segel, A theoretical study of receptor mechanisms in bacterial chemotaxis. SIAM J. Appl. Math. **32**, 653–665 (1977)
30. J.A. Sherratt, M. Smith, Periodic travelling waves in cyclic populations: field studies and reaction diffusion models. J. R. Soc. Interface **5**, 483–505 (2008)
31. G.W. Snedecor, W.G. Cochran, *Statistical Methods* (The Iowa State Iniversity Press, Ames, 1980)
32. V.M. Stern, Economic thresholds. Annu. Rev. Entomol. **18**, 259–280 (1973)
33. W.J. Sutherland (ed.), *Ecological Census Techniques: A Handbook* (Cambridge University Press, Cambridge, 1996)
34. L.R. Taylor, I.P. Woiwod, J.N. Perry, The negative binomial as a dynamic ecological model for aggregation, and the density dependence of k. J. Anim. Ecol. **48**, 289–304 (1979)
35. P. Turchin, *Quantitative Analysis of Movement: Measuring and Modeling Population Redistribution in Animals and Plants* (Sinauer, Sunderland, 1988)
36. L. Yaroslavsky, A. Moreno, J. Campos, Frequency responses and resolving power of numerical integration of sampled data. Opt. Express **13**(8), 2892–2905 (2005)

LECTURE NOTES IN MATHEMATICS

Edited by J.-M. Morel, B. Teissier; P.K. Maini

Editorial Policy (for Multi-Author Publications: Summer Schools / Intensive Courses)

1. Lecture Notes aim to report new developments in all areas of mathematics and their applications - quickly, informally and at a high level. Mathematical texts analysing new developments in modelling and numerical simulation are welcome. Manuscripts should be reasonably selfcontained and rounded off. Thus they may, and often will, present not only results of the author but also related work by other people. They should provide sufficient motivation, examples and applications. There should also be an introduction making the text comprehensible to a wider audience. This clearly distinguishes Lecture Notes from journal articles or technical reports which normally are very concise. Articles intended for a journal but too long to be accepted by most journals, usually do not have this "lecture notes" character.

2. In general SUMMER SCHOOLS and other similar INTENSIVE COURSES are held to present mathematical topics that are close to the frontiers of recent research to an audience at the beginning or intermediate graduate level, who may want to continue with this area of work, for a thesis or later. This makes demands on the didactic aspects of the presentation. Because the subjects of such schools are advanced, there often exists no textbook, and so ideally, the publication resulting from such a school could be a first approximation to such a textbook. Usually several authors are involved in the writing, so it is not always simple to obtain a unified approach to the presentation.

   For prospective publication in LNM, the resulting manuscript should not be just a collection of course notes, each of which has been developed by an individual author with little or no coordination with the others, and with little or no common concept. The subject matter should dictate the structure of the book, and the authorship of each part or chapter should take secondary importance. Of course the choice of authors is crucial to the quality of the material at the school and in the book, and the intention here is not to belittle their impact, but simply to say that the book should be planned to be written by these authors jointly, and not just assembled as a result of what these authors happen to submit.

   This represents considerable preparatory work (as it is imperative to ensure that the authors know these criteria before they invest work on a manuscript), and also considerable editing work afterwards, to get the book into final shape. Still it is the form that holds the most promise of a successful book that will be used by its intended audience, rather than yet another volume of proceedings for the library shelf.

3. Manuscripts should be submitted either online at www.editorialmanager.com/lnm/ to Springer's mathematics editorial, or to one of the series editors. Volume editors are expected to arrange for the refereeing, to the usual scientific standards, of the individual contributions. If the resulting reports can be forwarded to us (series editors or Springer) this is very helpful. If no reports are forwarded or if other questions remain unclear in respect of homogeneity etc, the series editors may wish to consult external referees for an overall evaluation of the volume. A final decision to publish can be made only on the basis of the complete manuscript; however a preliminary decision can be based on a pre-final or incomplete manuscript. The strict minimum amount of material that will be considered should include a detailed outline describing the planned contents of each chapter.

   Volume editors and authors should be aware that incomplete or insufficiently close to final manuscripts almost always result in longer evaluation times. They should also be aware that parallel submission of their manuscript to another publisher while under consideration for LNM will in general lead to immediate rejection.

4. Manuscripts should in general be submitted in English. Final manuscripts should contain at least 100 pages of mathematical text and should always include

   – a general table of contents;
   – an informative introduction, with adequate motivation and perhaps some historical remarks: it should be accessible to a reader not intimately familiar with the topic treated;
   – a global subject index: as a rule this is genuinely helpful for the reader.

   Lecture Notes volumes are, as a rule, printed digitally from the authors' files. We strongly recommend that all contributions in a volume be written in the same LaTeX version, preferably LaTeX2e. To ensure best results, authors are asked to use the LaTeX2e style files available from Springer's web-server at
   ftp://ftp.springer.de/pub/tex/latex/svmonot1/ (for monographs) and
   ftp://ftp.springer.de/pub/tex/latex/svmultt1/ (for summer schools/tutorials).
   Additional technical instructions, if necessary, are available on request from:
   lnm@springer.com.

5. Careful preparation of the manuscripts will help keep production time short besides ensuring satisfactory appearance of the finished book in print and online. After acceptance of the manuscript authors will be asked to prepare the final LaTeX source files and also the corresponding dvi-, pdf- or zipped ps-file. The LaTeX source files are essential for producing the full-text online version of the book. For the existing online volumes of LNM see:
   http://www.springerlink.com/openurl.asp?genre=journal&issn=0075-8434.
   The actual production of a Lecture Notes volume takes approximately 12 weeks.

6. Volume editors receive a total of 50 free copies of their volume to be shared with the authors, but no royalties. They and the authors are entitled to a discount of 33.3 % on the price of Springer books purchased for their personal use, if ordering directly from Springer.

7. Commitment to publish is made by letter of intent rather than by signing a formal contract. Springer-Verlag secures the copyright for each volume. Authors are free to reuse material contained in their LNM volumes in later publications: a brief written (or e-mail) request for formal permission is sufficient.

Addresses:
Professor J.-M. Morel, CMLA,
École Normale Supérieure de Cachan,
61 Avenue du Président Wilson, 94235 Cachan Cedex, France
E-mail: morel@cmla.ens-cachan.fr

Professor B. Teissier, Institut Mathématique de Jussieu,
UMR 7586 du CNRS, Équipe "Géométrie et Dynamique",
175 rue du Chevaleret,
75013 Paris, France
E-mail: teissier@math.jussieu.fr

For the "Mathematical Biosciences Subseries" of LNM:

Professor P. K. Maini, Center for Mathematical Biology,
Mathematical Institute, 24-29 St Giles,
Oxford OX1 3LP, UK
E-mail : maini@maths.ox.ac.uk

Springer, Mathematics Editorial I,
Tiergartenstr. 17,
69121 Heidelberg, Germany,
Tel.: +49 (6221) 4876-8259
Fax: +49 (6221) 4876-8259
E-mail: lnm@springer.com